Operational Amplifiers And Linear Integrated Circuits

OPERATIONAL AMPLIFIERS AND LINEAR INTEGRATED CIRCUITS

ROBERT F. COUGHLIN

and

FREDERICK F. DRISCOLL

Wentworth Institute

PRENTICE HALL, INC. *Englewood Cliffs, New Jersey*

Library of Congress Cataloging in Publication Data

COUGHLIN, ROBERT F
Operational amplifiers and linear integrated circuits.

Bibliography: p. 264
1. Operational amplifiers. 2. Linear integrated circuits. I. Driscoll, Frederick F., 1943– joint author. II. Title.
TK7871.58.06C68 621.3815'35 76-7539
ISBN 0-13-637850-1

10 9 8 7 6 5 4 3 2

Printed in the United States of America

PRENTICE-HALL INTERNATIONAL, INC., *London*
PRENTICE-HALL OF AUSTRALIA, PTY. LTD., *Sydney*
PRENTICE-HALL OF CANADA, LTD., *Toronto*
PRENTICE-HALL OF INDIA PRIVATE LIMITED, *New Delhi*
PRENTICE-HALL OF JAPAN, INC., *Tokyo*
PRENTICE-HALL OF SOUTHEAST ASIA PTE. LTD., *Singapore*

To Bae and Fred
and to
Millie and Wally

Contents

10 Band Width, Slew Rate, Noise, and Frequency Compensation 163

Preface

Op amps and other linear integrated circuits are both fun and easy to use, especially if the application does not require the devices to operate near their design limits. It is the purpose of this text to show just how easy they are to use in a variety of applications involving instrumentation, signal generation, and control.

When first learning about how to use an op amp, one should not be presented with a myriad op amps and asked to make an informed selection. Rather our introduction begins with an inexpensive, reliable op amp that forgives most mistakes in wiring, ignores long lead capacitance, and does not burnout too easily. Such an op amp is the 741, whose characteristics are documented in Appendix 1 and whose applications are sprinkled throughout the text.

If a slightly faster op amp is needed for a wider bandwidth, another inexpensive and widely used op amp is the 301. See Appendix 2 for its electrical characteristics and Chapter 10 to learn when one might prefer the 301 over the 741.

Chapters 1 through 8 show how the op amp can be used in a wide variety of applications. The limitations of op amps are not shown in Chapters 1 to 8. It is very important to gain confidence in using op amps before pushing performance to its limits. When studying transistors or other devices, we do not begin with their limitations. Regretably, much of the integrated circuit literature begins with the limitations of integrated circuits and thus obscures the inherent simplicity and overwhelming advantages of basic integrated circuits over basic transistor circuits. For these reasons, op amp limitations are not presented until Chapters 9 and 10. Furthermore, not all op amp limitations apply to every op amp circuit. For example, dc op amp limitations such as offset voltage are usually not important if the op amp is used in an ac amplifier circuit. Thus dc limitations (Chapter 9) are treated separately from ac limitations (Chapter 10).

A fascinating integrated circuit, the multiplier, is presented in Chapter 11, because it makes analysis and design of communication circuits very easy. Modulators, demodulators, frequency doublers, frequency shifters, and a host of other applications are performed by the multiplier, an op amp, and a few resistors.

In Chapter 12, the four basic types of active filters are shown: they are low-pass, high-pass, band-pass and band-reject filters. The Butterworth filters were selected because they are simple to design. If you want to design a three-pole (60 db/decade) Butterworth low- or high-pass filter, Chapter 12 tells you how to do it in four steps with a pencil and paper. No calculator or computer is required. Basic algebra is the only mathematics that is required throughout the text.

Chapter 13 is devoted to another class of ICs that is as widely used and as versatile as the op amp, namely IC timers. They are inexpensive and simple to apply.

This text has enough material for a one-semester course. All the circuits have been classroom and laboratory tested and have been selected with great care to illustrate (only partially) the tremendous range of applications where ICs provide inexpensive solutions to practical problems. The material is suitable for both non-electronics specialists who just want to learn something about linear ICs and electronics majors who wish to use linear ICs.

We thank Dean Charles M. Thomson for his encouragement to write this book; our students for insistance on relevant and timely subject matter in the field of semiconductors; Mary Hatfield, Barbara Baum, Judith Trebach, and Maureen Meehan for their skillful typing; our colleagues Robert S. Villanucci and William Megow for their design and construction of the breadboard systems in Chapter 1, and William Bradley and Joseph Rudis for the photographs.

R. F. COUGHLIN
F. F. DRISCOLL
Boston, Mass.

Operational Amplifiers And Linear Integrated Circuits

1

Breadboarding Op Amp Circuits

1-0 Introduction

Linear integrated circuits (ICs) are being incorporated almost daily into more and more electronic applications in such fields as audio and radio communications, medical technology, instrumentation, manufacturing controls, and automotive technology. The reaons for their growing use are small size, ease of use, reliability, and low cost. Circuits that just a few years ago required weeks of design time can now be purchased as ICs. This allows the ICs to be used as a functional block. With a few external components added to the IC, the design is completed. This functional block is also much simpler to analyze and trouble-shoot than its discrete-component equivalent.

There are two types of linear integrated circuits manufactured today, namely *monolithic integrated circuits* and *hybrid integrated circuits.* Monolithic ICs are manufactured on a single semiconductor substrate (usually silicon) along with some resistors and capacitors. Hybrid ICs are a combination of discrete components, passive elements, and monolithic ICs. Most circuits in this text use a monolithic IC, but some, such as the multipliers of Chapter 11, are of hybrid construction.

This book is an introductory text, but all the circuits have been laboratory tested. They have been chosen to demonstrate a wide variety of applications and to illustrate the basic characteristics being discussed at the time. Many readers of a text wish to set up the circuits to check results and formulas and to prepare for study in more depth. Section 1-1 shows two types of linear IC breadboards that allow linear IC circuits to be set up quickly and easily.

1-1 Linear Integrated-Circuit Breadboards

1-1.1 Breadboard Requirements. A breadboarding system should allow you to:

1. set up a circuit in a minimum amount of time (approximately two minutes for most of the circuits in this text)
2. change external elements easily for any design modifications
3. replace the IC without having to unsolder it
4. connect leads from audio oscillators, function generators, voltmeters, and oscilloscopes
5. have a positive and negative power supply

One type of breadboard that will be described uses two 9-V batteries for the power supply. A second type uses a plus/minus regulated power supply, the design of which is included in Chapter 7.

1-1.2 Battery-Operated IC Breadboard. Figure 1-1 shows an inexpensive IC breadboard that allows experimentation with a 741 operational amplifier. The op amp is soldered to the bottom side of the printed-circuit board as shown in Fig. 1-2. When the power switch is thrown to "on,"

Figure 1-1 Inexpensive breadboard for a 741 op amp.

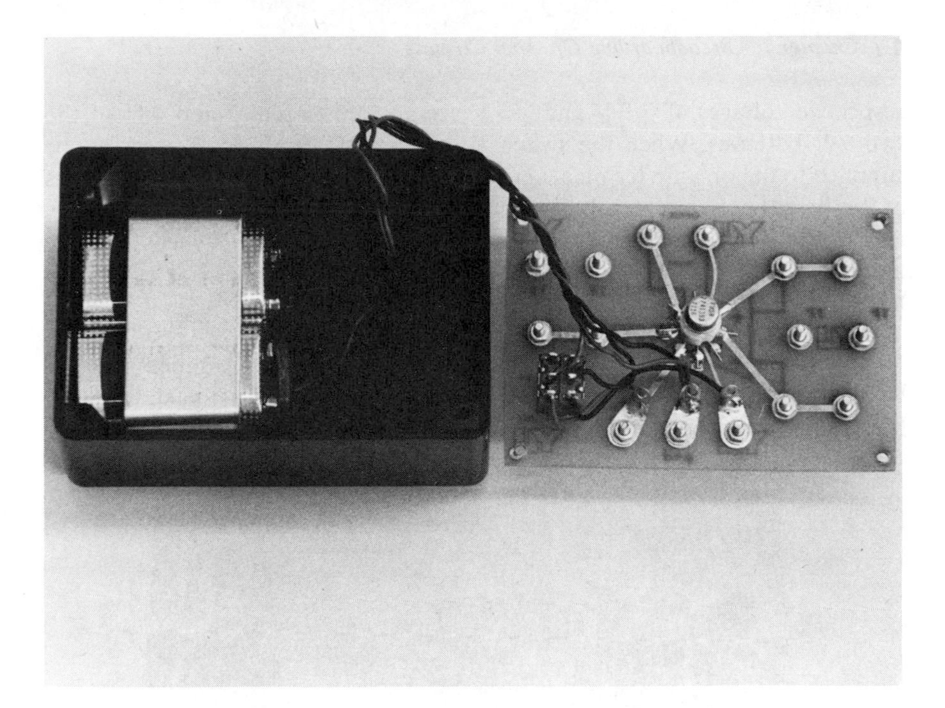

Figure 1-2 Bottom view of IC breadboard.

Figure 1-3 Linear IC breadboard.

respective voltages of +9 V and −9 V are applied to pins 7 and 4 from the two 9-V batteries. When the switch is in the "off" position, power can be supplied to the op amp terminal springs from an external supply. Advantages of this breadboard are portability, low cost, terminals that are easily available, and extra tie points. Its limitations are that it has been designed for only a particular IC (the 741 or 301 op amp) and that the batteries will eventually have to be replaced.

1-1.3 IC Breadboard with Regulated Power Supply. A more versatile linear IC breadboard is shown in Figs. 1-3 and 1-4. It consists simply of a box frame, two printed-circuit (pc) breadboards, and a ±15-V power supply.

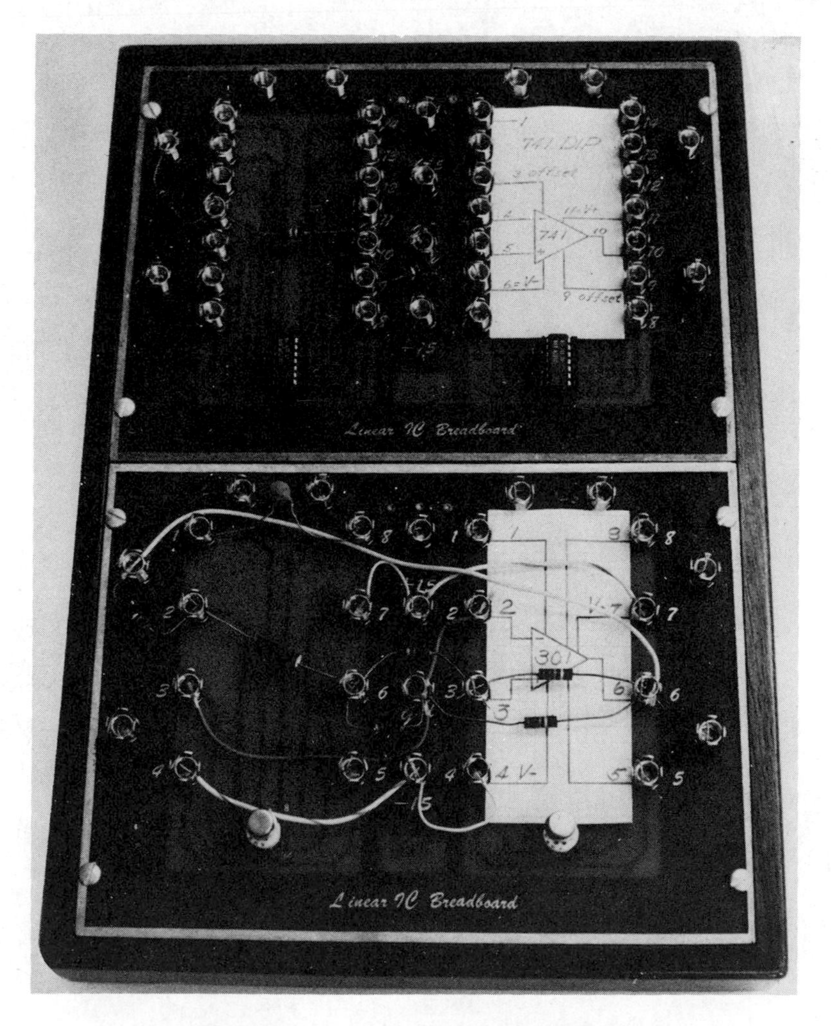

Figure 1-4 Using the linear IC breadboard.

The power supply is mounted on the inside rear wall as shown in Fig. 1-5. The undersides of the breadboards are shown in Fig. 1-6.

The three center terminals of the breadboards provide +15 V, ground, and −15 V. Figure 1-4 shows the breadboarding arrangement for three of the circuits studied in the text. The bottom two ICs are wired for the triangular-wave generator in Fig. 6-10. The upper left IC is the inverting amplifier of Fig. 3-1.

As shown in Fig. 1-4, a paper template can be made for the op amp pin connections to identify input, output, power supply, and other terminals. A template for a 741 dual-in-line package op amp is shown for the top right IC in Fig. 1-4; the bottom right template is for a 301. For more experienced users, the template can be omitted. Further details on this linear IC breadboard are provided in Section 1-2.

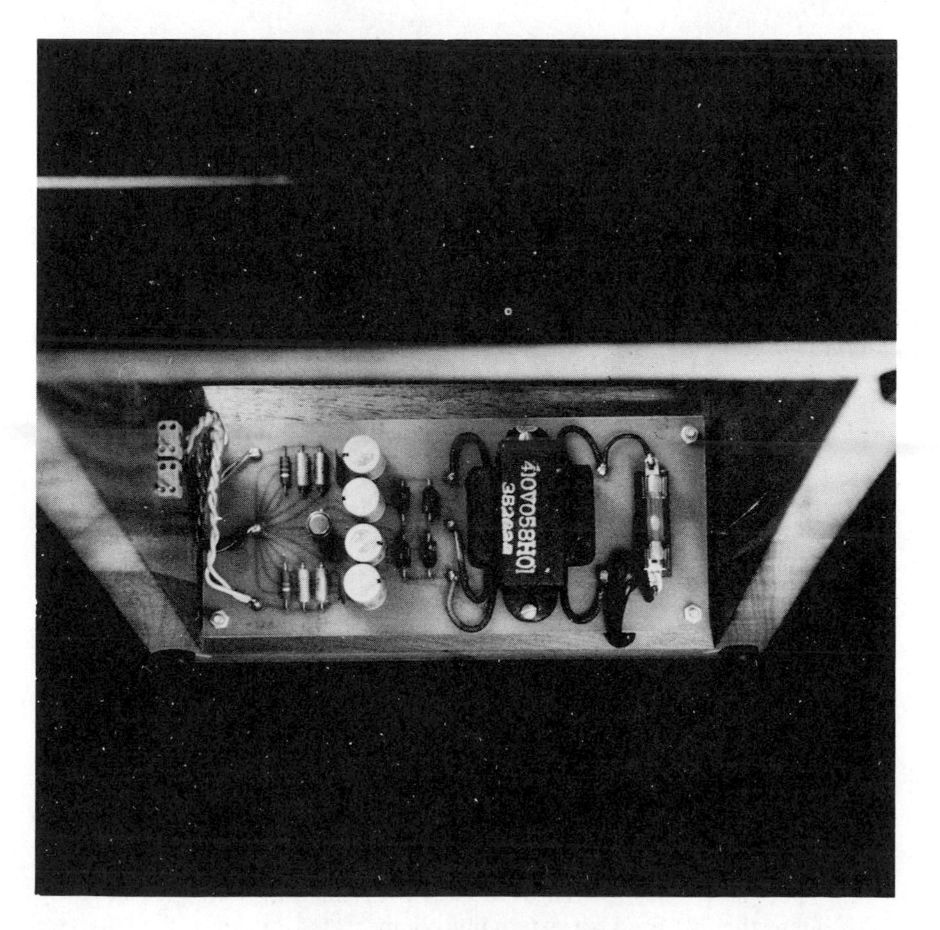

Figure 1-5 Power supply mounting.

Figure 1-6 Undersides of breadboards.

1-2 Hardware for the Linear Breadboard

1-2.1 ICs and IC Sockets. IC sockets can be purchased for all the standard ICs. The op amps chosen for most of the circuits in this text are the 741 and 301 (see Appendices 1 and 2). The reasons for choosing these two devices are

1. They are general-purpose op amps.
2. They show the differences between an internally compensated op amp, the 741, and an externally compensated op amp, the 301 (see Chapters 9 and 10).

3. They are inexpensive, costing less than $1.
4. Both are readily available in a 14-pin dual-in-line package (DIP) and an 8-pin round case (TO99). See Appendices 1, 2, and 3 for details on pin counting and package styles.

1-2.2 Spring-Clip Connectors. The spring clips (OMNI-GRIP®; U. S. Pat. Nos. 3, 150,911. Courtesy of Cosmic Voice, Inc.) Figs. 1-1 to 1-4 were chosen for the following reasons.

1. They accept most wire sizes used in electronic work.
2. Components with different lead thicknesses can be connected to the same spring clip.
3. They are large enough to work with easily.
4. They will accept alligator and spade connectors.
5. One spring clip can be chosen as the common or ground terminal, thus eliminating ground loops.

1-2.3 Breadboard Material List. The materials used for the linear IC breadboard in Figs. 1-3 to 1-5 are

1. *Box frame* or chassis, as available; these were made by stripping a vacuum-tube power supply and sawing the box in half at an angle
2. *OMNI-GRIP® spring clips*, 40 for DIP board and 28 for 8-pin metal can board. Available from General Electronics Associates, Inc., P.O. Box 156, Northfield, Ohio, 44067
3. *IC sockets*, four 14-pin for DIP board or four 8-pin for metal can board
4. *Mounting hardware* 4/40 nuts and bolts
5. G10 double-sided printed-circuit board
6. *Power Supply*

    Option 1: two 9-V batteries

    Option 2: buy a ±15-V, 50-mA power supply for approximately $40

    Option 3: build the supply of Fig. 1-5 from information given in Figs. 7-2 and 7-16 with the following equipment (approximate cost: $15)

        One Raytheon RC4195T Fixed ±15-V Dual tracking voltage regulator

        One 110-V to 30-VCT @ 1 A transformer

        Four 1-A diodes

        Four 500-μF electrolytic capacitors at 50 WV dc for unregulated supply

        Four 5-μF electrolytic capacitors, 35 WV dc tantalum, for output of the 4195

        Two 10-kΩ, $\frac{1}{4}$-W bleeder resistors, one from $+V_o$ to ground and one from $-V_o$ to ground of the 4195

One fuse holder and 0.5-A fuse
One line cord plus hookup wire

1-3 Using the IC Breadboard

1-3.1 Breadboarding an IC Circuit. After you have designed and analyzed a circuit containing an IC and obtained all the components, you are ready to build and test your design. With an IC breadboard similar to those described in Figs. 1-1 to 1-6, the time needed to build a prototype should be only a matter of minutes.

Insert the IC and, with power off, connect the +15-V and −15-V terminals on the breadboard to the $+V$ and $-V$ terminals of the IC. Connect all grounded leads to the single-ground spring clip, if possible. Complete the circuit by connecting all other components to the proper spring clips. Use tie points if necessary. Now, most importantly, *recheck* your entire layout; if you are like the rest of us, you have already made a mistake. Try to keep leads short.

1-3.2 Testing an IC Circuit. After you have checked your wiring layout, you are ready to proceed.

1. Connect the power by plugging in the line cord (not shown).
2. With a CRO on the 10 V/cm scale and dc coupled, or with a dc voltmeter, check to be sure that the $+V$ terminal is at $+V$ and the $-V$ terminal at $-V$.
3. Connect the input signal and measure it.
4. Measure the output voltage to be sure the output is neither in saturation (see Chapter 2) nor slew rate limited (see Chapter 10). If the input signal is ac, the output must be checked on an oscilloscope to be sure that the output is not overdriven or distorted.
5. Take all measurements with respect to ground. For example, if a resistor is connected between two terminals of an IC, do not connect either a meter or a CRO across the resistor; instead, measure the voltage on one side of the resistor and then on the other side and calculate the voltage across the resistor.
6. Unless otherwise instructed, avoid using ammeters. Measure the voltage as in step 5 and calculate current.
7. Load resistors should not be less than 2 kΩ; the reason for this is discussed in Section 2-1.2.
8. Disconnect the input signal before the dc power is removed. Otherwise, the IC may be destroyed.
9. These ICs will stand much abuse. But *never*
   a. reverse polarity of the power supplies,

   b. drive the op amp's input pins above or below the potentials at the $+V$ and $-V$ terminal, or
   c. leave an input signal connected with no power on the IC.
10. If unwanted oscillations appear at the output and the circuit connections check out,
   a. connect a 0.1-μF capacitor between the op amp's $+V$ pin and ground and another 0.1-μF capacitor between the op amp's $-V$ pin and ground,
   b. shorten your leads,
   c. check that test instrument, signal generator, load, and power supply ground leads come together in one spring clip (star grounding).
11. These same principles apply to all other linear ICs.

We now proceed to our first experience with an op amp.

2

First Experience With An Op Amp

2-0 Introduction

The name *operational amplifier* was given to early high-gain amplifiers designed to perform the mathematical operations of addition, subtraction, multiplication, and division. They worked with high voltages such as ± 300 V, but they could accomplish computations, like solving calculus problems, that were not economical to undertake prior to their invention.

The modern successor of those amplifiers is the *linear integrated-circuit op amp*. It inherits the name, works at lower voltages, and is at least as good. Today's op amp is so low in cost that millions are now used annually. Their low cost, versatility, and dependability have expanded their use far beyond applications envisioned by early designers. Some present-day uses for op amps are in the fields of process control, communications, computers, power and signal sources, displays, and testing or measuring systems. The op amp is still basically a very good high-gain dc amplifier.

One's first experience with a linear IC op amp should concentrate on its most important and fundamental properties. Accordingly, our objectives in this chapter will be to identify each terminal of the op amp and to learn its purpose, some of its electrical limitations, and how to apply it usefully.

Op amps have five basic terminals: two for supply power, two for input signals, and one for output. Internally they are complex, as shown by the schematic diagram in Fig. 2-1 and Appendix 1. It is not necessary to know anything about the internal operation of the op amp in order to use it. The people who design and build op amps have done such an outstanding job

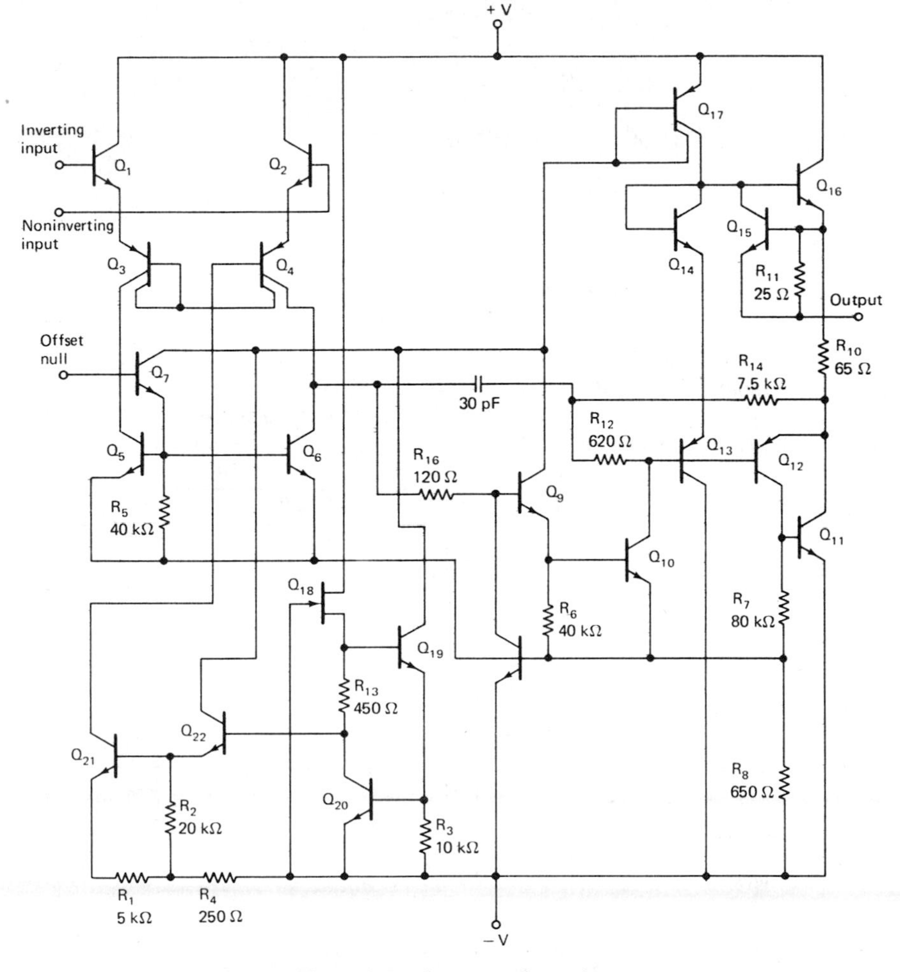

Figure 2-1 Op amp schematic.

that external components connected to the op amp, not the op amp itself, determine what it will do.

Three common packages are shown in Fig. 2-2, and more appear in Appendix 1. As viewed from the top, pins are counted in a counterclockwise direction, similar to vacuum tubes. Pin 1 is identified by a notch on the DIP of Fig. 2-2(b) and by a dot on the flat pack of Fig. 2-2(c). Pin 8 is identified by a metal tab on the metal can package of Fig. 2-2(a).

2-1 Op Amp Terminals

The circuit schematic for the op amp is an arrowhead, as shown in Fig. 2-2. The arrowhead symbolizes amplification and points from input to output.

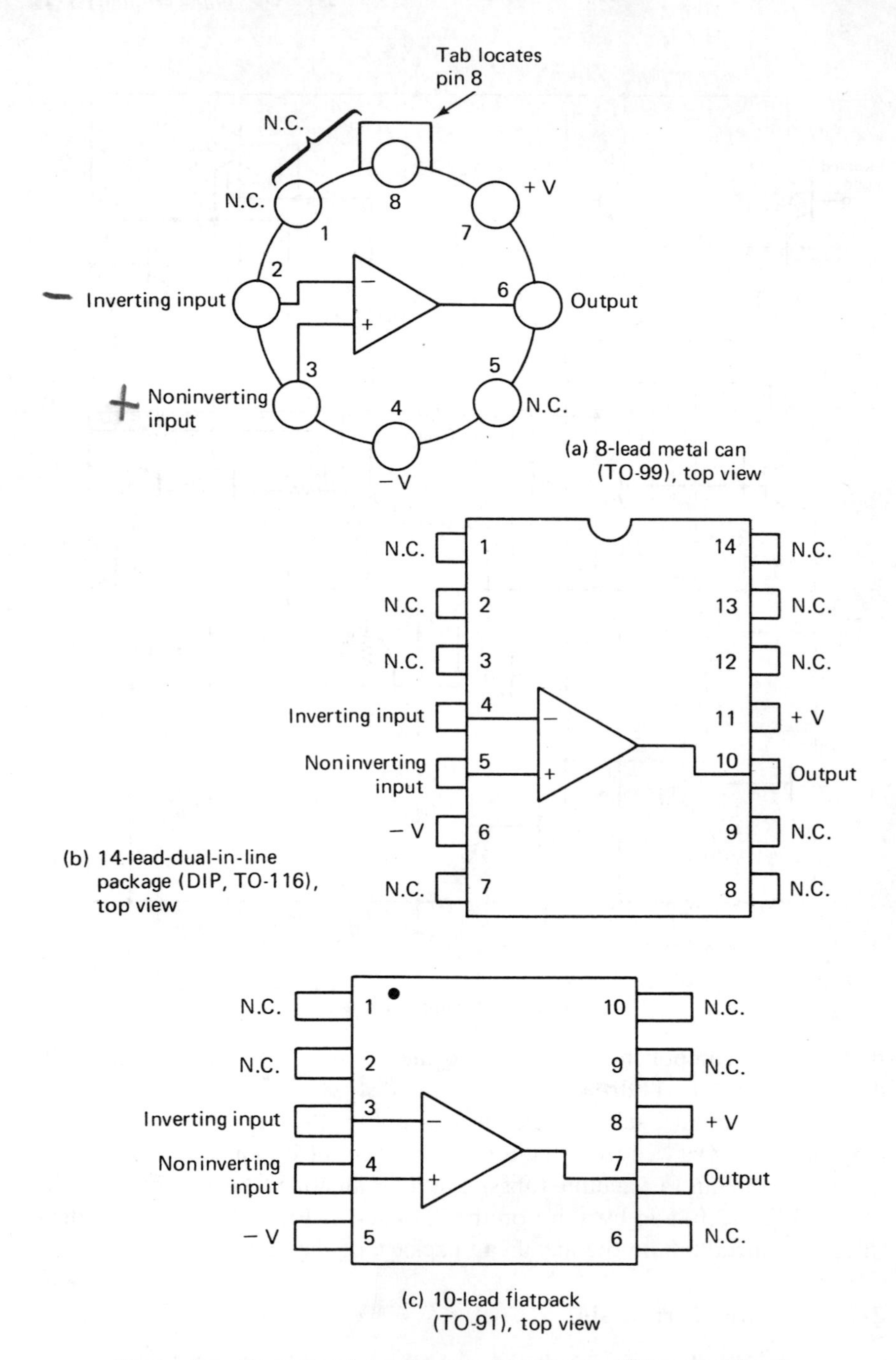

Figure 2-2 Operational amplifier and packages. (N.C. signifies "no connection.")

2-1.1 Power Supply Terminals. Op amp terminals labeled $+V$ and $-V$ identify those op amp terminals that must be connected to the power supply; see Fig. 2-3 and Appendices 1 and 2. Note that the power supply has *three* terminals (positive, negative, and common ground). This type of supply is called a *split supply*, with typical values of ± 15 V, ± 12 V, and ± 6 V. Special-purpose op amps may require nonsymmetrical supplies such as $+12$ V and -6 V, or even a single polarity supply such as $+30$ V and ground. Note that the ground is *not* wired to the op amp in Fig. 2-3. Currents returning to the supply from the op amp must return through external circuit elements such as the load resistor R_L.

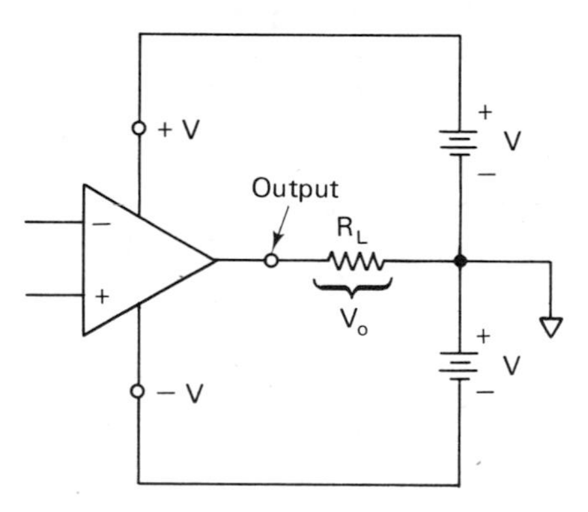

Figure 2-3 Wiring power and a load to an op amp.

The maximum supply voltage that can be applied between $+V$ and $-V$ is typically 36 V or ± 18 V.

2-1.2 Output Terminal. In Fig. 2-3 the op amp's output terminal is connected to one side of the load resistor R_L. The other side of R_L is wired to ground. Output voltage V_o is measured with respect to ground. Since there is only one output terminal in an op amp, it is called a *single-ended output*. There is a limit to the current that can be drawn from the output terminal of an op amp, usually on the order of 5 to 10 mA. There are also limits on the output terminal's voltage levels; these limits are set by the supply voltages and by output transistors Q_{16} and Q_{11} in Fig. 2-1. (Also see Appendix 1, "Output voltage as a function of supply voltage.") These transistors need about 1 V to 2 V from collector to emitter to insure that they are acting as amplifiers and not as switches. Thus the output terminal can rise to within 2 V of $+V$ and drop to within 2 V of $-V$. The upper limit of V_o is called the *positive saturation voltage*, $+V_{sat}$, and the lower limit is called the *negative saturation voltage*, $-V_{sat}$. For example, with a supply voltage of ± 15 V, $+V_{sat} = +13$ V and $-V_{sat} = -13$ V. Therefore, V_o is restricted to a peak-to-peak swing of ±13 V. Both current and voltage limits place a *minimum* value on the load resistance R_L of 2 kΩ.

Some op amps, such as the 741, have internal circuitry that automatically limits current drawn from the output terminal. Even with a short circuit for R_L, output current is limited to about 25 mA, as noted in Appendix 1. This feature prevents destruction of the op amp in the event of a short circuit. Normally, the complete output circuit showing the batteries is not drawn for every op amp circuit; a simplified diagram, as illustrated in Fig. 2-4, is used instead.

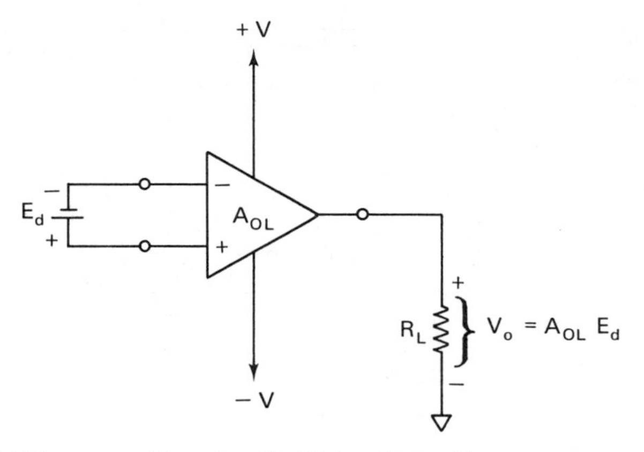

(a) V_o goes positive when the (+) input is positive with respect to the (−) input.

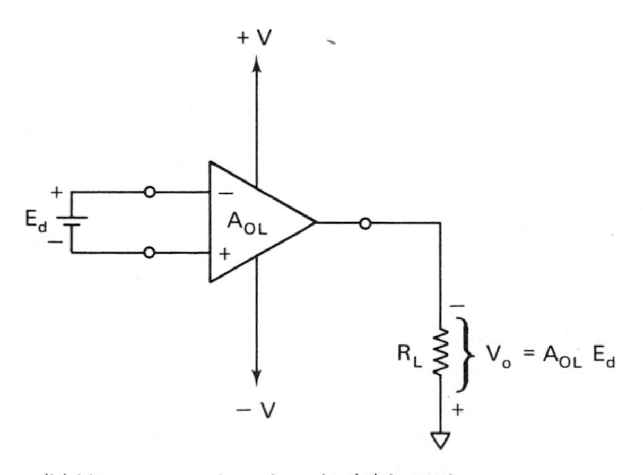

(b) V_o goes negative when the (+) input is negative with respect to the (−) input.

Figure 2-4 Polarity of V_o depends on the polarity of differential input voltage E_d.

2-1.3 Input Terminals. In Fig. 2-4 there are two input terminals, labeled − and +. They are called *differential input terminals* because output voltage V_o depends on the *difference* in voltage between them, E_d, and the gain of the amplifier, A_{OL}. As shown in Fig. 2-4(a), the output terminal is positive with respect to ground when the (+) input is positive with respect to the (−) input. When E_d is reversed in Fig. 2-4(b) to make the (+) input negative with respect to the (−) input, V_o becomes negative with respect to ground.

We conclude from Fig. 2-4 that the polarity of the output terminal is the same as the polarity of (+) input terminal. Moreover, the polarity of the output terminal is opposite or inverted from the polarity of the (−) input terminal. For these reasons the (−) input is designated the *inverting input* and the (+) input the *noninverting input*. See Appendix 1.

It is important to emphasize that the polarity of V_o depends only on the *difference* in voltage between inverting and noninverting inputs. This difference voltage can be found by

$$E_d = \text{voltage at the (+) input} - \text{voltage at the (−) input} \qquad (2\text{-}1)$$

Equation (2-1) assumes that the (+) input is positive with respect to the (−) input. The sign of E_d tells us (1) the polarity of the (+) input with respect to the (−) input and (2) the polarity of the output terminal with respect to ground. This equation holds if the inverting input is grounded, if the noninverting input is grounded, and even if both inputs are above or below ground potential. One other important characteristic of the input terminals is the high impedance between them and also between each input terminal and ground.

2-2 Open-Loop Voltage Gain

2-2.1 Definition. Refer to Fig. 2-4. If differential input voltage E_d is small enough, output voltage V_o will be determined by both E_d and the *open-loop voltage gain*, A_{OL}. A_{OL} is called open-loop voltage gain because possible feedback connections from output terminal to input terminals are left open. Accordingly, V_o would be ideally expressed by the simple relationship

$$\text{output voltage} = \text{differential input voltage} \times \text{open loop gain}$$

$$V_o = E_d \times A_{OL} \qquad (2\text{-}2)$$

2-2.2 Differential Input Voltage, E_d. The value of A_{OL} is extremely large, often 200,000 or more. Recall from section 2-1.2 that V_o can never exceed positive or negative saturation voltages $+V_{sat}$ and $-V_{sat}$. For a ±15-V supply, saturation voltages would be about ±13 V. Thus, for the op amp to act as an amplifier, E_d must be limited to a maximum voltage of ±65 μV. This conclusion is reached from Eq. (2-2).

$$E_{d\,\max} = \frac{+V_{\text{sat}}}{A_{OL}} = \frac{13\text{ V}}{200{,}000} = 65\ \mu\text{V}$$

$$-E_{d\,\max} = \frac{-V_{\text{sat}}}{A_{OL}} = \frac{-13\text{ V}}{200{,}000} = -65\ \mu\text{V}$$

In the laboratory or shop it is difficult to measure 65 μV, because induced noise, 60-Hz hum, and leakage currents on the typical test setup can easily generate a millivolt (1000 μV). Furthermore, it is difficult and inconvenient to measure very high gains. The op amp also has tiny internal unbalances that *act* as a small voltage that offsets, and even may exceed, E_d. This *offset voltage* will be discussed in Chapter 9.

2-2.3 Conclusions. There are three conclusions to be drawn from these brief comments. First, V_o in the circuit of Fig. 2-4 either will be at one of the limits $+V_{\text{sat}}$ and $-V_{\text{sat}}$ or will be oscillating between these limits. Don't be disturbed, because this behavior is what a high-gain amplifier usually does. Second, to maintain V_o between these limits we must go to a feedback type of circuit that forces V_o to depend on stable, precision elements such as resistors and signal generators rather than A_{OL} and E_d.

The last and most important conclusion is that *if E_d is so small that we cannot measure it easily, then for all practical purposes E_d equals 0V.* This conclusion is by no means trivial. We will repeatedly use the fact that $E_d \approx 0$ *if V_o lies between the saturation voltages.* To state this in another way, the (−) input will always be at about the same potential as the (+) input with respect to ground if V_o lies between saturation voltages. Without learning any more about the op amp, it is possible to understand basic comparator applications. In a comparator application, the op amp performs not as an amplifier but as a device that tells when an unknown voltage is below, above, or just equal to a known reference voltage. Before introducing the comparator in the next section, an example is given to illustrate ideas presented thus far.

Example 2-1: In Fig. 2-4, $+V = 15$ V, $-V = -15$ V, $+V_{\text{sat}} = +13$ V, $-V_{\text{sat}} = -13$ V, and gain $A_{OL} = 200{,}000$. Assuming ideal conditions, find the magnitude and polarity of V_o for each of the following input voltages, given with respect to ground.

	Voltage at (−) input	*Voltage at (+) input*
(a)	−10 μV	−15 μV
(b)	−10 μV	+15 μV
(c)	−10 μV	−5 μV
(d)	+1.000001 V	+1.000000 V
(e)	+5 mV	0 V
(f)	0 V	+5 mV

Solution: The polarity of V_o is the same as the polarity of the (+) input with respect to the (−) input. The (+) input is more negative than the (−) input in (a), (d), and (e). This is shown by Eq. (2-1), and therefore V_o will go negative. From Eq. (2-2), the magnitude of V_o is A_{OL} times the difference, E_d, between voltages at the (+) and (−) inputs. But if $A_{OL} \times E_d$ exceeds $+V$ or $-V$, then V_o must stop at $+V_{sat}$ or $-V_{sat}$ as in (e) and (f). Calculations are summarized as follows.

	E_d *(using Eq. (2-1))*	*Polarity of (+) input with respect to (−) input*	V_o *(from Eq. (2-2))*	*Polarity of output terminal with respect to ground*
(a)	$-5\ \mu V$	−	$5\ \mu V \times 200{,}000 = -1.0$ V	−
(b)	$25\ \mu V$	+	$25\ \mu V \times 200{,}000 = 5.0$ V	+
(c)	$5\ \mu V$	+	$5\ \mu V \times 200{,}000 = 1.0$ V	+
(d)	$-1\ \mu V$	−	$1\ \mu V \times 200{,}000 = -0.2$ V	−
(e)	-5 mV	−	$-13\ V = -V_{sat}$	−
(f)	5 mV	+	$13\ V = +V_{sat}$	+

2-3 Introduction to the Ideal Comparator

2-3.1 Voltage Sensing with the (+) Input. In Fig. 2-5(a), a ground or 0-V reference is applied to the (−) input. The voltage to be sensed, E_i, is applied to the (+) input. The comparator's input circuit compares E_i to the 0 reference voltage, and the comparator's output tells whether E_i is positive or negative with respect to the 0 reference voltage.

As shown in Fig. 2-5(b), E_i is positive during time 0 to A. The (+) input will be positive with respect to the (−) input, so V_o will be driven to $+V_{sat}$. When E_i is negative, during time A to B, V_o is driven to $-V_{sat}$ because the (+) input is negative with respect to the (−) input. Thus V_o tells us when E_i is above (more positive than) or below (more negative than) the 0-V reference. Furthermore, the transition in V_o tells *when* E_i crossed the reference voltage and in *what direction.* When E_i crossed the 0-V reference going positive (as at times 0 and B), V_o went positive. When V_o goes negative (as at time A), it means that E_i just crossed the reference voltage going negative.

2-3.2 Voltage Sensing with the (−) Input. In Fig. 2-6(a), the 0 reference voltage (ground) is applied to the (+) input. E_i is applied to the (−) input. The comparator's input compares E_i to the 0 reference just as in Fig. 2-5(a). However, as shown in Fig. 2-6(b), V_o is driven to $-V_{sat}$ when E_i is above (more positive than) 0 V, during time 0 to A. During time A to B, when E_i is below (more negative than) the 0-V reference, V_o is at $+V_{sat}$.

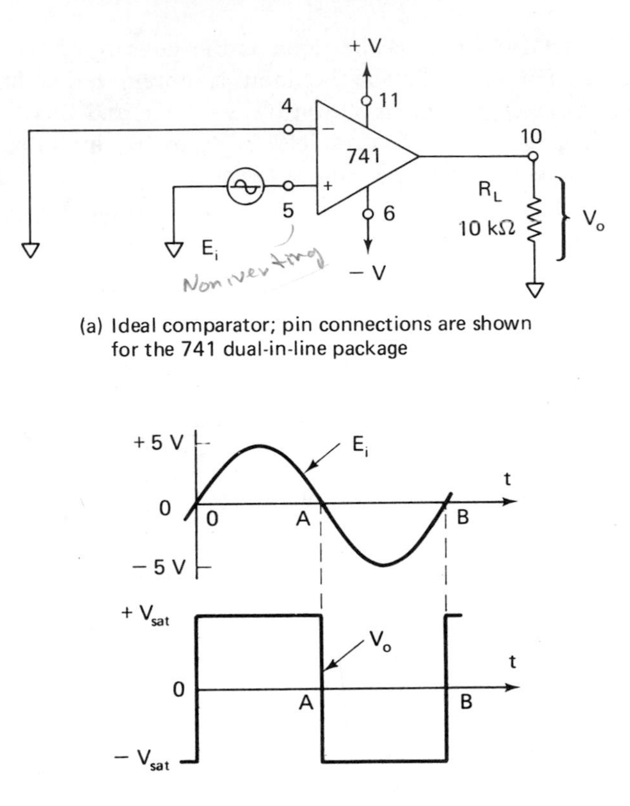

(a) Ideal comparator; pin connections are shown for the 741 dual-in-line package

(b) Input voltage E_i and output voltage V_o

Figure 2-5 A comparator sensing for 0V with the (+) input.

Summary: When $V_o = -V_{sat}$, E_i is greater than 0 V; when $V_o = +V_{sat}$, E_i is less than 0 V. The comparator circuits of Figs. 2-5 and 2-6 are called *zero crossing detectors.*

2-4 Nonzero-Level Detectors

2-4.1 Voltage-Level Detector. The comparator circuits of Figs. 2-5 and 2-6 may be expanded to detect voltages other than 0. In Fig. 2-7(a), a voltage reference of +2 V is applied to the (−) input. When input voltage E_i is less than V_{ref}, V_o is at $-V_{sat}$ because the (+) input is negative with respect to the (−) input. When E_i is greater than V_{ref}, the (+) input becomes positive with respect to the (−) input and V_o goes to $+V_{sat}$. As shown in Fig. 2-7(b), V_o tells whether E_i is greater or less than V_{ref}; when V_o changes from one saturation voltage to another, E_i equals V_{ref}.

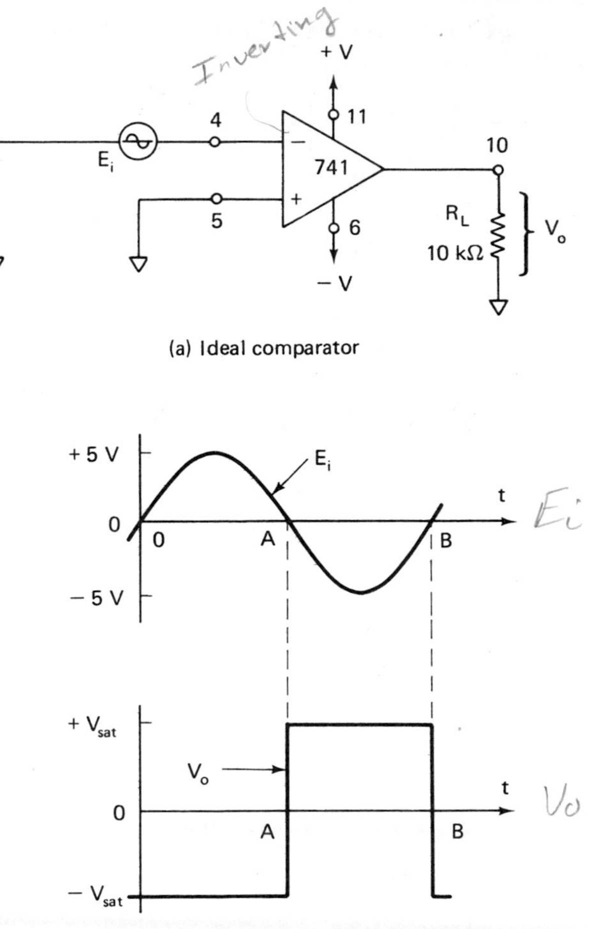

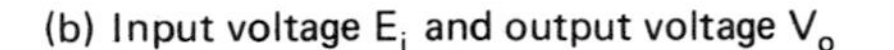
(b) Input voltage E_i and output voltage V_o

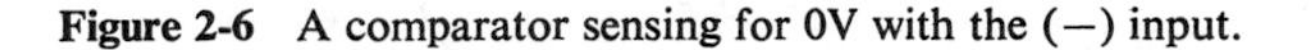
Figure 2-6 A comparator sensing for 0V with the (−) input.

Example 2-2: If the polarity of V_{ref} is reversed in Fig. 2-7(a), find the resulting shape of V_o if E_i is the same triangular wave as shown in Fig. 2-7(b).
Solution: When E_i is less than V_{ref}, the (−) input is more positive than the (+) input and $V_o = -V_{sat}$. As E_i crosses V_{ref} (−2 V) going positive, V_o switches to $+V_{sat}$ as shown in Fig. 2-8. When E_i crosses V_{ref} going negative, V_o switches to $-V_{sat}$. Thus V_o tells whether E_i is above, below, or at V_{ref}.

2-4.2 Voltage-Level Detector with Light-Emitting Diodes. A modification of the voltage-level detector of Fig. 2-7 is shown in Fig. 2-9. When E_i is above V_{ref}, V_o is positive and forward-biases the green light-emitting diode (LED). This condition could indicate an *on* or *go* situation. If E_i is

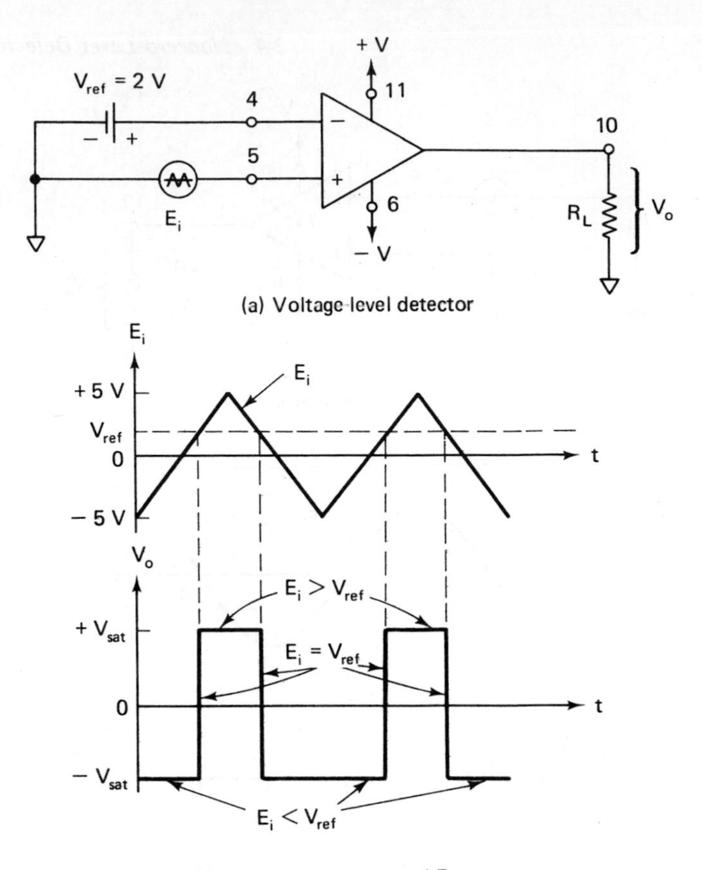

(a) Voltage-level detector

(b) Wave shapes for V_o and E_i

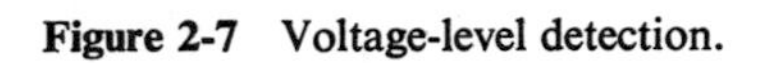

Figure 2-7 Voltage-level detection.

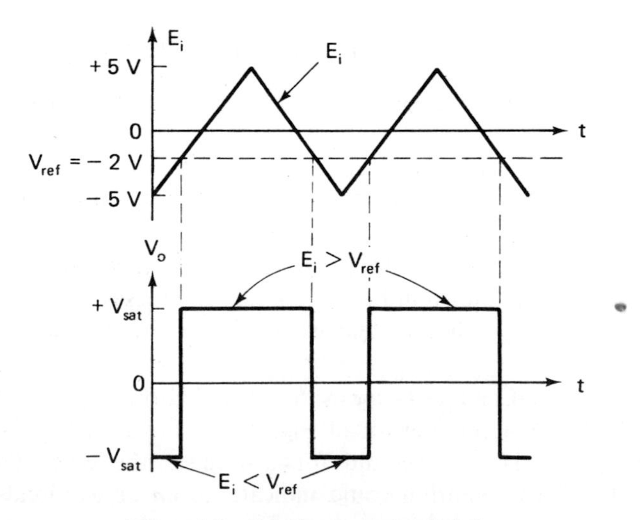

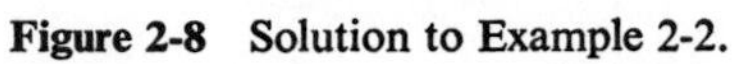

Figure 2-8 Solution to Example 2-2.

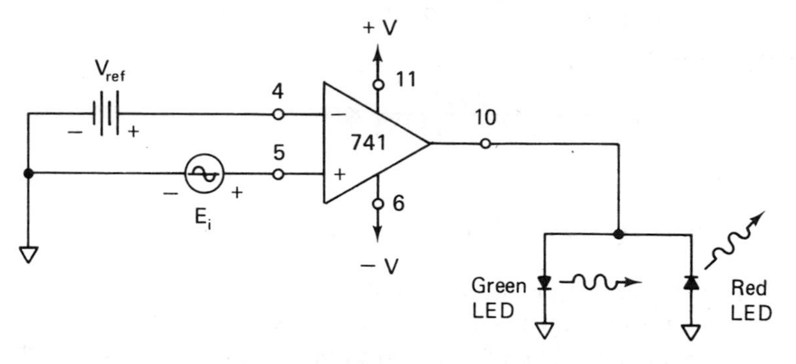

Figure 2-9 Voltage-level detector using LEDs at the output.

below V_{ref}, then V_o is negative and the red LED is lit. This could correspond to the *off* or *no-go* condition. The output saturation current of a 741 is approximately 25 mA. This value of current is sufficient to make an LED glow quite brightly. Green and red LEDs are available in a single package for such an application.

2-5 Practical Reference Voltages

Since op amps are usually powered by well regulated supply voltages, we can take advantage of this fact to make an inexpensive reference voltage. Resistors R_1 and R_2 in Fig. 2-10 divide the supply voltage in accordance with the voltage divider:

$$V_{ref} = \frac{R_2}{R_1 + R_2}(+V) \tag{2-3}$$

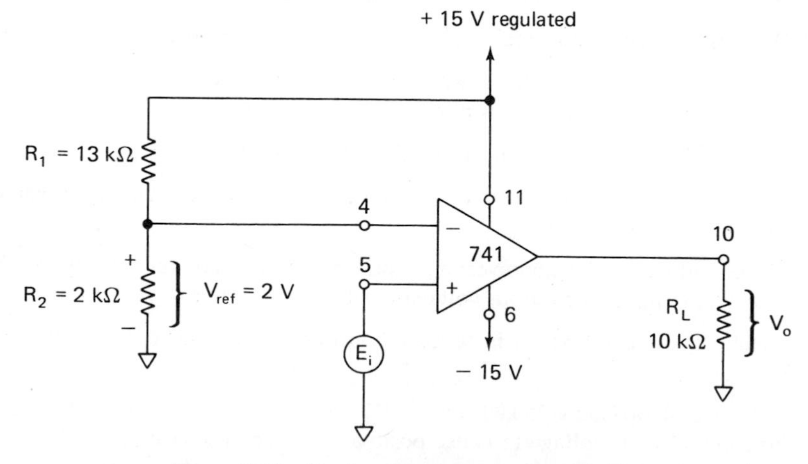

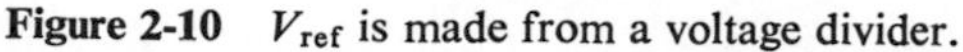
Figure 2-10 V_{ref} is made from a voltage divider.

Example 2-3: Evaluate V_{ref} in Fig. 2-10.
Solution: From Eq. (2-2),

$$V_{ref} = \frac{2\ \text{k}\Omega}{2\ \text{k}\Omega + 13\ \text{k}\Omega} \times 15\ \text{V} = 2\ \text{V}$$

2-6 Practical Considerations

When a comparator circuit is actually constructed, we learn that there are several differences between actual performance of a real op amp and the ideal situation presented thus far. First, the output voltage V_o does not change instantly from $-V_{sat}$ to $+V_{sat}$. It takes a measureable amount of time to make the change. Second, when differential input voltage E_d is 0, V_o is not precisely 0. Finally, the op amp output voltage may even oscillate unpredictably, either just as E_d approaches 0 V or continuously. This points up the need to learn about a few more of the op amp's characteristics and how to minimize differences between ideal and actual performance. More advanced applications of the comparator will be studied in Chapter 4, but first we need to learn about the effect of feedback connections from the output to the (—) input; this will be discussed in Chapter 3.

Problems

2-1 List the basic terminals of an op amp and draw the circuit schematic.

2-2 What are the three common types of packages used to house op amps?

2-3 Is the input of an op amp a differential input or a single-ended input?

2-4 Is the output of an op amp a differential output or a single-ended output?

2-5 What is the maximum supply voltage that can be applied to most op amps?

2-6 How much current can be drawn from an op amp output terminal without having it go into current saturation?

2-7 What is the short-circuit current of the 741 op amp?

2-8 If an op amp is powered from a $\pm$10-V supply, what would be typical values of $+V_{sat}$ and $-V_{sat}$?

2-9 If the voltage at the noninverting input is greater than the voltage at the inverting input, is the output voltage positive or negative?

2-10 Define A_{OL} and express it in terms of output voltage and differential input voltage.

2-11 If the open-loop gain is 250,000 and $\pm V_{sat} = \pm 10$ V, determine the maximum differential input voltage to cause positive and negative saturation.

2-12 For the op amp of Problem 2-11, if the voltage at (−) input is 3.02 mV and the voltage at (+) input is 3.015 mV, find the output voltage.

2-13 In Fig. 2-5, if the peak value of E_i is 10 V, does the output waveform change? Explain.

2-14 $V_{ref} = 1$ V in Fig. 2-7. Draw the output waveform.

2-15 If E_i and V_{ref} are swapped in Fig. 2-7, draw the output waveform.

2-16 In Fig. 2-10, if $R_1 = 24\ k\Omega$ and $R_2 = 6\ k\Omega$, calculate V_{ref}.

2-17 If R_1 in Fig. 2-10, instead of being connected to +15 V, is connected to −15 V, find the value of V_{ref}.

3

Inverting and Noninverting Amplifiers

3-0 Introduction

This chapter uses the op amp in one of its most important applications, namely making an amplifier. An *amplifier* is a circuit that receives a signal at its input and delivers an undistorted larger version of the signal at its output. All circuits in this chapter have one feature in common: An external feedback resistor is connected between the output terminal and (—) input terminal. This type of circuit is called a *negative feedback circuit.*

There are many advantages obtained with negative feedback, all based on the fact that circuit performance no longer depends on open-loop gain of the op amp, A_{OL}. By adding the feedback resistor we form a loop from output to (—) input. The resulting circuit now has a *closed-loop gain* or *amplifier* gain, A_{CL}, which is independent of A_{OL}.

As will be shown, closed-loop gain, A_{CL}, depends only on external resistors. For best results 1% resistors should be used, and A_{CL} will be known within about 1%. Note that adding external resistors does not change open-loop gain A_{OL}. A_{OL} still varies from op amp to op amp. So adding negative feedback will allow us to ignore changes in A_{OL} as long as A_{OL} is large. We begin with the inverting amplifier to show that A_{CL} depends simply on the ratio of two resistors.

3-1 The Inverting Amplifier

3-1.1 Introduction. The circuit of Fig. 3-1 is the most widely used op amp circuit. It is an amplifier whose closed-loop gain from E_i to V_o is set by

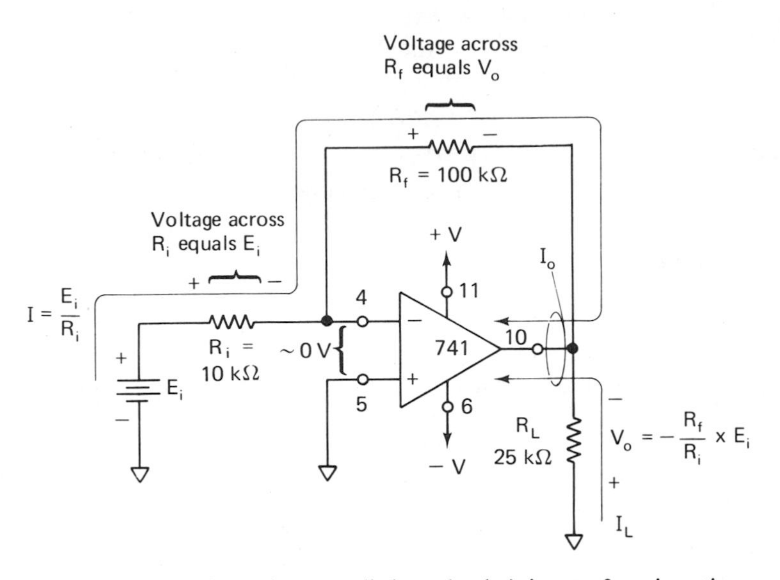

Figure 3-1 Positive voltage applied to the (−) input of an inverting amplifier. Numbers next to the small circles are pin connections for the 14-pin dual-in-line package.

R_f and R_i. It can amplify ac or dc signals. To understand how this circuit operates, we make two realistic simplifying assumptions that were introduced in Chapter 2.

1. The voltage E_d between (+) and (−) inputs is essentially 0.
2. The current drawn by either the (+) or the (−) input terminal is negligible.

3-1.2 Positive Voltage Applied to the Inverting Input. In Fig. 3-1, positive voltage E_i is applied through input resistor R_i to the op amp's (−) input. Negative feedback is provided by feedback resistor R_f. The voltage between (+) and (−) inputs is essentially equal to 0 V. Therefore, the (−) input terminal is also at 0 V, so ground potential is at the (−) input. For this reason, the (−) input is said to be at *virtual* ground.

Since one side of R_i is at E_i and the other is at 0 V, the voltage drop across R_i is E_i. The current I through R_i is found from Ohm's Law:

$$I = \frac{E_i}{R_i} \tag{3-1a}$$

R_i includes the resistance of the signal generator.

All of the input current I flows through R_f, since a negligible amount is drawn by the (−) input terminal. Note that the current through R_f is set by R_i and E_i; not by R_f, V_o, or the op amp.

The voltage drop across R_f is simply $I(R_f)$, or

$$V_{R_f} = I(R_f) = \frac{E_i}{R_i}R_f \tag{3-1b}$$

But as shown in Fig. 3-1, one side of R_f and one side of load R_L are connected. The voltage from this connection to ground is V_o. The other sides of R_f and of R_L are at ground potential. Therefore, V_o equals V_{R_f} (the voltage across R_f). To obtain the polarity of V_o, note that the left side of R_f is at ground potential. The current direction established by E_i forces the right side of R_f to go negative. Therefore, V_o is negative when E_i is positive. Now, equating V_o with V_{R_f} and adding a minus sign to signify that V_o goes negative when E_i goes positive, we have

$$V_o = -E_i\frac{R_f}{R_i} \tag{3-2a}$$

Now, introducing the definition that the closed-loop gain of the amplifier is A_{CL}, we rewrite Eq. (3-2a) as

$$A_{CL} = \frac{V_o}{E_i} = -\frac{R_f}{R_i} \tag{3-2b}$$

The minus sign in Eq. (3-2b) shows that the polarity of the output V_o is inverted with respect to E_i. For this reason, the circuit of Fig. 3-1 is called an *inverting amplifier*.

3-1.3 Load and Output Currents. The load current, I_L that flows through R_L is determined only by R_L and V_o and is furnished from the op amp's output terminal. Thus $I_L = V_o/R_L$. The current I through R_f must also be furnished by the output terminal. Therefore the op amp output current I_o is

$$I_o = I + I_L \tag{3-3}$$

The maximum value of I_o is set by the op amp; it is usually between 5 and 10 mA.

Example 3-1: For Fig. 3-1, let $R_f = 100\text{ k}\Omega$, $R_i = 10\text{ k}\Omega$, and $E_i = 1\text{ V}$. Calculate (a) I, (b) V_o, and (c) A_{CL}
Solution: (a) From Eq. (3-1a),

$$I = \frac{E_i}{R_i} = \frac{1\text{ V}}{10\text{ k}\Omega} = 0.1\text{ mA}$$

(b) From Eq. (3-2a),

$$V_o = -\frac{R_f}{R_i} \times E_i = -\frac{100\text{ k}\Omega}{10\text{ k}\Omega}(1\text{ V}) = -10\text{ V}$$

(c) Using Eq. (3-2b),

$$A_{CL} = -\frac{R_f}{R_i} = -\frac{100\ \text{k}\Omega}{10\ \text{k}\Omega} = -10$$

This answer may be checked by taking the ratio of V_o to E_i:

$$A_v = \frac{V_o}{E_i} = \frac{-10\ \text{V}}{1\ \text{V}} = -10$$

Example 3-2: Using the values given in Example 3-1 and $R_L = 25\ \text{k}\Omega$, determine (a) I_L and (b) the total current into the output pin of the op amp.
Solution: (a) Using the value of V_o calculated in Example 3-1,

$$I_L = \frac{V_o}{R_L} = \frac{10\ \text{V}}{25\ \text{k}\Omega} = 0.4\ \text{mA}$$

The direction of current is shown in Fig. 3-1.
(b) Using Eq. (3-3) and the value of I from Example 3-1,

$$I_o = I + I_L = 0.1\ \text{mA} + 0.4\ \text{mA} = 0.5\ \text{mA}$$

The input resistance seen by E_i is R_i. One of the reasons for using the op amp is its high input resistance. In order to keep input resistance of the *circuit* high, R_i should be equal to or greater than 10 kΩ.

3-1.4 Negative Voltage Applied to the Inverting Input. Fig. 3-2 shows a negative voltage, E_i, applied via R_i to the inverting input. All the principles

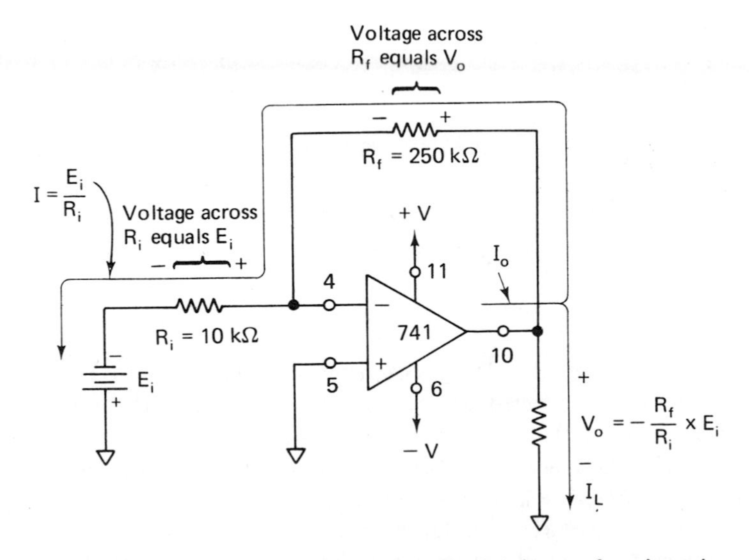

Figure 3-2 Negative voltage applied to the (−) input of an inverting amplifier.

and equations of Sections 3-1.1 to 3-1.3 still apply. The only difference between Figs. 3-1 and 3-2 is the direction of the currents. Reversing the polarity of the input voltage, E_i reverses the direction of all currents and the voltage polarities. Now the output of the amplifier will go positive when E_i goes negative.

Example 3-3: For Fig. 3-2, let $R_f = 250\ \text{k}\Omega$, $R_i = 10\ \text{k}\Omega$, and $E_i = 0.5$ V. Calculate (a) I, (b) the voltage across R_f, (c) V_o, using Eq. (3-2a).
Solution: (a) From Eq. (3-1a),

$$I = \frac{E_i}{R_i} = \frac{0.5\ \text{V}}{10\ \text{k}\Omega} = 50\ \mu\text{A} = 0.05\ \text{mA}$$

(b) From Eq. (3-1b),

$$V_{R_f} = I \times R_f$$
$$= (50\ \mu\text{A})(250\ \text{k}\Omega) = 12.5\ \text{V}$$

(c) From Eq. (3-2a),

$$V_o = -\frac{R_f}{R_i} \times E_i = -\frac{250\ \text{k}\Omega}{10\ \text{k}\Omega}(0.5\ \text{V}) = -12.5\ \text{V}$$

Thus the output voltage does equal the voltage across R_f, and $A_{CL} = -25$.

Example 3-4: Using the values in Example 3-3, determine (a) R_L for a load current of 2 mA, (b) I_o, (c) the circuit's input resistance.
Solution: (a) Using Ohm's Law and V_o from Example (3-3),

$$R_L = \frac{V_o}{I_L} = \frac{12.5\ \text{V}}{2\ \text{mA}} = 6.25\ \text{k}\Omega$$

(b) From Eq. (3-3) and Example 3-3,

$$I_o = I + I_L = 0.05\ \text{mA} + 2\ \text{mA} = 2.05\ \text{mA}$$

(c) The circuit's input resistance, or the resistance seen by E_i is $R_i = 10\ \text{k}\Omega$.

3-1.5 AC Voltage Applied to the Inverting Input. Figure 3-3 shows an ac signal voltage E_i applied to the inverting input. For the positive half-signal, the voltage polarities and the direction of currents are the same as in Fig. 3-1. For the negative half-signal voltage, the polarities and direction of currents are the same as in Fig. 3-2. The output waveform is the negative (or 180° out of phase) of the input wave as shown in Fig. 3-3. That is, when E_i is positive, V_o is negative; and vice versa. The equations developed in Section 3-1.2 are applicable to Fig. 3-3 for ac voltages.

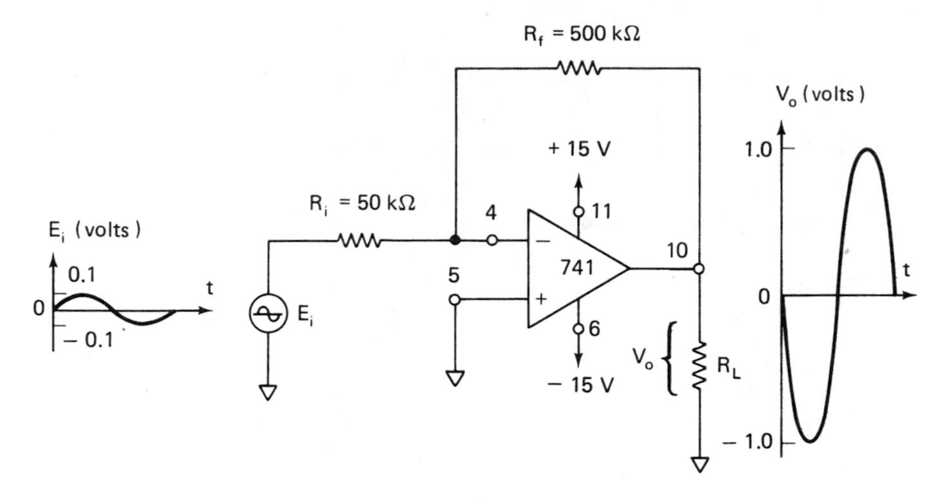

Figure 3-3 Inverting amplifier with ac input signal.

Example 3-5: For the circuit of Fig. 3-3, $R_f = 500\ \text{k}\Omega$, and $R_i = 50\ \text{k}\Omega$, calculate the voltage gain A_{CL}.
Solution: From Eq. (3-2b),

$$A_{CL} = -\frac{R_f}{R_i} = \frac{-500\ \text{k}\Omega}{50\ \text{k}\Omega} = -10$$

Example 3-6: If the peak input voltage in Example 3-5 is 0.1 V, determine the peak output voltage.
Solution: Using Eq. (3-2a),

$$\text{Peak } V_o = \frac{-R_f}{R_i} \times E_i = A_{CL}E_i = (-10)(0.1\ \text{V}) = -1.0\ \text{V}$$

3-2 Inverting Adder and Audio Mixer

3-2.1 Inverting Adder. In the circuit of Fig. 3-4, V_o equals the sum of the input voltages with polarity reversed. Mathematically,

$$V_o = -(E_1 + E_2 + E_3) \tag{3-4}$$

Circuit operation is explained by noting that the summing point S and the (—) input are at ground potential. Current I_1 is set by E_1 and R, I_2 by E_2 and R, and I_3 by E_3 and R. Expressed mathematically,

$$I_1 = \frac{E_1}{R}, \qquad I_2 = \frac{E_2}{R}, \qquad I_3 = \frac{E_3}{R} \tag{3-5}$$

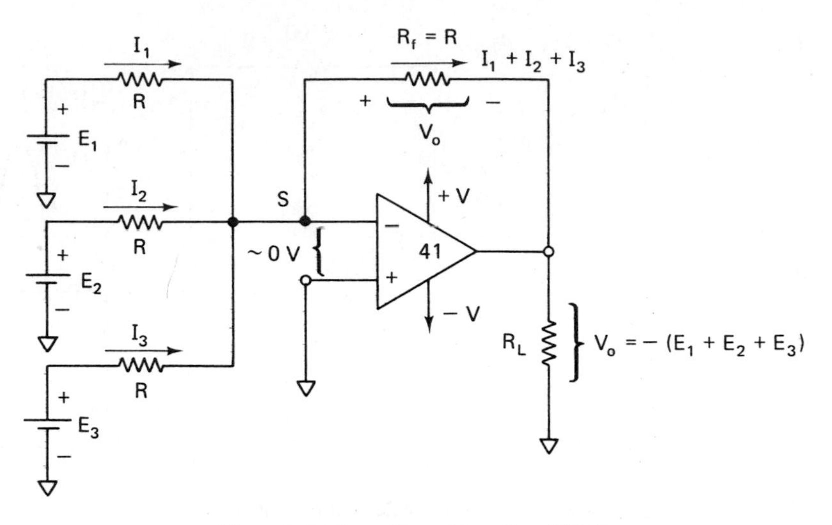

Figure 3-4 Inverting adder, $R = 10\ k\Omega$.

Since the (—) input draws negligible current, I_1, I_2, and I_3 all flow through R_f. That is, the sum of the input currents flows through R_f and sets up a voltage drop across R_f equal to V_o, or

$$V_o = -(I_1 + I_2 + I_3)R_f$$

Substituting for the currents from Eq. (3-5) and substituting R for R_f, we obtain Eq. (3-4)

$$V_o = -\left(\frac{E_1}{R} + \frac{E_2}{R} + \frac{E_3}{R}\right)R = -(E_1 + E_2 + E_3)$$

Example 3-7: In Fig. 3-4, $E_1 = 2$ V, $E_2 = 3$ V, $E_3 = 1$ V, and all resistors are 10 kΩ. Evaluate V_o.
Solution: From Eq. (3-4), $V_o = -(2\text{ V} + 3\text{ V} + 1\text{ V}) = -6$ V

Example 3-8: If the polarity of E_3 is reversed in Fig. 3-4 but values are the same as in Example 3-7, find V_o.
Solution: From Eq. (3-4), $V_o = -(2\text{ V} + 3\text{ V} - 1\text{ V}) = -4$ V

If only two input signals E_1 and E_2 are needed, simply replace E_3 with a short circuit to ground. If four signals must be added, simply add another equal resistor R between the fourth signal and summing point S. Equation (3-4) can be changed to include any number of input voltages.

3-2.2 Audio Mixer. In the adder of Fig. 3-4, all of the input currents flow through feedback resistor R_f. This means that I_1 does not affect I_2

or I_3. More generally, the input currents do not affect one another because each sees ground potential at the summing node. Therefore, the input currents—and consequently the input voltages E_1, E_2, and E_3—do *not* interact.

This feature is especially desirable in an audio mixer. For example, let E_1, E_2, and E_3 be replaced by microphones. The ac voltages from each microphone will be added or mixed at every instant. Then if one microphone is carrying guitar music, it won't come out of a second microphone facing the singer. If a 100-kΩ volume control is installed between each microphone and associated input resistor, their relative volumes can be adjusted and added. A weak singer can then be heard above a very loud guitar.

3-3 Inverting Adder with Gain

The three-input multiplying inverting adder in Fig. 3-5 is similar to the inverting amplifier except that each input voltage can be multipled by a constant voltage gain and the results added. Just as in the adder, each input current is set by its input voltage and input resistance.

$$I_1 = \frac{E_1}{R_1}, \qquad I_2 = \frac{E_2}{R_2}, \qquad I_3 = \frac{E_3}{R_3} \tag{3-6}$$

Likewise, all input currents add together in R_f to generate an output voltage equal to R_f times the current sum, or

$$V_o = -(I_1 + I_2 + I_3)R_f = -\left(E_1\frac{R_f}{R_1} + E_2\frac{R_f}{R_2} + E_3\frac{R_f}{R_3}\right) \tag{3-7}$$

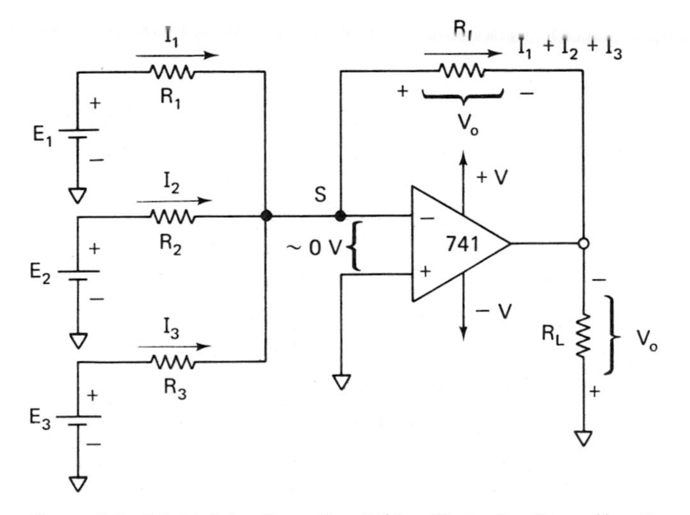

Figure 3-5 Multiplying Inverting Adder, $R_f > R_1$, R_2 and/or R_3.

Equation (3-7) shows that the gain for each input may be adjusted individually by choosing the desired ratio between R_f and each corresponding input resistor.

Example 3-9: In Fig. 3-5, $R_f = 100\text{ k}\Omega$, $R_1 = 10\text{ k}\Omega$, $R_2 = 20\text{ k}\Omega$, and $R_3 = 50\text{ k}\Omega$. Find (a) the *magnitude* of voltage gain applied to each input voltage and (b) the output voltage if $E_1 = E_2 = 0.1$ V, $E_3 = -0.1$ V.
Solution: (a) From Eq. (3-7) we can deduce the closed loop gain A_{CL} for each input. For E_1,

$$|A_{CL_1}| = \frac{R_f}{R_1} = \frac{100\text{ k}\Omega}{10\text{ k}\Omega} = 10$$

For E_2,

$$|A_{CL_2}| = \frac{R_f}{R_2} = \frac{100\text{ k}\Omega}{20\text{ k}\Omega} = 5$$

For E_3,

$$|A_{CL_3}| = \frac{R_f}{R_3} = \frac{100\text{ k}\Omega}{50\text{ k}\Omega} = 2$$

(b) From Eq. (3-7),

$$\begin{aligned} V_o &= -(0.1(10) + 0.1(5) + (-0.1)2) \\ &= -(1.0 + 0.5 - 0.2) = -1.3\text{ V} \end{aligned}$$

3-4 Inverting Averaging Amplifier

An *averaging amplifier* gives an output voltage proportional to the average of all the input voltages. If there are three input voltages, the averager should add the input voltages and divide the sum by three. The averager is the same circuit arrangement as the inverting adder in Fig. 3-4 or the inverting adder with gain in Fig. 3-5. The difference is that the input resistors are made equal to some convenient value R and the feedback resistor is made equal to R divided by the number of inputs. Let n equal the number of inputs. Then for a three-input averager, $n = 3$ and $R_f = R/3$. Proof is found by substituting into Eq. (3-7), for $R_f = R/3$ and $R_1 = R_2 = R_3 = R$ to show that

$$V_o = -\left(\frac{E_1 + E_2 + E_3}{n}\right) \tag{3-8}$$

Example 3-10: In Fig. 3-5, $R_1 = R_2 = R_3 = R = 100\text{ k}\Omega$, and $R_f = 100\text{ k}\Omega/3 = 33\text{ k}\Omega$. If $E_1 = +5$ V, $E_2 = +5$ V, and $E_3 = -1$ V, find V_o.

Solution: Since $R_f = R/3$, the amplifier is an averager, and from Eq. (3-8) with $n = 3$ we have

$$V_o = -\left(\frac{5\text{ V} + 5\text{ V} + (-1\text{ V})}{3}\right) = -\left(\frac{9\text{ V}}{3}\right) = -3\text{ V}$$

3-5 Voltage Follower

The circuit of Fig. 3-6 is called a *voltage follower*, but it is also referred to as a *source follower, unity gain amplifier, buffer amplifier*, or *isolation*

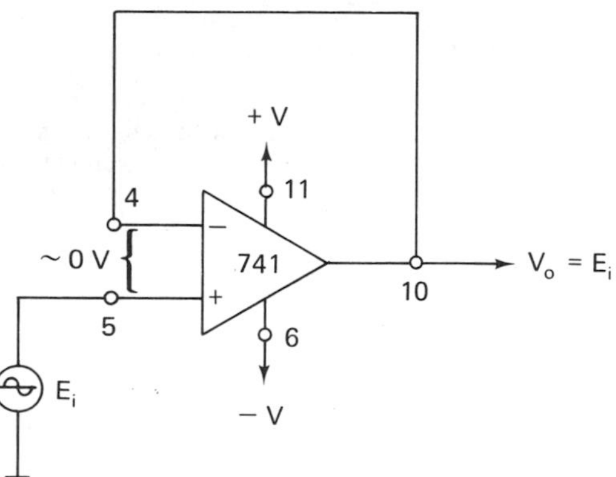

Figure 3-6 Voltage follower.

amplifier. The input voltage, E_i, is applied directly to the (+) input. Since the voltage between (+) and (−) pins of the op amp can be considered 0,

$$V_o = E_i \tag{3-9a}$$

Note that the output voltage equals the input voltage in both magnitude and sign. Therefore, as the name of the circuit implies, the output voltage *follows* the input or source voltage. The voltage gain is 1 (or unity), as shown by

$$A_{CL} = \frac{V_o}{E_i} = 1 \tag{3-9b}$$

The input impedance looking into the (+) input is very high, on the order of several megohms. Therefore, the input and output voltages are isolated or buffered from one another.

Example 3-11: For Fig. 3-7(a), determine (a) V_o, (b) I_L, and (c) I_o.
Solution: (a) From Eq. (3-9a),

$$V_o = E_i = 4\text{ V}$$

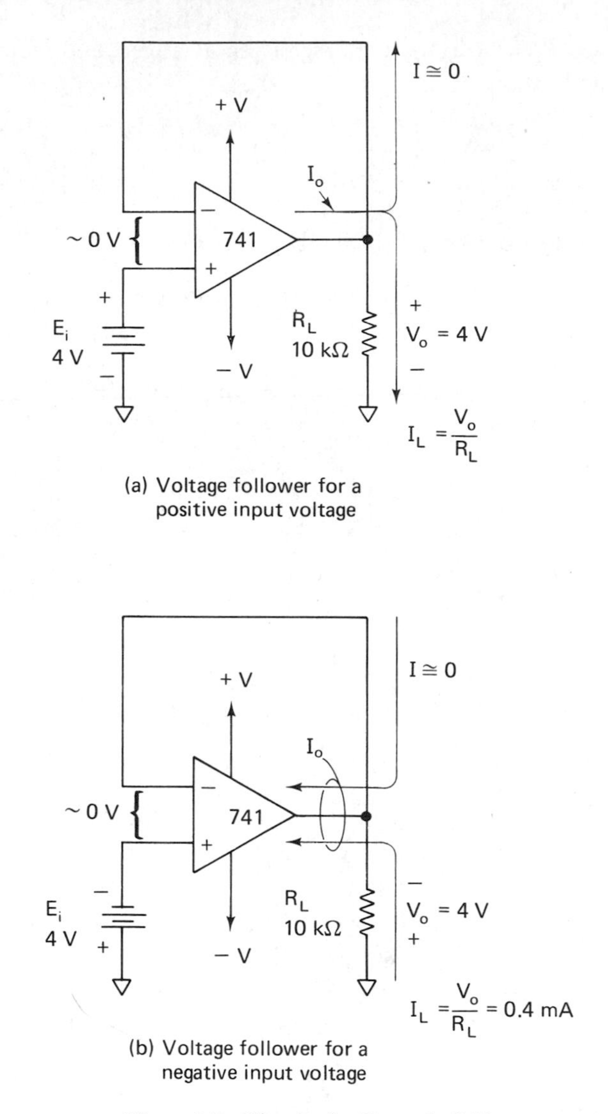

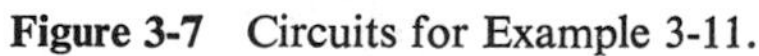

Figure 3-7 Circuits for Example 3-11.

(b) From Ohm's Law,

$$I_L = \frac{V_o}{R_L} = \frac{4\text{ V}}{10\text{ k}\Omega} = 0.4\text{ mA}$$

(c) From Eq. (3-3),

$$I_o = I + I_L$$

But $I \approx 0$, since input terminals of op amps draw negligible current; therefore,

$$I_o = 0 + 0.4 \text{ mA} = 0.4 \text{ mA}$$

If E_i were reversed, the polarity of V_o and the direction of currents would be reversed, as shown in Fig. 3-7(b).

3-6 Noninverting Amplifier

Figures 3-8(a) and (b) are noninverting amplifiers; that is, the output voltage, V_o, is of the same polarity as the input voltage, E_i. The input resistance of the inverting amplifier (Section 3-1) is R_i, but the input resistance of the noninverting amplifier is the input resistance of the op amp, which is extremely large, typically exceeding 100 MΩ. Since there is practically 0 voltage between the (+) and (−) pins of the op amp, both pins are at the same potential E_i. Therefore, E_i appears across R_1. E_i causes current I to flow as given by

$$I = \frac{E_i}{R_1} \tag{3-10a}$$

The direction of I depends on the polarity of E_i. Compare Figs. 3-8(a) and (b). The input current to the op amp's (−) terminal is negligible. Therefore, I flows through R_f and the voltage drop across R_f is represented by V_{R_f} and expressed as

$$V_{R_f} = I(R_f) = \frac{R_f}{R_1} \times E_i \tag{3-10b}$$

Equations (3-10a) and (3-10b) are similar to Eqs. (3-1a) and (3-1b).

The output voltage V_o is found by adding the voltage drop across R_1, which is E_i, to the voltage across R_f, which is V_{R_f}

$$V_o = E_i + \frac{R_f}{R_1} E_i$$

or

$$V_o = \left(1 + \frac{R_f}{R_1}\right) E_i \tag{3-11a}$$

Rearranging Eq. (3-11a) to express voltage gain, we get

$$A_{CL} = \frac{V_o}{E_i} = 1 + \frac{R_f}{R_1} \tag{3-11b}$$

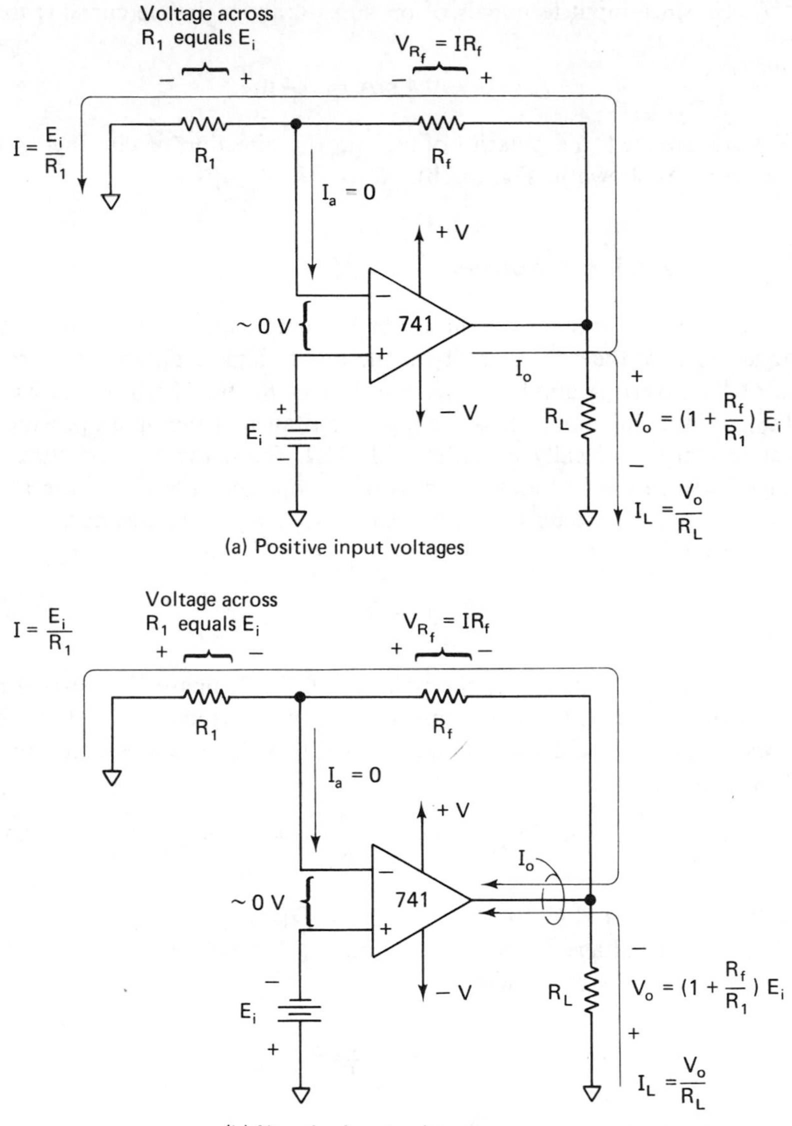

Figure 3-8 Voltage polarities and direction of currents for noninverting amplifiers.

Equation (3-11b) shows that the voltage gain of a noninverting amplifier equals the *magnitude* of the gain of an inverting amplifier (R_f/R_1) plus 1.

The load current I_L is given by V_o/R_L and therefore depends only on V_o and R_L. I_o, the current drawn from the output pin of the op amp, is given by Eq. (3-3).

Example 3-12: For the circuit of Fig. 3-8(a), let $R_1 = 5\ \text{k}\Omega$, $R_f = 20\ \text{k}\Omega$, and $E_i = 2$ V. Calculate (a) V_o and (b) A_{CL}.
Solution: (a) From Eq. (3-11a),

$$V_o = \left(1 + \frac{20\ \text{k}\Omega}{5\ \text{k}\Omega}\right)(2\ \text{V}) = 10\ \text{V}$$

(b) Using Eq. (3-11b),

$$A_{CL} = \frac{V_o}{E_i} = \frac{10\ \text{V}}{2\ \text{V}} = 5$$

or

$$A_{CL} = 1 + \frac{R_F}{R_1} = 1 + \frac{20\ \text{k}\Omega}{5\ \text{k}\Omega} = 1 + 4 = 5$$

Example 3-13: Using the circuit values of Example 3-12 and $R_L = 5\ \text{k}\Omega$, calculate (a) the load current, I_L, and (b) the output op amp current, I_o.
Solution: (a) Since $V_o = 10$ V in Example 3-12,

$$I_L = \frac{V_o}{R_L} = \frac{10\ \text{V}}{5\ \text{k}\Omega} = 2\ \text{mA}$$

(b) Applying Eq. (3-3) and the value of $I = 2\ \text{V}/5\ \text{k}\Omega = 0.4$ mA,

$$I_o = I + I_L = 0.4\ \text{mA} + 2\ \text{mA} = 2.4\ \text{mA}$$

Example 3-14: The circuit of Fig. 3-8(b) is to be designed for $A_{CL} = 16$, $E_i = 0.2$ V, and $R_1 = 2\ \text{k}\Omega$; calculate R_f.
Solution: (a) Rearranging Eq. (3-11b), we have

$$\frac{R_f}{R_1} = A_{CL} - 1 = 16 - 1 = 15$$

Then

$$R_f = 15R_1 = 15(2\ \text{k}\Omega) = 30\ \text{k}\Omega$$

3-7 Noninverting Adder

3-7.1 Two-Input Noninverting Adder. A two-input noninverting adder is shown in Fig. 3-9(a). All resistors are equal. To find the voltage, E_i, applied to the (+) input, refer to Fig. 3-9(b). The difference between E_1 and E_2 divides equally between the input resistors R, so $E_i = (E_1 + E_2)/2$. The simplified (Thévenin) equivalent input circuit is shown in Fig. 3-9(c) as a noninverting amplifier with a gain of 2. (See Section 3-6.) The amplifier multiplies $(E_1 + E_2)/2$ by 2 to give an output V_o of

$$V_o = E_1 + E_2 \tag{3-12}$$

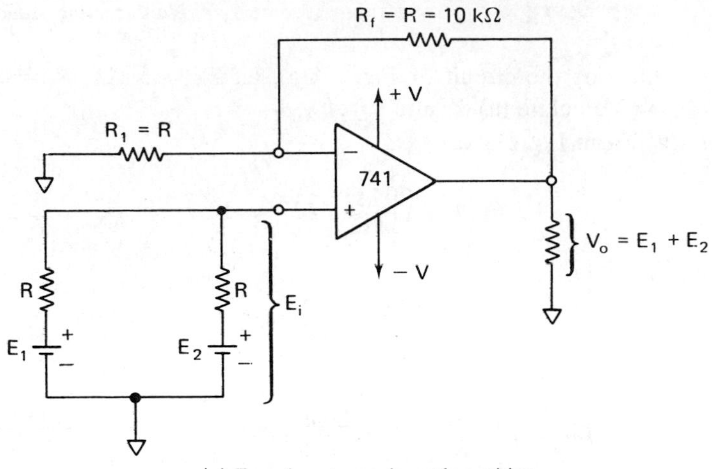

(a) Two-input non-inverting odder.

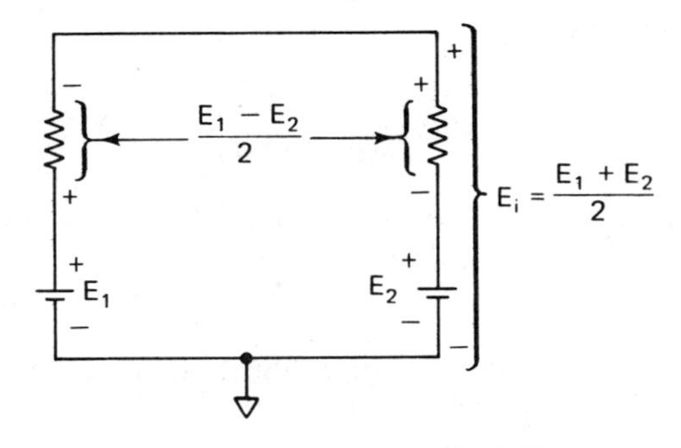

(b) Circuit to calculate E_i at (+) input, $E_1 > E_2$

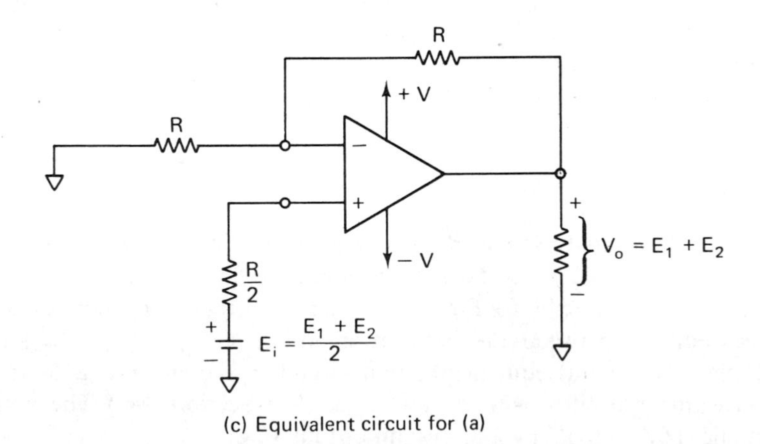

(c) Equivalent circuit for (a)

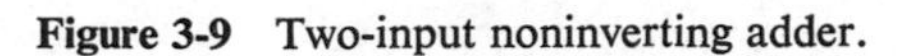

Figure 3-9 Two-input noninverting adder.

3-7.2 *N*-input Noninverting Adder. If more than two input signals are to be added, we make all resistors equal except the feedback resistor R_f. For the three-input noninverting adder in Fig. 3-10, R_f is made equal to

$$R_f = (n - 1)R \tag{3-13}$$

where n = number of inputs. Now E_i is the sum of the input voltages divided by the number of inputs (the average input voltage). The gain of the amplifier is then set to equal the number of inputs. Therefore, V_o simply adds the input voltages.

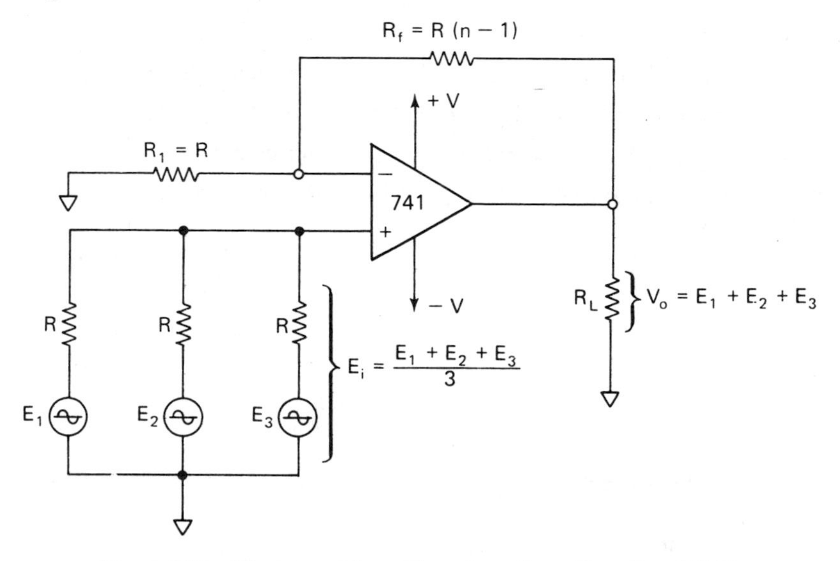

Figure 3-10 Three-input ($n = 3$) noninverting adder; $R = 10\ \text{k}\Omega$.

Example 3-15: In Fig. 3-10, $E_1 = E_2 = 2$ V and $E_3 = -1$ V. If $R_1 = R = 10\ \text{k}\Omega$, find (a) n, (b) R_f and (c) V_o.
Solution: (a) $n = 3$; (b) by Eq. (3-13), $R_f = (3 - 1)10\ \text{k}\Omega = 20\ \text{k}\Omega$; (c) $V_o = E_1 + E_2 + E_3 = 2\text{ V} + 2\text{ V} - 1\text{ V} = 3\text{ V}$.

3-8 Difference Between Measured and Calculated Values

If the circuits of this chapter are tested, there may be differences between measured and calculated values of output voltage. These differences will be due to the fact that unavoidable op amp limitations have not been accounted for. Mention of such limitations in this chapter would divert attention from and needlessly complicate understanding of circuit operation. These limitations will, however, be covered thoroughly in Chapters 10 and 11.

Problems

3-1 What type of feedback is applied to an op amp when an external component is connected between the output terminal and the inverting input?

3-2 If the open-loop gain is very large, does the closed-loop gain depend on the external components or the op amp?

3-3 What are the two assumptions that have been used to analyze the circuits of this chapter?

3-4 Repeat Example 3-1 with $R_i = 50\ k\Omega$.

3-5 In Fig. 3-1, if $R_L = 10\ k\Omega$, determine (a) I_L and (b) I_o. $E_i = 1$ V.

3-6 If $R_i = 50\ k\Omega$ in Fig. 3-1, what is the input resistance as seen by E_i?

3-7 In Example 3-3, if $E_i = 0.05$ V, calculate (a) I, (b) V_{R_f}, (c) V_o.

3-8 In Fig. 3-2, if $E_i = 0.4$ V and $R_L = 5\ k\Omega$, determine (a) I_L and (b) I_o.

3-9 If $R_f = 100\ k\Omega$ and $R_i = 20\ k\Omega$, will the closed-loop gain be the same for both dc and ac input signals?

3-10 For Fig. 3-3, if $\pm V_{sat} = \pm 15$ V, what peak-to-peak value of E_i begins to cause the output to saturate?

3-11 Calculate V_o for Fig. 3-4 if (a) $E_1 = -2$ V, $E_2 = -1$ V, and $E_3 = +0.5$ V, (b) $E_1 = 2$ V, $E_2 = -3$ V, and $E_3 = 1$ V. All resistors equal 20 kΩ.

3-12 Using the values in Example 3-7, determine (a) the load current and (b) the current into the output terminal of the op amp. $R_L = 10\ k\Omega$.

3-13 Repeat Example 3-9 for $R_f = 50\ k\Omega$.

3-14 In Fig. 3-5, if $E_3 = 0$ but E_1 and E_2 are not 0, why doesn't $I_1 + I_2$ flow through R_3 to ground instead of through R_f.

3-15 If $E_1 = -8$ V, $E_2 = +2$ V, and $E_3 = 0$ in Example 3-10, find V_o.

3-16 In Fig. 3-7(a), if $E_i = +10$ V, calculate (a) V_o, (b) I_L (c) I_o.

3-17 Give three other names for the voltage-follower circuit of Fig. 3-6.

3-18 Let $R_1 = 25\ k\Omega$, $R_f = 100\ k\Omega$, and $E_i = 2.5$ V in Fig. 3-8. Determine (a) A_{CL} and (b) V_o.

3-19 For the values given in Problem 3-18, determine (a) I_L and (b) I_o. $R_L = 20\ k\Omega$.

3-20 Design the circuit of Fig. 3-8(b) for a closed-loop gain of 10. $R_1 = 22\ k\Omega$.

3-21 For the values given in Problem 3-20, calculate V_{R_f} if $E_i = 440$ mV.

3-22 In Fig. 3-9(a), let all resistors equal 10 kΩ, $E_1 = 5$ V, and $E_2 = -3$ V. Determine (a) E_i and (b) V_o.

3-23 Repeat Problem 3-22 with all resistors equal to 10 kΩ except for $R_f = 50\ k\Omega$.

3-24 In Fig. 3-10, if a fourth input E_4 is added, determine R_f. Let $R = 20\ k\Omega$.

3-25 Refer to Example 3-15. Determine (a) the load current if $R_L = 6\ k\Omega$ and (b) the current through R_f.

4

Comparators

4-0 Introduction

A comparator compares a signal voltage on one input with a reference voltage on the other input. Chapter 2 introduced this idea and showed that the reference voltage may be positive, negative, or 0. If the comparator is a general-purpose op amp, the output will be a positive or negative saturation voltage depending on which input voltage is higher. Comparators are used in circuits such as the following:

1. *Schmitt trigger* or squaring circuit, a circuit that converts an irregular-shaped wave to a square wave or pulse
2. 0-*crossing detector*, a circuit that indicates when and in what direction an input signal crosses 0 V
3. *Voltage-level detector*, a circuit that indicates when the input voltage reaches a given reference voltage
4. *Oscillator*, a circuit that generates triangular or square waves

An introduction to a basic comparator was given in Sections 2-4 and 2-5 to show an immediate application for an op amp. The versatility and reliability of the basic comparator can be extended by adding a few external components to a general-purpose op amp. To obtain far better performance, we will look at two types of integrated circuits designed specifically as comparators.

A resistor network will be added to the basic op amp comparator so that there will be a connection between the output terminal and the (+) input.

This is *positive feedback*. Ordinarily, positive feedback leads to instability in an amplifier. However, under suitable conditions, positive feedback can be controlled to yield better performance in the basic comparator as a square-wave generator (multivibrator), a single-pulse (one-shot) generator, or a voltage-level detector. This chapter will begin with improving the op amp comparator. Chapter 6 will discuss multivibrators and one-shot generators.

4-1 Need for Improving the Op Amp Comparator

Figure 4-1 shows a triangular voltage E_i applied to the (−) input. The (+) input is connected to ground. The circuit is the same as the 0-V sensing circuit of Fig. 2-6(a), but now a triangular input signal is applied. The ideal analysis is the same: when E_i is less than 0, V_o is at $+V_{sat}$, and when E_i is greater than 0, V_o is at $-V_{sat}$. As E_i crosses the 0 reference going positive, V_o switches from $+V_{sat}$ to $-V_{sat}$. V_o switches from $-V_{sat}$ to $+V_{sat}$ when E_i crosses 0 going negative, as shown in Fig. 4-1.

In some practical applications, E_i may approach the 0 reference very slowly and actually hover close to 0. Then V_o either would not switch quickly from one saturation voltage to the other or would oscillate between $+V_{sat}$ and $-V_{sat}$. This oscillation is quite likely because of the inevitable presence

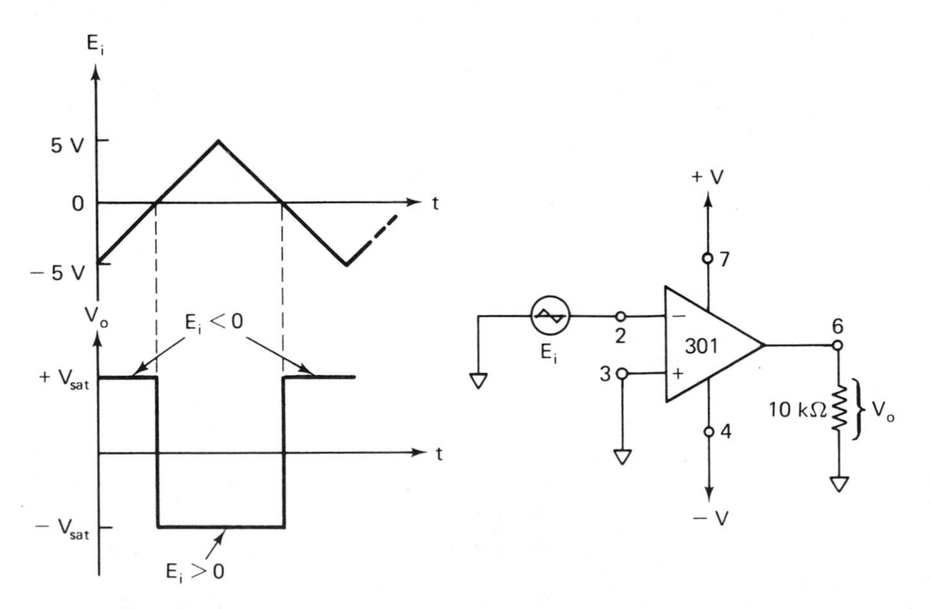

Figure 4-1 Zero crossing detector with a triangular input wave and corresponding output wave.

of noise on wires leading to the op amp's input terminals. For simplicity, noise voltage is shown in Fig. 4-2(a) as a sine wave in series with the signal voltage E_i. Figure 4-2(b) shows the noise voltage and the signal voltage added together to give the total instantaneous voltage at the (—) input of the op amp. Note in Fig. 4-2(b) how the noise voltage causes this total input voltage to cross the 0 reference at several points. For each 0 crossing V_o changes as shown in Fig. 4-2(c). Therefore, V_o is now detecting 0 reference crossings

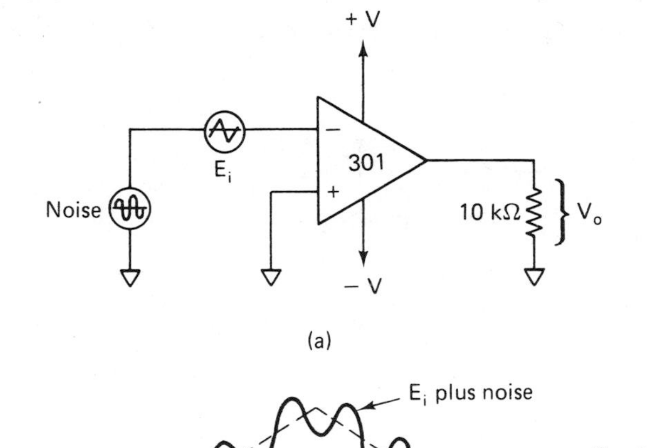

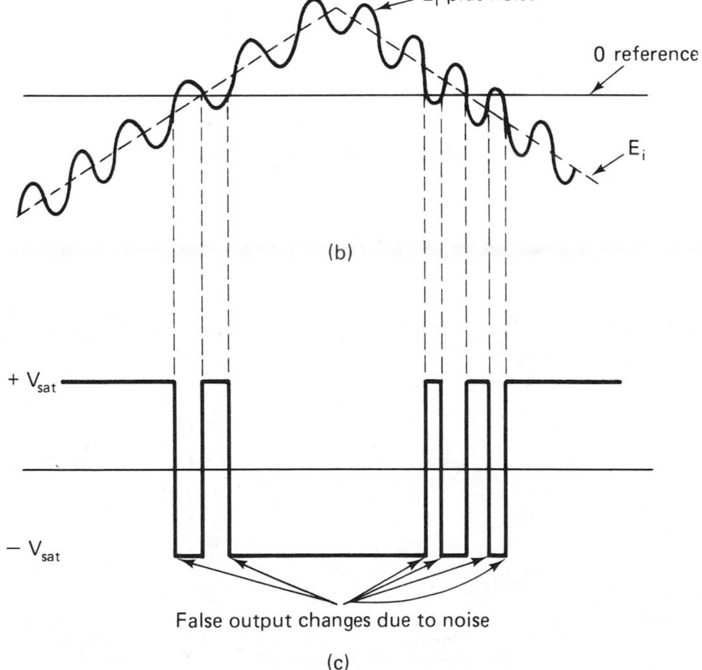

Figure 4-2 Effects of noise in a 0-crossing detector. (a) Circuit diagram; (b) total input voltage; (c) output voltage wave form.

for both noise voltages and E_i. Although the noise voltage can probably not be eliminated, we must prevent the output voltage from sensing these false crossings. Their elimination is accomplished by positive feedback.

4-2 Positive Feedback

4-2.1 Introduction. Positive feedback is accomplished by taking a fraction of the output voltage V_o and applying it to the (+) input. In Fig. 4-3(a), output voltage V_o divides between R_1 and R_2. A fraction of V_o is

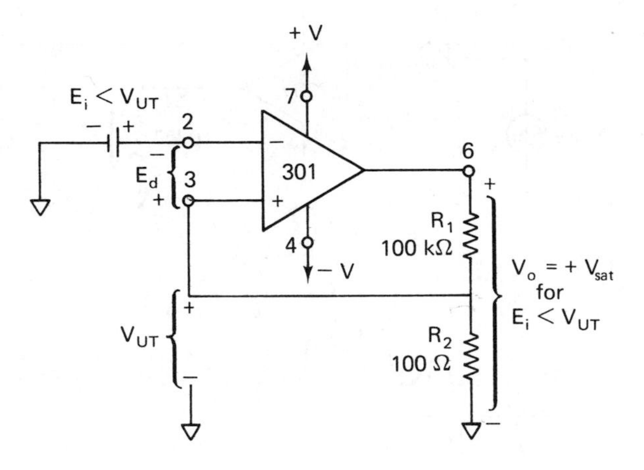

(a) Upper-threshold voltage, V_{UT}

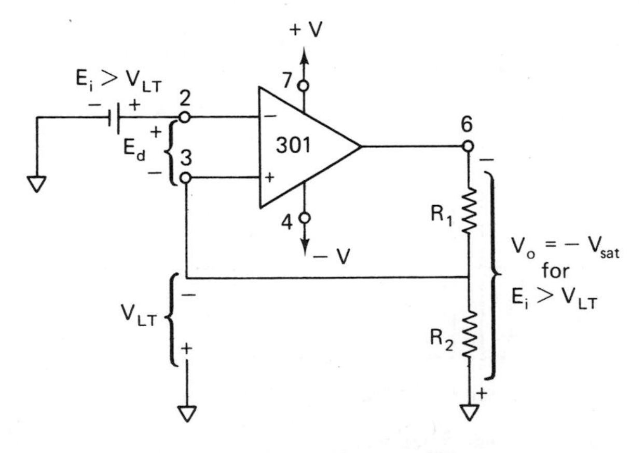

(b) Lower-threshold voltage, V_{LT}

Figure 4-3 R_1 and R_2 feed back a positive voltage from output to (+) input terminal.

fed back to the (+) input and creates a variable reference voltage. The idea of a reference voltage was introduced in Chapter 2, where it was always a fixed voltage. Now, this feedback voltage is variable because it depends on V_o. We will now study positive feedback and how it can be used to eliminate false output changes due to noise.

4-2.2 Upper-Threshold Voltage. In Fig. 4-3(a), output voltage V_o divides between R_1 and R_2. A fraction of V_o is fed back to the (+) input and is called *upper-threshold voltage*, V_{UT}. V_{UT} is expressed from the voltage divider as

$$V_{UT} = \frac{R_2}{R_1 + R_2}(+V_{sat}) \tag{4-1}$$

For E_i values less than V_{UT}, the voltage at the (+) input is greater than the voltage at the (−) input. Therefore, V_o is locked at $+V_{sat}$.

If E_i is made slightly more positive than V_{UT}, the polarity of E_d, as shown, reverses and V_o begins to drop in value. Now the fraction of V_o fed back to the positive input is smaller, so E_d becomes larger. V_o then drops even faster and is driven quickly to $-V_{sat}$. The circuit is then stable at the condition shown in Fig. 4-3(b).

4-2.3 Lower-Threshold Voltage. When V_o is at $-V_{sat}$, the voltage fed back to the (+) input is called *lower-threshold voltage*, V_{LT}, and is given by

$$V_{LT} = \frac{R_2}{R_1 + R_2}(-V_{sat}) \tag{4-2}$$

Note that V_{LT} is negative with respect to ground. Therefore, V_o will stay at $-V_{sat}$ as long as E_i is above, or positive with respect to, V_{LT}. V_o will switch back to $+V_{sat}$ if E_i goes more negative than, or below, V_{LT}.

We conclude that positive feedback induces a snap action to switch V_o faster from one limit to the other. Once V_o begins to change, it causes a regenerative action that makes V_o change even faster. If the threshold voltages are larger than the peak noise voltages, then positive feedback will eliminate false output transitions. This principle is investigated in the following examples.

Example 4-1: If $+V_{sat} = 14$ V in Fig. 4-3(a), find V_{UT}.
Solution: By Eq. (4-1),

$$V_{UT} = \frac{100\ \Omega}{100{,}100\ \Omega}(14\ \text{V}) \approx 14\ \text{mV}$$

Example 4-2: If $-V_{sat} = -13$ V in Fig. 4-3(b), find V_{LT}.

Solution: By Eq. (4-2),

$$V_{\text{LT}} = \frac{100\ \Omega}{100{,}100\ \Omega}(-13\ \text{V}) \approx -13\ \text{mV}$$

Example 4-3: In Fig. 4-4, E_i is a triangular wave applied to the (—) input in Figure 4-3(a). Find the resultant output voltage.

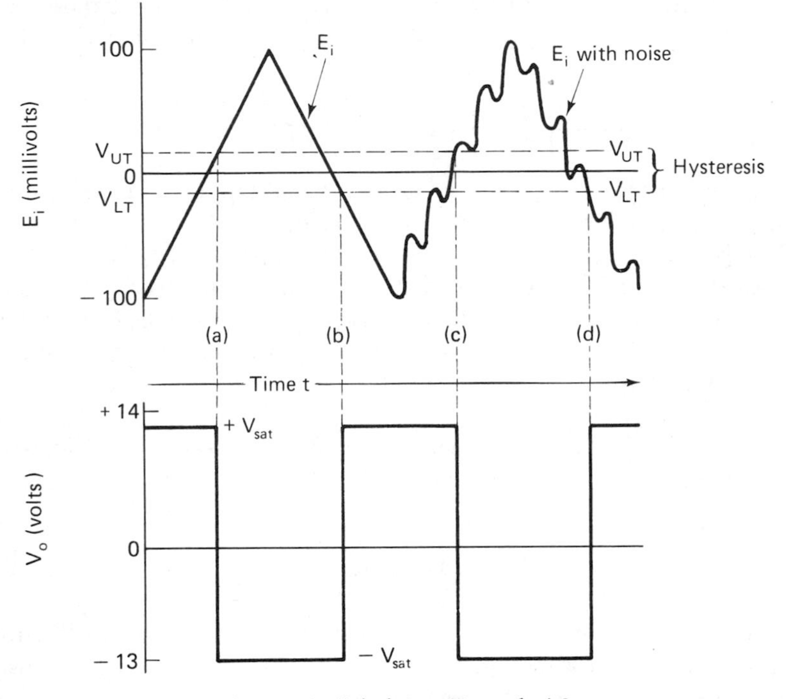

Figure 4-4 Solution to Example 4-3.

Solution: The dashed lines drawn on E_i in Fig. 4-4, locate V_{UT} and V_{LT}. At time $t = 0$, E_i is less than V_{LT}, so V_o is at $+V_{\text{sat}}$ (as in Fig. 4-4). When E_i becomes greater than V_{UT}, at times (a) and (c), V_o switches quickly to $-V_{\text{sat}}$. When E_i becomes less than V_{LT}, at times (b) and (d), V_o switches quickly to $+V_{\text{sat}}$. Observe how positive feedback has eliminated the false crossings.

4-3 Hysteresis

There is a standard technique of showing comparator performance on one graph instead of the two graphs, as in Fig. 4-4. By plotting E_i on the horizontal axis and V_o on the vertical axis, we obtain the output–input

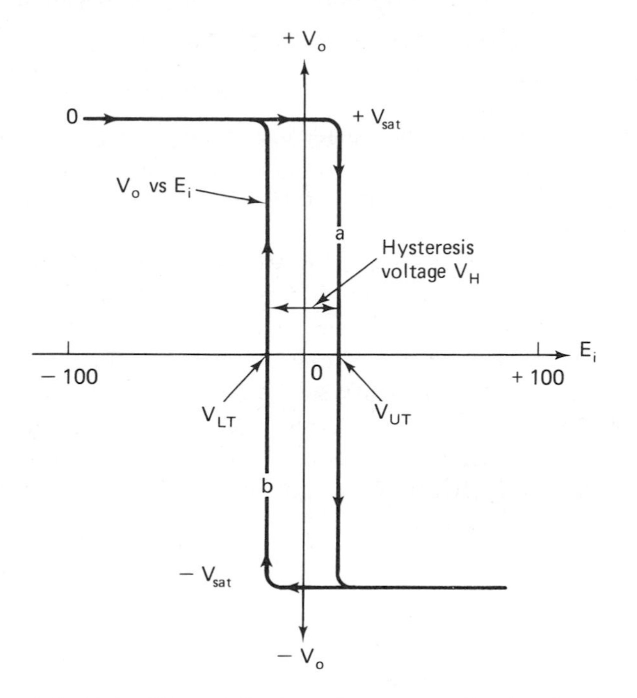

Figure 4-5 Plot of V_o vs E_i illustrates the amount of hysteresis voltage in a comparator circuit.

voltage characteristic, as in Fig. 4-5. For E_i less than V_{UT}, $V_o = +V_{sat}$. The vertical line (a) shows V_o going from $+V_{sat}$ to $-V_{sat}$ as E_i becomes greater than V_{UT}. Vertical line (b) shows V_o changing from $-V_{sat}$ to $+V_{sat}$ when E_i becomes less than V_{LT}. The difference in voltage between V_{UT} and V_{LT} is called the *hysteresis voltage*, V_H.

Whenever any circuit changes from one state to a second state at some input signal and then reverts from the second to the first state at a *different* input signal, the circuit is said to exhibit *hysteresis*. For the positive-feedback comparator, the difference in input signals is

$$V_H = V_{UT} - V_{LT} \tag{4-3}$$

For Examples 4-1 and 4-2, the hysteresis voltage is 14 mV − (−13 mV) = 27 mV.

Since the comparator converts a sine wave input or a triangular wave input to square wave output, it is called a *squaring circuit*. A circuit that triggers an output voltage change when E_i crosses a threshold voltage is called a *Schmitt trigger* circuit.

4-4 Limitations of the 741 and 301 Op Amps as Comparators

Even though positive feedback made the op amp's output terminal switch faster from one saturation voltage to another, the transition time is relatively slow. The time required to change from $+V_{sat}$ to $-V_{sat}$ or vice versa is typically several microseconds.

Moreover, the output voltage levels are fixed at either $+V_{sat}$ or $-V_{sat}$ and depend on supply voltage. They may not be compatible with voltage levels required by a particular load. For example, transistor–transistor logic (TTL) requires input voltages that are either approximately +5 V or 0 V. Clearly, the +14 V and −13 V saturation voltages exceed these limits. The process of imposing limits on these output voltages is called *bounding*. The principles will be shown in Section 4-5. For faster switching time, we must use op amps designed specifically for fast comparator applications. Two such comparators will be studied in Sections 4-6 and 4-7.

4-5 Bounding or Limiting Output Voltage

4-5.1 Single-Zener Bounding. In Fig. 4-6, a zener diode is connected from output to (−) input. E_i drives the (−) input negative, forcing V_o to go positive. When V_o reaches the zener's breakdown voltage, V_Z, the zener

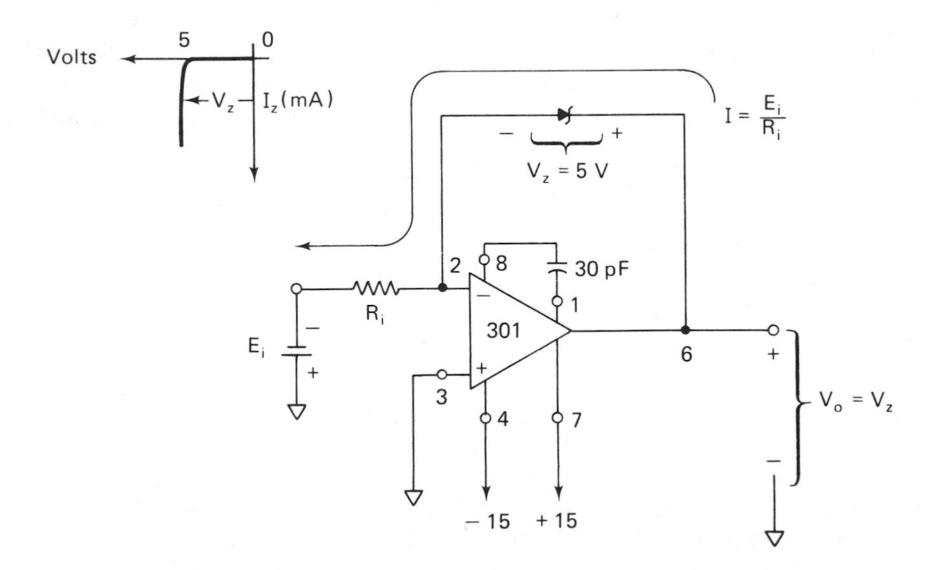

Figure 4-6 Output V_o is bounded by V_Z to +5 V instead of $+V_{sat}$.

conducts. The zener terminal voltage then stays constant at the rated value of V_Z. In Fig. 4-6, the zener voltage $V_Z = 5$ V (1N5231). The differential input voltage E_d is approximately 0 V so pin 2 is at ground potential. Thus E_i and R_i determine the current through R_i and the zener (since negligible current is drawn by the (—) input). Pin 2 puts one end of the zener at ground potential; the other end of the zener holds output terminal pin 6 at $V_Z = 5$ V.

If E_i is varied, the current through the zener will change. However, as shown in Fig. 4-6, a zener diode holds its terminal voltage fairly constant no matter what current it conducts. Therefore, V_Z, and consequently V_o, will be independent of E_i. We conclude that for all negative values of E_i, between a few millivolts and negative supply voltage, V_o will be *limited* or *bounded* at $V_Z = 5$ *V* and not at $+V_{sat}$.

The zener voltage must be less than the power supply voltage. The circuit of Fig. 4-6 makes an excellent voltage reference. Any load connected to pin 6 draws current from the op amp and not the zener, and the load will see a constant voltage.

4-5.2 Negative-Output Bounding. If E_i in Fig. 4-6 is reversed in polarity, the current direction through R_i is reversed, as in Fig. 4-7. The zener diode is forward-biased and acts as any other forward-biased silicon diode does. That is, the diode voltage stays at about 0.6 V no matter what current it conducts. The diode current is set by E_i/R_i. The (—) input holds one side of the diode at ground potential, and the other side of the diode limits V_o to -0.6 V instead of $-V_{sat}$.

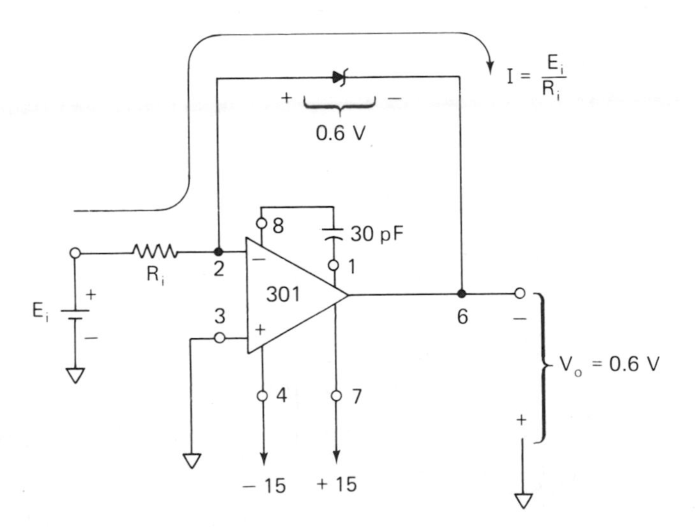

Figure 4-7 Output V_o is bounded to one diode drop of 0.6 V instead of $-V_{sat}$.

Example 4-4: Given $R_i = 1\ k\Omega$ and $E_i = 2$ V in both Figs. 4-6 and 4-7. The zener voltage is 4.5 V. Find V_o and the zener current in both (a) Fig. 4-6 and (b) Fig. 4-7.
Solution: For both figures, the zener current is $I = E_i/R_i = 2\ V/1\ k\Omega = 2$ mA. (a) For Fig. 4-6, $V_o = V_Z = 4.5$ V. (b) For Fig. 4-7, $V_o = -0.6$ V.

4-5.3 Symmetrical-Output Voltage Bounding. In Fig. 4-8(a) Z_1 and Z_2 are identical zener diodes. When E_i goes positive, V_o is forced to go nega-

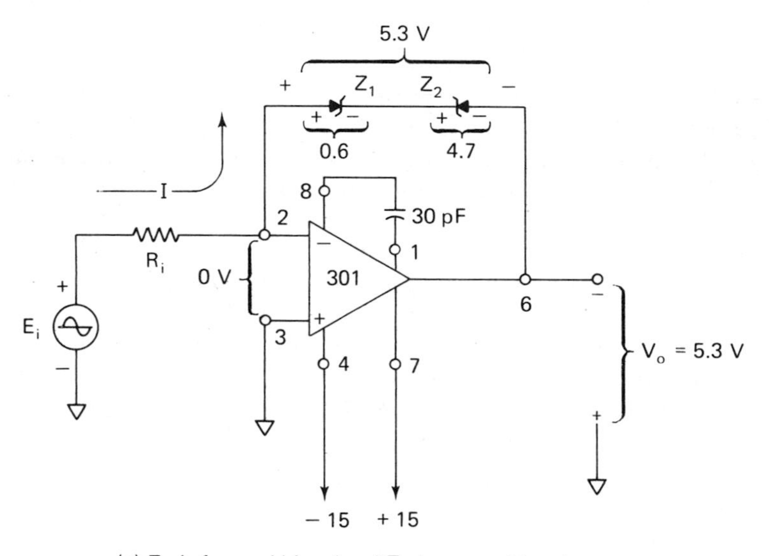

(a) Z_1 is forward-biased, and Z_2 is reverse-biased

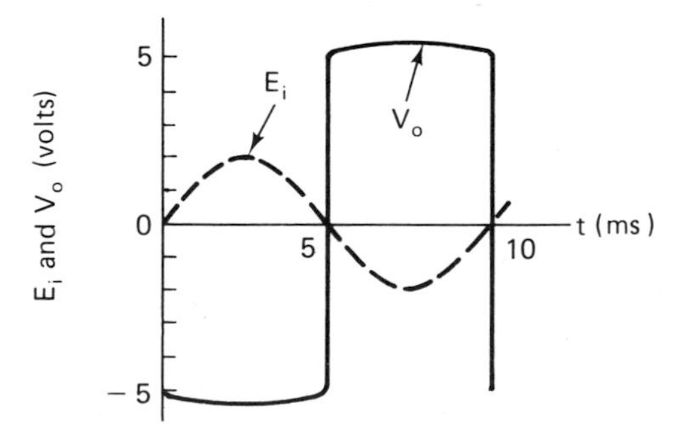

(b) Output and input voltage wave shapes

Figure 4-8 Comparator output voltage symmetrically bounded to 5.3 V.

tive. When V_o gets to -5.3 V, Z_1 is forward-biased and Z_2 is reverse-biased into zener breakdown. Since pin 2 is held at 0 V, the total voltage drop across both zeners bounds V_o at -5.3 V instead of $-V_{sat}$. This condition is shown in Fig. 4-8(b) to occur between 0 and 5 ms. When E_i goes negative (time 5 to 10 ms), V_o is bounded at $+5.3$ V. Z_1 becomes reverse-biased at 4.7 V, and Z_2 becomes forward-biased at 0.6 V. Thus V_o is symmetrically bounded at -5.3 V and $+5.3$ V for positive and negative inputs, respectively.

4-6 The 710 IC Comparator

4-6.1 Electrical Characteristics. The 710 is a popular fast comparator. Pin connections for the metal can T0 99 case are shown in Fig. 4-9(a). Key electrical specifications are as follows:

1. *Supply voltages:* +12 V on pin 8, −6 V on pin 4, and ground on pin 1. Note that an asymmetrical supply is needed. This may be a disadvantage in an overall design.
2. *Output voltage V_o:* V_o has an upper bound at +3.2 V and a lower bound at −0.5 V, so the output can drive digital integrated logic circuits such as transistor–transistor logic (TTL).
3. *Differential inputs:* If (+) input pin 2 is positive with respect to (−) input pin 3, V_o will go positive to +3.2 V. Reversing the differential input voltage will drive V_o negative to −0.5 V. Voltage between pins 2 and 3 should never exceed 5 V of either polarity. Voltage between pin 2 and ground or between pin 3 and ground should not exceed ±7 V.
4. *Response time:* Remove all resistors in Fig. 4-9(a), ground pin 2, and apply +100 mV to pin 3. V_o will go to −0.5 V. Then apply −120 mV to pin 3. The time interval for V_o to rise to +3.2 V is defined as response time and is less than 40 ns. To understand how fast the 710 responds, we contrast it with a 741 general-purpose op amp that would require about 8000 ns to do the same job.
5. *Input bias currents:* Input bias currents are large, typically 16 μA, to allow fast response time. Therefore, input or source resistance should be small, less than 1 kΩ if possible. Input bias currents will be covered in Chapter 10.

4-6.2 Voltage-Level Detection with the 710 Comparator. Figure 4-9(a) shows the 710 comparator used as a voltage-level detector. The circuit employs both hysteresis and a reference voltage, and the 710 does its own output bounding. The 100 kΩ and 470 Ω resistors give a hysteresis voltage of about 15 mV (see Section 4-3). The voltage at pin 2 is always very nearly

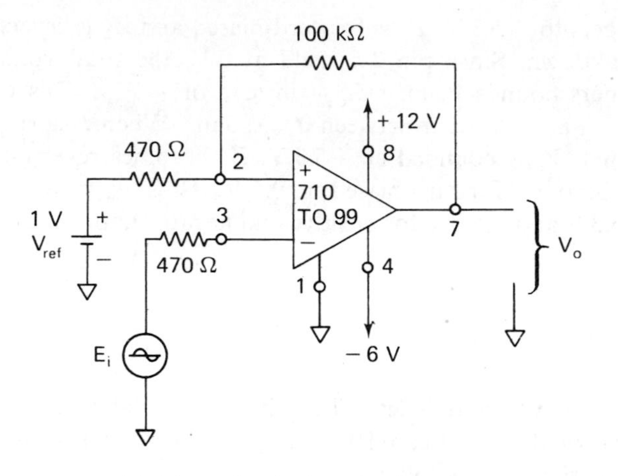

(a) Schematic of voltage-level detector

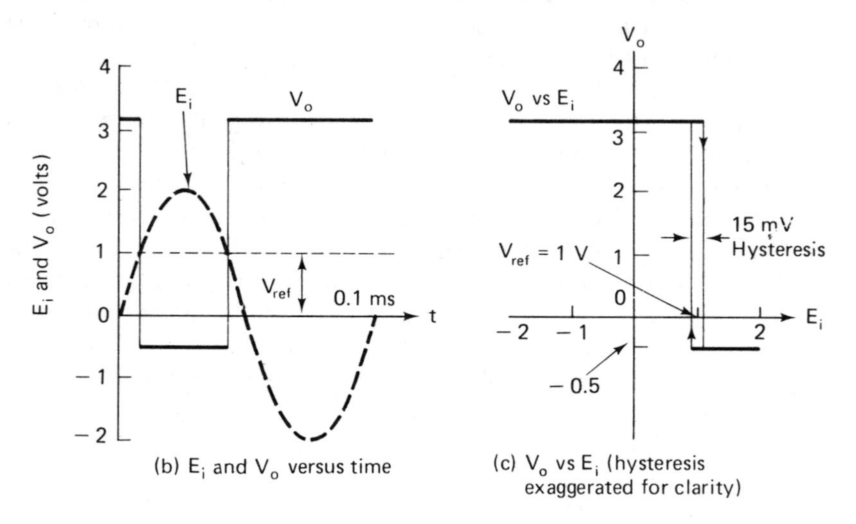

(b) E_i and V_o versus time

(c) V_o vs E_i (hysteresis exaggerated for clarity)

Figure 4-9 The 710 comparator as a voltage-level detector.

equal to V_{ref} or 1.0 V. This is because voltage drop across the 470 Ω resistor never exceeds about 15 mV.

E_i and V_o are plotted against time in Fig. 4-9(b). For all values of E_i less than V_{ref}, V_o is at +3.2 V. When E_i exceeds V_{ref}, V_o is at −0.5 V. Observe that when V_o goes negative it means that E_i just crossed V_{ref} in a positive direction. V_o is plotted against E_i in Fig. 4-9(c) for comparison with the 0-crossing detector of Fig. 4-5.

4-7 IC Precision Comparator, 111/311

4-7.1 Electrical Characteristics. The 111 (military) or 311 (commercial) comparator (see Appendix 3) is much more versatile than the 710 but has a slower response time (200 ns as compared to 40 ns for the 710). Advantages of the 111 are seen by a comparison of typical electrical characteristics as follows.

Characteristic	*111/311*	*710*
Supply voltage	±15 V or 0 to +5 V	+12 V, −6 V
Input bias current	100 nA	13 μA
Differential input voltage	±30 V	±5 V
Input voltage	±14 V	±5 V
Voltage gain	200,000	800
Output voltage	see Sec. 4-7.2	+3.2 V to −0.5 V
Strobe capability	yes	no

4-7.2 Output Terminal Operation. A simplified model of the 311 in Fig. 4-10(a) shows that its output behaves like a switch Sw connected between output pin 7 and pin 1. Pin 7 can be wired to any voltage V^{++} with magnitudes up to 40 V more positive than the $-V$ supply terminal (pin 4). When (+) input pin 2 is more positive than (−) input pin 3, the 311's equivalent output switch is open. V_o is then determined by V^{++} and is +5 V.

When the (+) input is less positive than (below) the (−) input, the 311's equivalent output switch closes and extends the ground on pin 1 to output pin 7. R_f and R_i add about 50 mV of hysteresis to minimize noise effects so that pin 2 is essentially at 0 V. Wave shapes for V_o and E_i are shown in Fig. 4-10(b). V_o is 0 V (switch closed) for positive half-cycles of E_i. V_o is +5 V (switch open) for negative half-cycles of E_i. This is a typical interface circuit; that is, voltages may vary between levels of +15 V and −15 V, but V_o is restrained between +5 V and 0 V, which are typical digital signal levels. So the 311 can be used for converting analog voltage levels to digital voltage levels (interfacing).

4-7.3 Strobe Terminal Operation. The strobe terminal of the 311 is pin 6. (See also Appendix 3.) This strobe feature allows the comparator output either to respond to input signals or to be independent of input signals. Fig. 4-11 uses the 311 comparator as a 0-crossing detector. A 10 kΩ resistor is connected to the strobe terminal. The other side of the resistor is connected to a switch. With the strobe switch open, the 311 operates normally. That is, the output voltage is at V^{++} for negative values of E_i and 0 for positive values of E_i. When the strobe switch is closed (connecting the 10 kΩ to ground), the output voltage goes to V^{++} regardless of the input

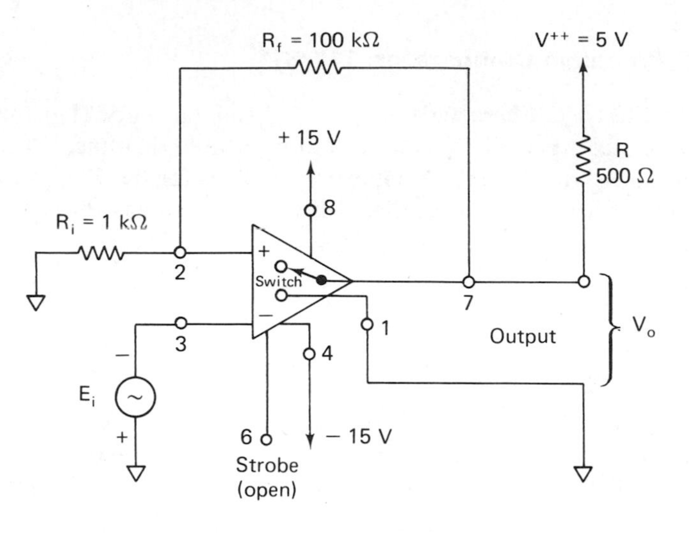

(a) 311 0-crossing detector with hysteresis

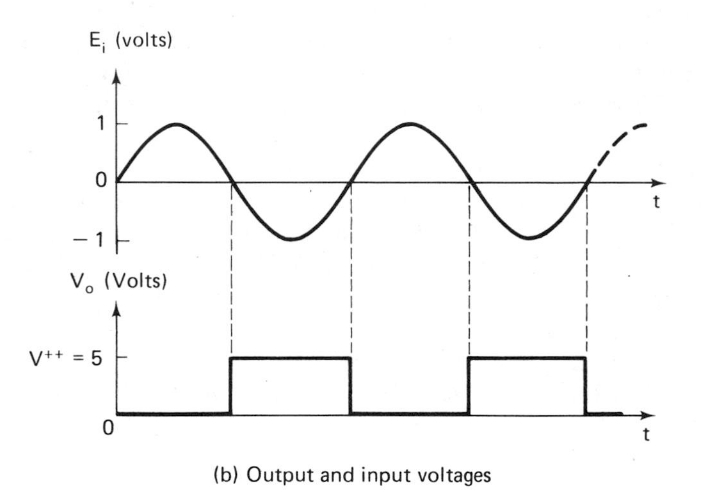

(b) Output and input voltages

Figure 4-10 Simplified model of the 311 comparator with input and output voltage waveforms.

signal. V_o will stay at V^{++} as long as the strobe switch is closed. See Fig. 4-11(b). The output is then independent of the inputs until the strobe switch is again opened.

The strobe feature is useful when a comparator is used to determine what type of signal is to be read out of a computer memory. The strobe switch is closed to ignore extraneous input signals that may occur up until the readout

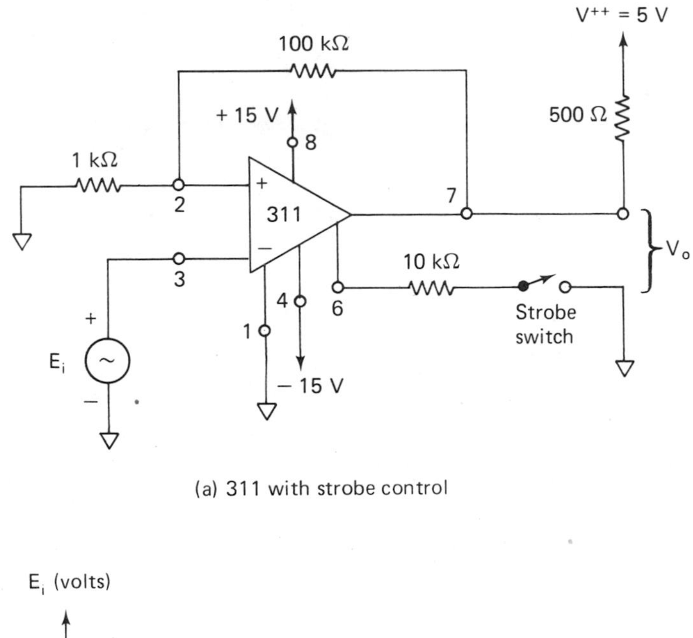

(a) 311 with strobe control

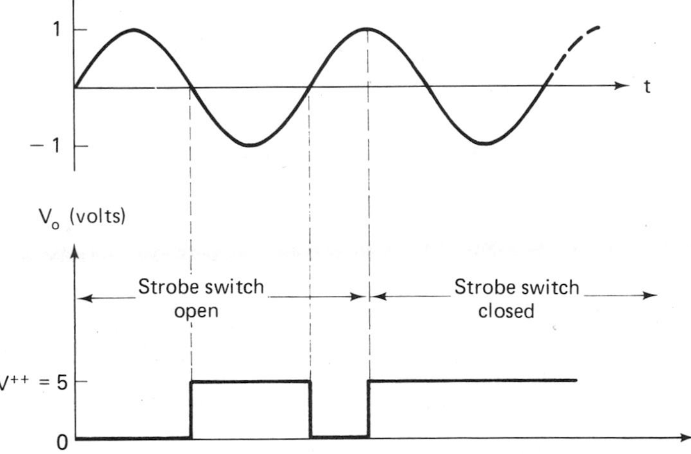

(b) When the strobe switch is closed, $V_o = V^{++}$

Figure 4-11 Operation of the strobe terminal.

is due. Then during the readout time the switch is opened, and the 311 performs as a regular comparator. Current from the strobe terminal should be limited to about 3 mA. If the strobe feature is not to be used, the strobe terminal is left open or wired to $+V$. See Appendix 3.

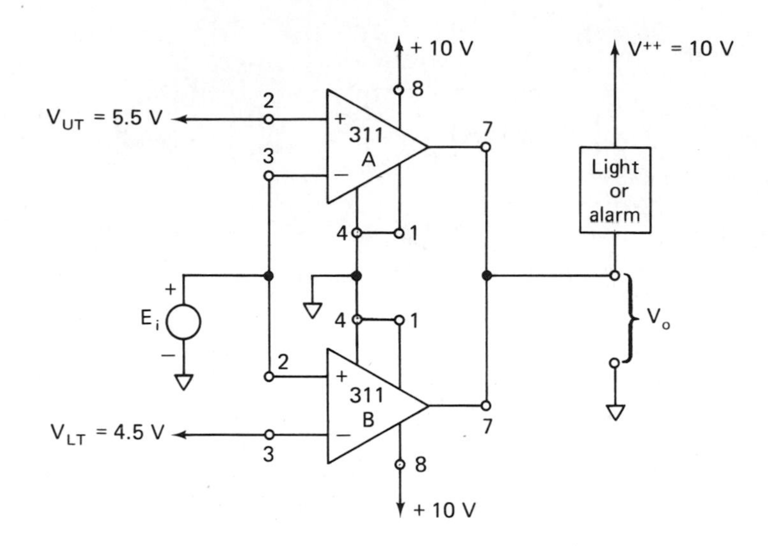

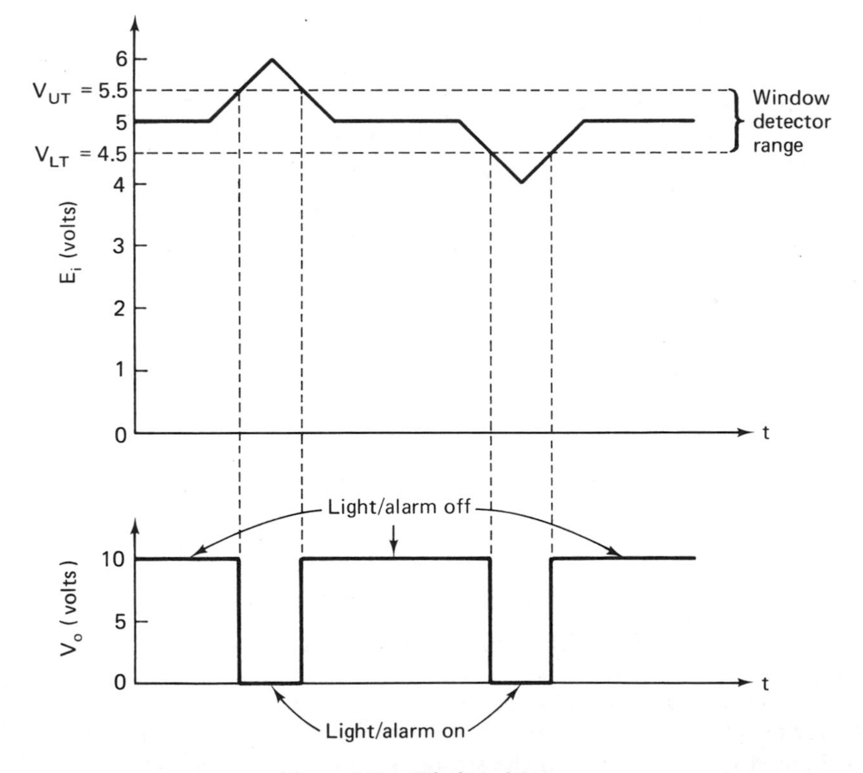

Figure 4-12 Window detector.

4-8 Window Detector

4-8.1 Introduction. The circuit of Fig. 4-12 is designed to monitor an input voltage and indicate when this voltage goes either above or below prescribed limits. For example, IC logic power supplies for TTL must be regulated to 5.0 V. If the supply voltage should exceed 5.5 V, the logic may be damaged, and if the supply voltage should drop below 4.5 V, the logic may exhibit marginal operation. Therefore, the limits for TTL power supplies are 4.5 V and 5.5 V. The power supply should be looking through a window whose limits are 4.5 V and 5.5 V, hence the name *window detector*. This circuit is sometimes called a *double-ended limit detector*.

In Fig. 4-12, input voltage E_i is connected to the (−) input of comparator A and the (+) input of comparator B. Upper limit V_{UT} is applied to the (+) input of A, while lower limit V_{LT} is applied to the (−) input of B. When E_i lies between V_{LT} and V_{UT}, the light/alarm is off. But when E_i drops below V_{LT} or goes above V_{UT} the light/alarm goes on to signify that E_i is not between the prescribed limits.

4-8.2 Circuit Operation. Circuit operation is as follows. Assume $E_i = 5$ V. Since E_i is greater than V_{LT} and less than V_{UT}, the output voltage of both comparators is at V^{++}. The lamp/alarm is off. Next, assume that $E_i = 6.0$ V or $E_i > V_{UT}$. The input at pin 3 of A is more positive than at pin 2, so the A output is at the potential of pin 1 or ground. This ground lights the lamp, and $V_o = 0$ V. Now assume that E_i drops to 4.0 V or $E_i < V_{LT}$. The (+) input of B is less than its (−) input, so the B output goes to 0 V (the voltage at its pin 1). Once again this ground causes the lamp/alarm to light. Note that this application shows that output pins of the 311 can be connected together and the output is at V^{++} only when the output of each comparator is at V^{++}.

Problems

4-1 What is the name given to a circuit that converts an irregular-shaped wave to a square wave or pulse?

4-2 For what applications would a 0-crossing detector be used?

4-3 What is the name of a circuit that indicates when an input voltage reaches a specified reference voltage?

4-4 What type of feedback occurs when a resistor is connected between the output terminal and the (+) input?

4-5 In Fig. 4-3(a), if $R_2 = 200\ \Omega$ and $V_{sat} = \pm 10$ V, what is the value for (a) the upper-threshold voltage (b) the lower-threshold voltage?

4-6 For the values given in Problem 4-5, determine the hysteresis voltage.

4-7 Calculate R_1 in Fig. 4-3(b) so that the circuit has a hysteresis voltage of 0.1 V. $R_2 = 100\ \Omega$ and $\pm V_{sat} = \pm 15$ V.

4-8 What is placing limits on the output voltage called?

4-9 If $R_i = 2\ \text{k}\Omega$ and $E_i = 10$ V in Fig. 4-6, what is (a) V_o and (b) the zener current?

4-10 If the zener voltage for both zeners is 8 V, the output voltage is bounded between what two limits?

4-11 Name a disadvantage of the 710 comparator as compared to the 311.

4-12 If V^{++} in Fig. 4-10(a) is +10 V, draw the output waveform.

5

Selected Applications of Op Amps

5-0 Introduction

Why is the op amp such a popular device? This chapter attempts to answer that question by presenting a wide selection of applications. They were selected to show that the op amp can perform as a very nearly ideal device. Moreover the diversity of operations that the op amp can perform is almost without limit. In fact, applications that are normally very difficult, such as measuring short-circuit current, are rendered simple by the op amp. Together with a few resistors and a power supply, the op amp can, for example, measure the output from photodetectors, give audio tone control, equalize tones of different amplitudes, control high currents, and allow matching of semiconductor device characteristics, We begin with selecting an op amp circuit to make a high-resistance dc and ac voltmeter.

5-1 High-Resistance DC Voltmeter

5-1.1 Basic Voltage-Measuring Circuit. Figure 5-1 shows a simple but very effective high input-resistance dc voltmeter. The voltage to be measured, E_i, is applied to the (+) input terminal. Since the differential input voltage is 0 V, E_i is developed across R_i. The meter current, I_m, is set by E_i and R_i just as in the noninverting amplifier.

$$I_m = \frac{E_i}{R_i} \tag{5-1}$$

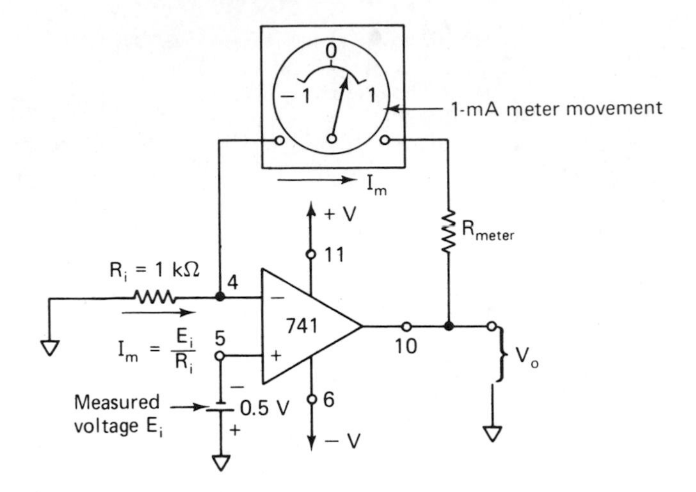

Figure 5-1 High input-resistance dc voltmeter.

If R_i is 1 kΩ, then 1 mA of meter current will flow for $E_i = 1$ V dc. Therefore, the milliameter can be calibrated directly in volts. As shown, this circuit can measure any dc voltage from −1 V to +1 V.

Example 5-1: Find I_m in Fig. 5-1.
Solution: From Eq. (5-1), $I_m = 0.5$ V/1 kΩ $= 0.5$ mA. The needle is deflected halfway between 0 and +1 mA.

One advantage of Fig. 5-1 is that E_i sees the very high input impedance of the (+) input. Since the (+) input draws negligible current, it will not load down or change the voltage being measured. Another advantage of placing the meter in the feedback loop is that if the meter resistance should vary, it will have no effect on meter current. Even if we added a resistor in series with the meter, within the feedback loop, it would not effect I_m. The reason is that I_m is set only by E_i and R_i. The output voltage will change if meter resistance changes, but in this circuit we are not concerned with V_o. This circuit is sometimes called a *voltage-to-current* converter.

5-1.2 Voltmeter Scale Changing. Since the input voltage in Fig. 5-1 must be less than the power supply voltages (±15 V), a convenient maximum limit to impose on E_i is ±10 V. The simplest way to convert Fig. 5-1 from a ±1 V voltmeter to a ±10 V voltmeter is to change R_i to 10 kΩ. In other words, pick R_i so that the full-scale input voltage E_{FS} equals R_i times the full-scale meter current I_{FS} or

$$R_i = \frac{E_{FS}}{I_{FS}} \tag{5-2}$$

Example 5-2: A microammeter with 50 μA $= I_{FS}$ is to be used in Fig. 5-1. Calculate R_i for $E_{FS} = 5$ V.
Solution: By Eq. (5-2), $R_i = 5\text{ V}/50\ \mu\text{A} = 100\ \text{k}\Omega$. Before measuring higher-input voltages, use a voltage-divider circuit. The output of the divider is applied to the (+) input.

5-2 High-Resistance AC Voltmeter

The circuit of Fig. 5-2 is a high-resistance ac voltmeter. The ac voltage E_i and resistance R_i determine the instantaneous value of the current through R_i and the feedback loop. The four diodes form a full-wave bridge rectifier around the meter. They are shown connected so that the meter current flows only in the direction (+) to (−) through the meter. Thus the meter current I_m is a rectified version of the ac current through R_i. The dc meter movement *averages* the rectified current. If E_i is a sine wave, the meter current I_m will be

$$I_m = 0.636\frac{E_{i_p}}{R_i} = 0.90\frac{E_{rms}}{R_i} \tag{5-3}$$

where E_{i_p} = the peak value of E_i and E_{rms} = rms value of E_i.

Example 5-3: In Fig. 5-2, I_m is a dc meter movement of 100 μA full scale. It should read full scale when $E_i = 1$ V rms. Find R_i so that the meter face can be calibrated for a full-scale deflection of 1 V.

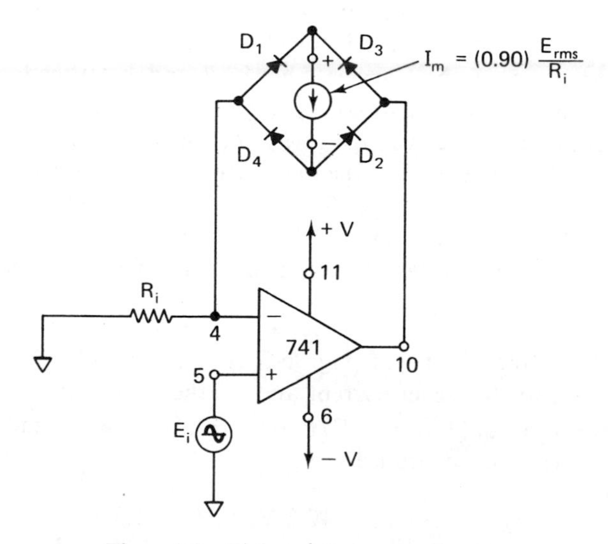

Figure 5-2 High resistance ac voltmeter.

Solution: Rewriting Eq. (5-3),

$$R_i = 0.90\frac{E_{rms}}{I_m} = 0.90\frac{1}{100\ \mu A} = 9\ k\Omega$$

It is emphasized that the diode resistance does not affect I_m. Only R_i and E_i are the circuit values that determine I_m. This circuit is sometimes called an ac voltage to dc current converter. The reader may have noticed that the high-resistance 1-V ac voltmeter will perform as a high-resistance 1-V dc voltmeter if we merely change R_i in Example 5-3 to 10 kΩ. Thus, connecting 1-kΩ in series with the 9-kΩ resistor is all that is needed to change the voltmeter from ac to dc.

5-3 Voltage-to-Current Converters: Floating Loads

5-3.1 Voltage Control of Load Current. From Sections 5-1 and 5-2 we learned not just how to make a voltmeter but that current in the feedback loop depends on the input voltage and R_i. There are applications where we need to pass a constant current through a load and hold it constant despite any changes in load resistance or load voltage. If the load does not have to be grounded, we simply place the load in the feedback loop and control both input and load current by the same principle developed in Section 5-1.

5-3.2 Zener Diode Tester. Suppose we have to test the breakdown voltage of a number of zener diodes at a current of precisely 5 mA. If we connect the zener in the feedback loop as in Fig. 5-3(a), our voltmeter circuit of Fig. 5-1 becomes a zener diode tester. That is, E_i and R_i set the load or zener current at a constant value. E_i forces V_o to go negative until the zener breaks down and clamps the zener voltage at V_Z. R_i converts E_i to a current, and so long as R_i and E_i are constant, the load current will be constant regardless of the value of the zener voltage. Since the zener is driven by a constant current, this circuit is called a *constant current source*. However, it is basically a voltage-to-current converter. Zener breakdown voltage can be calculated from V_o and E_i as $V_Z = V_o - E_i$.

Example 5-4: In the circuit of Fig. 5-3(a), $V_o = 10.3$ V, $E_i = 5$ V, and $R_i = 1$ kΩ. Find (a) the zener current and (b) the zener voltage.
Solution: (a) From Eq. (3-1), $I = E_i/R_i$ or $I = 5\ V/1\ k\Omega = 5$ mA. (b) From Fig. 5-3(a), rewrite the equation for V_o.

$$V_Z = V_o - E_i = 10.3\ V - 5\ V = 5.3\ V$$

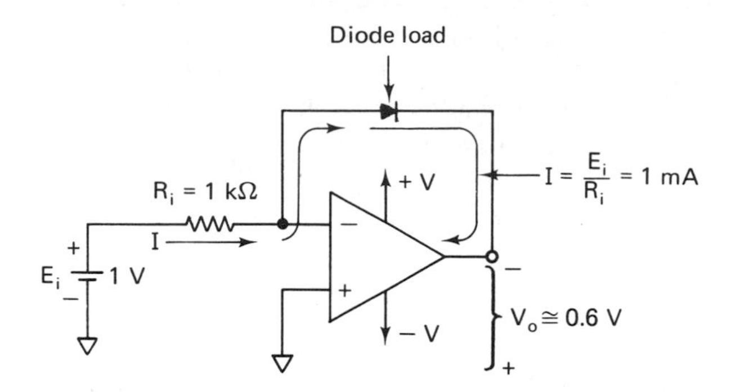

(b) Load current equals input current

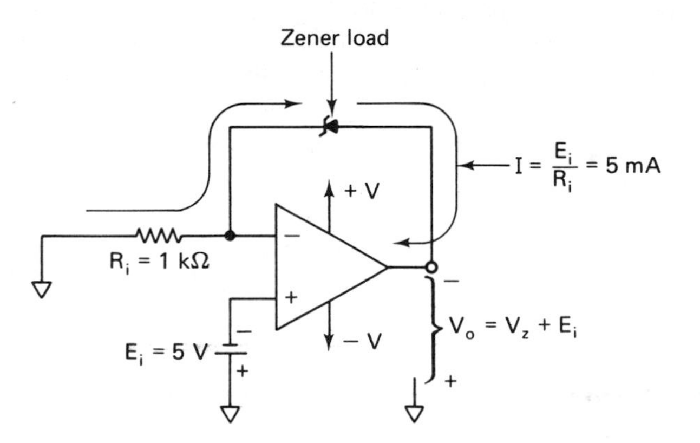

(a) Negligible current drawn from E_i, load current furnished by op amp

Figure 5-3 Voltage-controlled load currents with loads in feedback loop.

5-3.3 Diode Tester. Suppose we needed to select diodes from a production batch and find pairs with matching voltage drops at a particular value of diode current. Place the diode in the feedback loop as shown in Fig. 5-3(b). E_i and R_i will set the value of I. The (−) input draws negligible current, so I passes through the diode. As long as E_i and R_i are constant, current through the diode I will be constant at $I = E_i/R_i$. V_o will equal the diode voltage for the same reasons that V_o was equal to V_{R_f} in the inverting amplifier (see Section 3-1).

Example 5-5: $E_i = 1$ V, $R_i = 1$ kΩ, and $V_o = 0.6$ V in Fig. 5-3(b). Find (a) the diode current and (b) the voltage drop across the diode.
Solution: (a) $I = E_i/R_i = 1$ V/1 kΩ $= 1$ mA. (b) $V_{\text{diode}} = V_o = 0.6$ V.

There is one disadvantage with the circuit of Fig. 5-3(b): E_i must be able to furnish the current. Both circuits in Fig. 5-3 can only furnish currents up to 10 mA because of the op amp's output current limitation. Higher-load currents must be furnished from the power supply terminal as shown in Section 5-4.

5-4 Light-Emitting Diode Tester

The circuit of Fig. 5-4 converts E_i to a 20-mA load current based on the same principles discussed in Sections 5-1 to 5-3. Since the 741's output terminal can only supply about 5 mA to 10 mA, we cannot use the circuits of Figs. 5-1 to 5-3 for higher load currents. But if we add a transistor as in Fig. 5-4, load current is furnished from the negative supply voltage. The op amp's output terminal is required to furnish only base current, which is typically 1/100 of the load current. The factor 1/100 comes from assuming that the transistor's beta equals 100. Since the op amp can furnish an output current of up to 5 mA into the transistor's base this circuit can supply a maximum load current of $5\text{ mA} \times 100 = 0.5\text{ A}$.

A light-emitting diode such as the MLED50 is specified to have a typical brightness of 750 fL provided that the forward diode current is 20 mA. E_i and R_i will set the diode current I_L equal to $E_i/R_i = 2\text{ V}/100\ \Omega = 20\text{ mA}$. Now brightness of LEDs can be measured easily one after another for test or matching purposes, because the current through each diode will be exactly 20 mA regardless of the LED's forward voltage.

It is worthwhile to note that a load of two LEDs can be connected in series with the feedback loop and both would conduct 20 mA. The load could

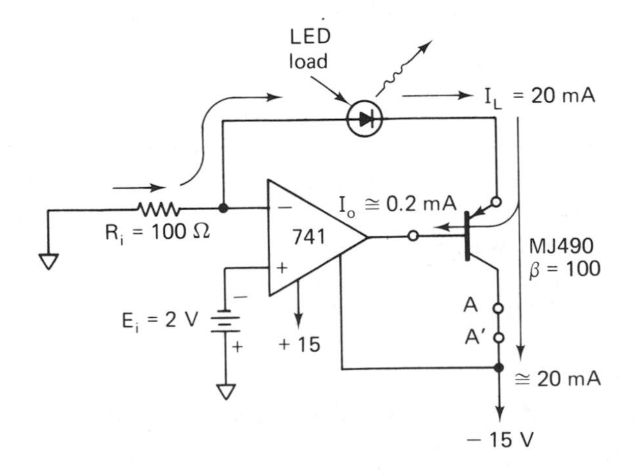

Figure 5-4 Voltage to high current converter.

also be connected in Fig. 5-4 between points AA' which is in series with the transistor's collector and still conduct about 20 mA. This is because the collector and emitter currents of a transistor are essentially equal. A load in the feedback loop is called a *floating load.* If one side of the load is grounded, it is a *grounded load.* To supply a constant current to a grounded load, another type of circuit must be selected, as shown in Section 5-5.

5-5 Furnishing a Constant Current to a Grounded Load

5-5.1 Introduction. In some applications, one terminal of the load must be grounded and load current controlled by an input voltage. The load current must depend *not* on load resistance but only on the input voltage.

A very flexible circuit that will accomplish this task is shown in Fig. 5-5. Two inputs labeled (+) input and (−) input are available to apply two control voltages, either one at a time or both together. The labels were chosen to correspond with the nearest op amp input terminal.

5-5.2 Controlling Load Current with the (+) Input. Analysis of Fig. 5-5 is accomplished by observing that V_o must divide equally between the two top resistors R. This division places the op amp's (−) input at $V_o/2$ with respect to ground. Since $E_d \approx 0$ V, the load voltage $V_L = V_o/2$. The

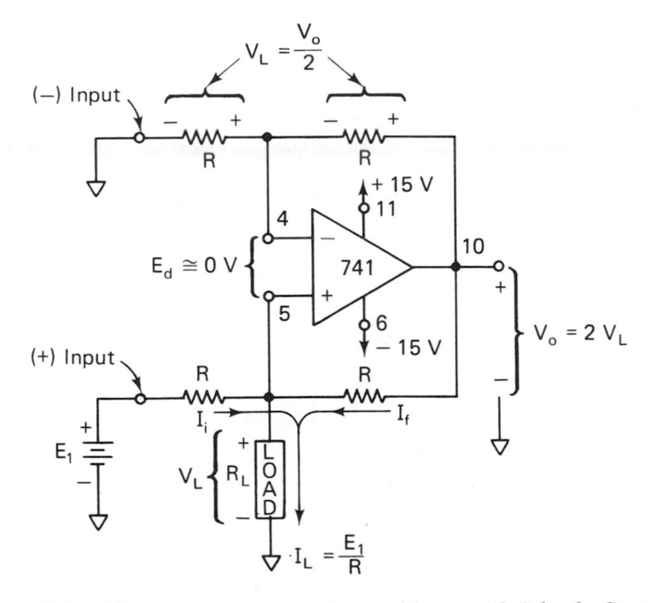

Figure 5-5 Voltage-to-current converter with grounded load. Control voltage on (+) input. $R = 10\text{ k}\Omega$.

current I_i is furnished from the source E_1 and I_f from V_o. I_i and I_f add to furnish the load current as follows:

$$I_L = I_i + I_f = \frac{E_1 - V_L}{R} + \frac{V_o - V_L}{R}$$

Since $V_o = 2\ V_L$, this equation simplifies to

$$I_L = \frac{E_1}{R} \tag{5-4}$$

Example 5-6: If $E_1 = 2$ V and $R = 1$ kΩ in Fig. 5-5, find I_L.
Solution: By Eq. (5-4), $I_L = 2$ V/1 kΩ = 2 mA.

Example 5-7: Find V_L and V_o for values given in Example 5-6, with $R_L =$ 500 Ω.
Solution: $V_L = I_L R_L = 2$ mA × 500 Ω = 1.0 V; $V_o = 2V_L = 2$ V.

5-5.3 Controlling Load Current with the (−) Input. The control voltage may be transferred from the (+) input in Fig. 5-5 to the (−) input, as shown in Fig. 5-6. The load current is still independent of the load and is expressed by

$$I_L = \frac{E_2}{R} \tag{5-5}$$

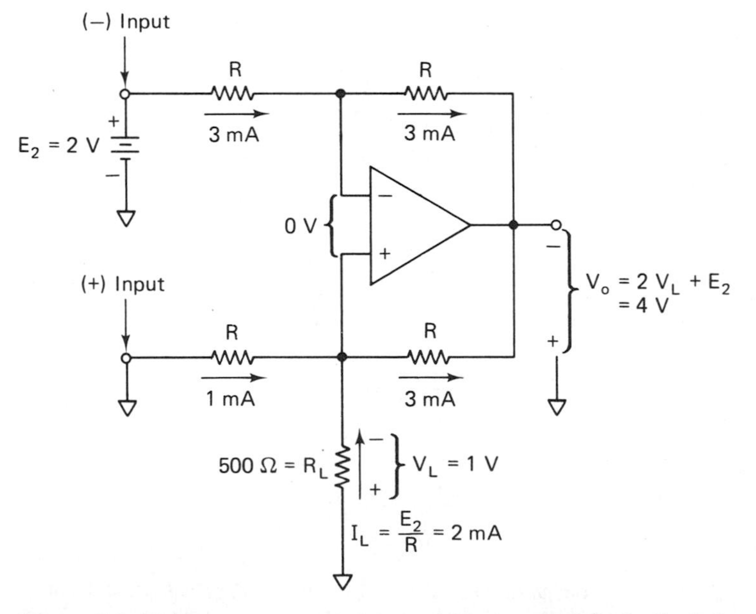

Figure 5-6 Voltage-to-current converter with grounded load. Control voltage on (−) input.

Note in Fig. 5-6 that (1) the load current direction, (2) the polarity of V_L, and (3) the polarity of V_o are all reversed from those in Fig. 5-5.

Example 5-8: In Fig. 5-6, $R_L = 500\ \Omega$, $E_2 = 2$ V, and $R = 1$ kΩ. Evaluate (a) I_L, (b) V_L, and (c) V_o.
Solution: (a) By Eq. (5-5), $I_L = 2\text{ V}/1\text{ k}\Omega = 2\text{ mA}$. (b) $V_L = I_L R_L = 2\text{ mA} \times 500\ \Omega = 1\text{ V}$.(c) From Fig. 5-6, $V_o = 2V_L + E_2 = 2(1\text{ V}) + 2\text{ V} = 4\text{ V}$.

5-5.4 Current Control with a Differential Voltage. It is possible and may even be desirable to have a current controlled by the difference between two voltages. This task can be accomplished by connecting *both* E_1 and E_2 to the (+) and (−) inputs, respectively, of Fig. 5-5 or 5-6. The load current is determined from

$$I_L = \frac{E_1 - E_2}{R} \qquad (5\text{-}6a)$$

and V_o from

$$V_o = 2V_L - E_2. \qquad (5\text{-}6b)$$

Reversing the polarity of E_1 will reverse I_L in Fig. 5-5. If E_2 is reversed in Fig. 5-6, I_L will reverse. When using Eqs. (5-6a) and (5-6b), if E_1 is greater than E_2, then the current directions and voltage polarities are as shown in Fig. 5-5. If E_2 is greater than E_1, the current directions and voltage polarities are as shown in Fig. 5-6.

5-5.5 Constant High Current Source, Grounded Load. In certain applications, such as electroplating, it is desirable to furnish a high current, of constant value, to a grounded load. The circuit of Fig. 5-7 will furnish con-

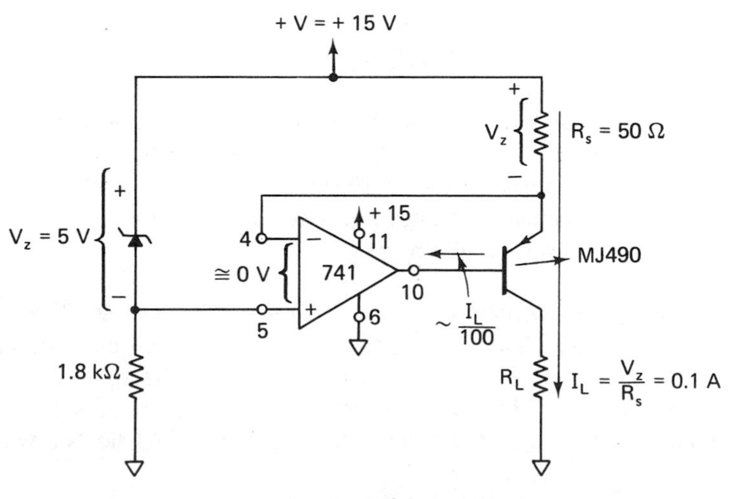

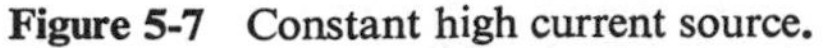

Figure 5-7 Constant high current source.

stant currents above 500 mA provided that the transistor is heat-sinked properly (above 5 W) and has a high beta ($\beta > 100$). The circuit operates as follows. The zener diode voltage is applied to one end of current sense resistor R_s and the op amp's positive input. Since the differential input voltage is 0 V, the zener voltage is developed across R_s. R_s and V_Z set the emitter current, I_E, constant at V_Z/R_s. The emitter and collector currents of a bipolar junction transistor are essentially equal. Since the collector current is load current I_L and $I_L \approx I_E$, the load current I_L is set by V_Z and R_s.

If the op amp can furnish a base current drive of over 5 mA and if the beta of the transistor is greater than 100, then I_L can exceed 5 mA $\times$ 100 = 500 mA. The voltage across the load must not exceed the difference between the supply and the zener voltage, otherwise the transistor and the op amp will go into saturation.

5-6 Short-Circuit Current Measurements

5-6.1 Introduction. Transducers such as phonograph pickups convert some physical quantity into electrical signals. For convenience, the transducers may be modeled by a signal generator as in Fig. 5-8(a). It is usually desirable to measure their maximum output current under short-circuit conditions; that is, we should place a short circuit across the output terminals and measure current through the short circuit. This technique is particularly suited to signal sources with very high internal resistance. For example, in Fig. 5-8(a), the short-circuit current I_{SC} should be 2.5 V/50 kΩ = 50 μA. However, if we place a microammeter across the output terminals of the generator, we no longer have a short circuit but a 5000 Ω resistance. The meter indication is

$$\frac{2.5\text{ V}}{50\text{ k}\Omega + 5\text{ k}\Omega} \cong 45\ \mu\text{A}$$

High-resistance sources are better modeled by an equivalent Norton circuit. This model is simply the ideal short-circuit current, I_{SC}, in parallel with its own internal resistance as in Fig. 5-8(b). This figure shows how I_{SC} splits between its internal resistance and the meter resistance. To eliminate this current split, we will use the op amp.

5-6.2 Using the Op Amp to Measure Short-Circuit Current. The op amp circuit of Fig. 5-8(c) effectively places a short circuit around the current source. The (−) input is at virtual ground because the differential input voltage is almost 0 V. The current source sees ground potential at both of its terminals, or the equivalent of a short circuit. *All* of I_{SC} flows toward the

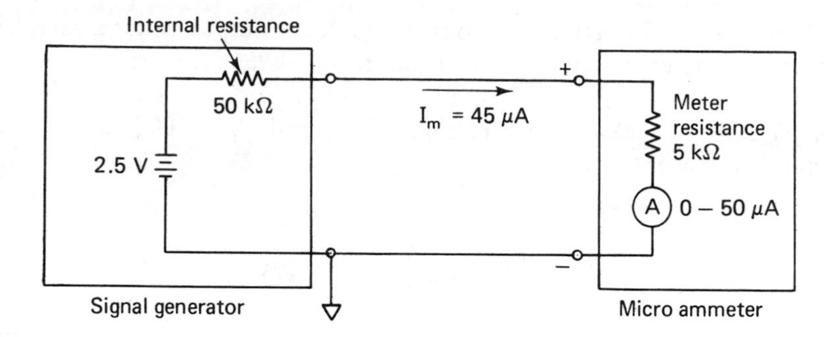

(a) Ammeter resistance reduces short-circuit current from the signal generator

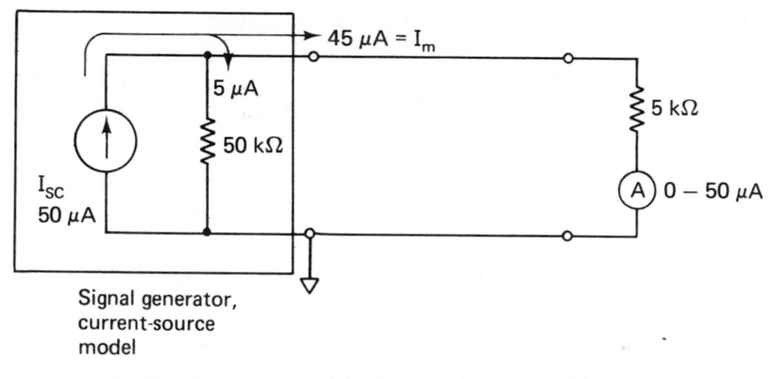

(b) Current-source model of signal generator in (a)

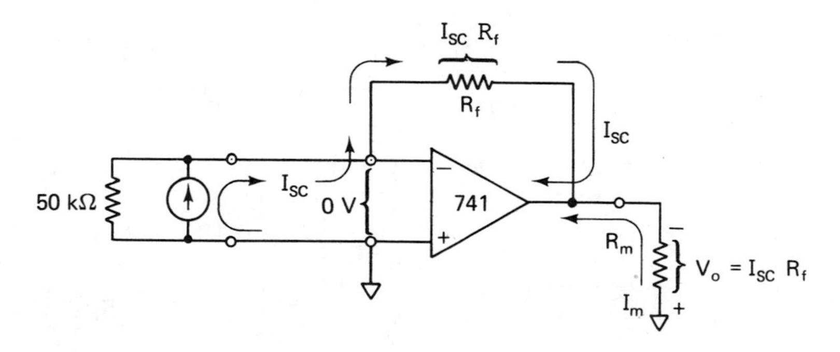

(c) Current-to-voltage converter

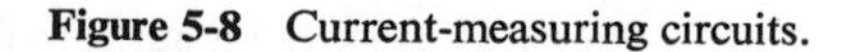

Figure 5-8 Current-measuring circuits.

(−) input and on through R_f. R_f converts I_{SC} to an output voltage, revealing the basic nature of this circuit to be a *current-to-voltage converter.*

Example 5-9: V_o measures 5 V in Fig. 5-8(c), and $R_f = 100\ \text{k}\Omega$. Find the short-circuit current. I_{SC}.
Solution: From Fig. 5-8(c),

$$I_{SC} = \frac{V_o}{R_f} = \frac{5\ \text{V}}{100\ \text{k}\Omega} = 50\ \mu\text{A}$$

The resistance R_m is the resistance of either the voltmeter or the CRO. The current I_m needed to drive either instrument comes from the op amp and not from I_{SC}.

5-7 Measuring Current from Photodetectors

5-7.1 Photoconductive Cell. With the switch at position 1 in Fig. 5-9, a photoconductive cell, sometimes called a light-sensitive resistor (LSR), is connected in series with the (−) input and E_i. The resistance of a photoconductive cell is very high in darkness and much lower when illuminated. Typically its dark resistance is greater than 500 kΩ and its light resistance in bright sun is approximately 5 kΩ. If $E_i = 5$ V, then current through the photoconductive cell, I, would be 5 V/500 kΩ = 10 μA in darkness and 5 V/5 kΩ = 1 mA in sunlight.

Example 5-10: In Fig. 5-9 the switch is in position 1 and $R_f = 10\ \text{k}\Omega$. If the current through the photoconductive cell is 10 μA in darkness and 1 mA in sunlight, find V_o for (a) the dark condition, (b) the light condition.

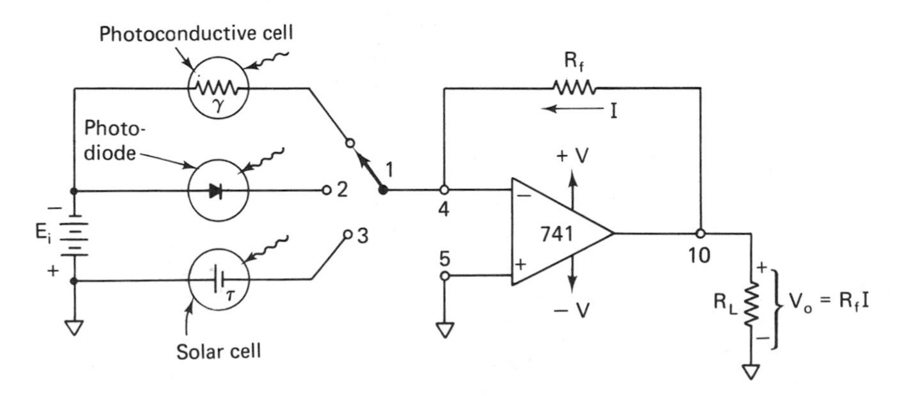

Figure 5-9 Using the op amp to measure output current from photodetectors.

Solution: From Fig. 5-9, $V_o = R_f I$.
(a) $V_o = 10\ \text{k}\Omega \times 10\ \mu\text{A} = 0.1\ \text{V}$; (b) $V_o = 10\ \text{k}\Omega \times 1\ \text{mA} = 10\ \text{V}$. Thus the circuit of Fig. 5-9 converts the output current from the photoconductive cell into an output voltage (a current-to-voltage converter).

5-7.2 Photodiode. When the switch is in position 2 in Fig. 5-9, E_i is on one side of the photodiode and virtual ground on the other. The photodiode is reverse-biased, as it must be for normal operation. In darkness the photodiode conducts a small leakage current on the order of nanoamperes. But depending on the radiant energy striking the diode, it will conduct 50 μA or more. Therefore, current I depends only on the energy striking the photodiode and not on E_i. This current is converted to a voltage by R_f.

Example 5-11: With the switch in position 2 in Fig. 5-9 and $R_f = 100\ \text{k}\Omega$, find V_o as the light changes photodiode current from (a) 1 μA to (b) 50 μA.
Solution: From $V_o = R_f I_L$,
(a) $V_o = 100\ \text{k}\Omega \times 1\ \mu\text{A} = 0.1\ \text{V}$; (b) $V_o = 100\ \text{k}\Omega \times 50\ \mu\text{A} = 5.0\ \text{V}$.

5-7.3 Solar Cell. The solar cell is a photovoltaic cell. It converts light energy directly into electrical energy. When the switch is in position 3 in Fig. 5-9, the solar cell sees essentially a short circuit because of the virtual ground at the (−) input. Under this short-circuit condition the solar cell generates a current proportional to the radiant energy striking its surface. This current is converted to voltage in the same manner as detailed in Sections 5-7.1 and 5-7.2. All circuits in Section 5-7 can be used not only to determine current but also to give an output voltage proportional to radiant intensity.

5-8 Current Amplifier

Characteristics of high-resistance signal sources were introduced in Section 5-6.1. There is no point in converting a current into an equal current, but a circuit that converts a current into a larger current can be very useful. The circuit of Fig. 5-10 is a current multiplier or current amplifier (technically a current-to-current converter). The signal current source I_{SC} is effectively short-circuited by the input terminals of the op amps. All of I_{SC} flows through resistor mR, and the voltage across it is mRI_{SC}. (Resistor mR is known as a multiplying resistor and m the multiplier.) Since R and mR are in parallel, the voltage across R is also mRI_{SC}. Therefore, the current through R must be mI_{SC}. Both currents add to form load current I_L. I_L is an amplified version of I_{SC} and is found simply from

$$I_L = (1 + m)I_{SC} \tag{5-7}$$

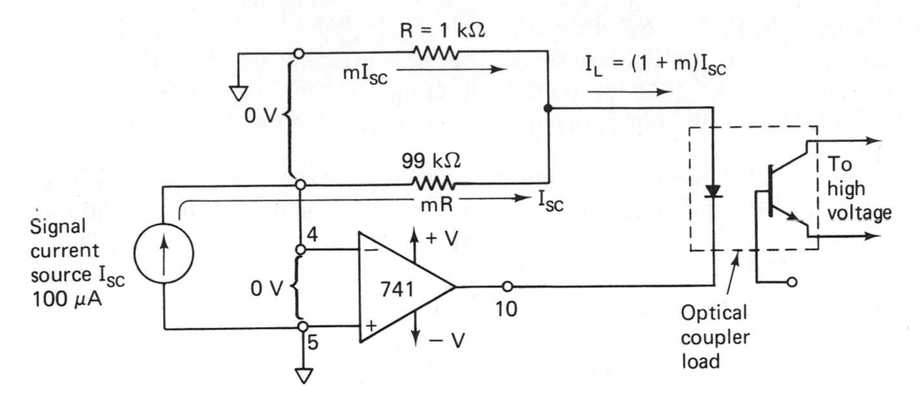

Figure 5-10 Current amplifier with optical coupler load.

Example 5-12: In Fig. 5-10, $R = 1\ \text{k}\Omega$ and $mR = 99\ \text{k}\Omega$. Therefore $m = 99\ \text{k}\Omega/1\ \text{k}\Omega = 99$. Find the current I_L through the emitting diode of the optical coupler.
Solution: By Eq. (5-7), $I_L = (1 + 99)(100\ \mu\text{A}) = 10\ \text{mA}$.

It is important to note that the load does not determine load current. Only the multiplier m and I_{SC} determine load current. For variable current gain, mR and R can be replaced by a single 100 kΩ potentiometer. The wiper goes to the emitting diode, one end to ground and the other end to the (−) input. The optical coupler isolates the op amp circuit from any high voltage load.

5-9 Phase Shifter

5-9.1 Introduction. An ideal phase-shifting circuit should transmit a wave without changing its amplitude but changing its phase angle by a preset amount. For example, a sine wave E_i with a frequency of 1 kHz and peak value of 1 V is the input of the phase shifter in Fig. 5-11(a). The output V_o has the same frequency and amplitude but lags E_i by 90°. That is, V_o goes through 0 V 90° *after* E_i goes through 0 V. Mathematically, V_o can be expressed by $V_o = E_i \angle -90°$. A general expression for the output voltage of the phase-shifter circuit in Fig. 5-11(b) given by

$$V_o = E_i \angle \theta \tag{5-8}$$

where θ is the phase angle and will be found from Eq. (5-9a).

5-9.2 Phase-Shifter Circuit. One op amp, three resistors, and one capacitor are all that is required as shown in Fig. 5-11(b) to make an excellent

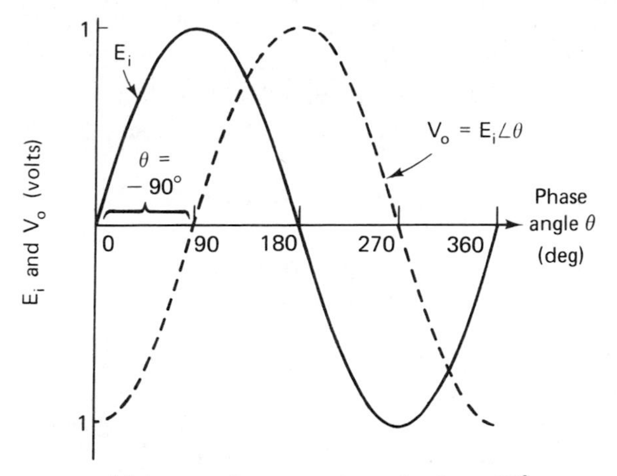

(a) Input and output voltages for $\theta = -90°$

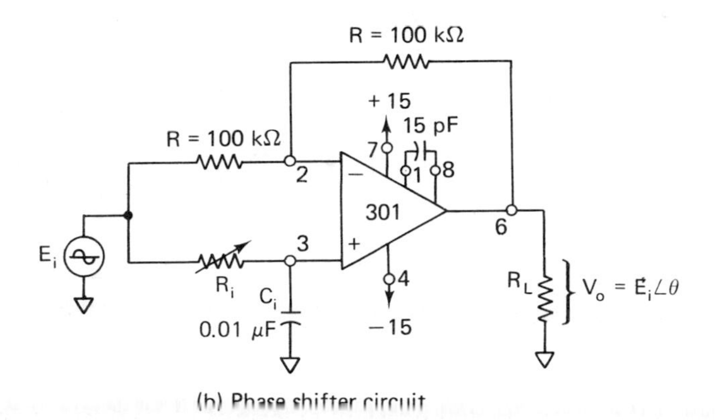

(b) Phase shifter circuit

Figure 5-11 Phase shifter.

phase shifter. The resistors R must be equal, and any convenient value from 10 kΩ to 220 kΩ may be used. Phase angle θ depends only on R_i, C_i, and the frequency f of E_i. The relationship is

$$\theta \approx -2 \arctan 2\pi f R_i C_i \tag{5-9a}$$

where θ is in degrees, f in Hertz, R_i in ohms, and C_i in farads. Equation (5-9a) is useful to find the phase angle if f, R_i, and C_i are known. If the desired phase angle is known, choose a value for C_i and solve for R_i:

$$R_i = -\frac{\tan \dfrac{\theta}{2}}{2\pi f C_i} \tag{5-9b}$$

Example 5-12: Find R_i in Fig. 5-11(b) so that V_o will lag E_i by 90°. The frequency of E_i is 1 kHz.
Solution: Since $\theta = -90°$, $\tan(-90°/2) = \tan(-45°) = -1$; from Eq. (5-9b),

$$R_i = \frac{-(-1)}{2\pi \times 1000 \times 0.01 \times 10^{-6}} = 15.9\ \text{k}\Omega$$

With $R_i = 15.9\ \text{k}\Omega$, V_o will have the phase angle shown in Fig. 5-11(a). This wave form is a negative cosine wave.

Example 5-13: If $R_i = 100\ \text{k}\Omega$ in Fig. 5-11(b), find the phase angle θ.
Solution: From Eq. (5-9a),

$$\begin{aligned}\theta &= -2 \arctan(2\pi)(1 \times 10^3)(100 \times 10^3)(0.01 \times 10^{-6}) \\ &= -2 \arctan 6.28 \\ &= -2 \times 81° = -162°\end{aligned}$$

It can be shown from Eq. (5-9a) that $\theta = -90°$ when R_i equals the reactance of C_i, or $1/(2\pi f C_i)$. As R_i is varied from 1 kΩ to 100 kΩ, θ varies from approximately −12° to −168°. Thus, the phase shifter can shift phase angles over a range approaching 180°. If R_i and C_i are interchanged in Fig. 5-11(b), the phase angle is positive.

5-10 Phonograph Preamplifier

5-10.1 Record-Cutting Equalization. When a phonograph record is cut, the cutting stylus moves with a constant velocity for a particular amplitude of voltage. If the side-to-side excursions are set to cut almost one groove width at 1 kHz, lower frequencies will drive the cutter into adjacent grooves. Higher frequencies will cause smaller cutter motion. Excursions of the cutter must be equalized. That is, signal frequencies from 500 Hz down to 50 Hz must be attenuated. Frequencies from 2 kHz to 20 kHz must be amplified. Otherwise, higher frequencies will be lost in the noise. This process is called recording equalization.

5-10.2 RIAA Playback Equalization. When a record is played back, a preamplifier circuit must reverse the recording equalization process. The frequency response curve required for the preamplifier is shown for a magnetic or reluctance phono pickup in Fig. 5-12(a). The ideal curve, shown by dashed lines, is called the *RIAA playback equalization curve*. (This equalization curve is the standard playback characteristic curve approved by the

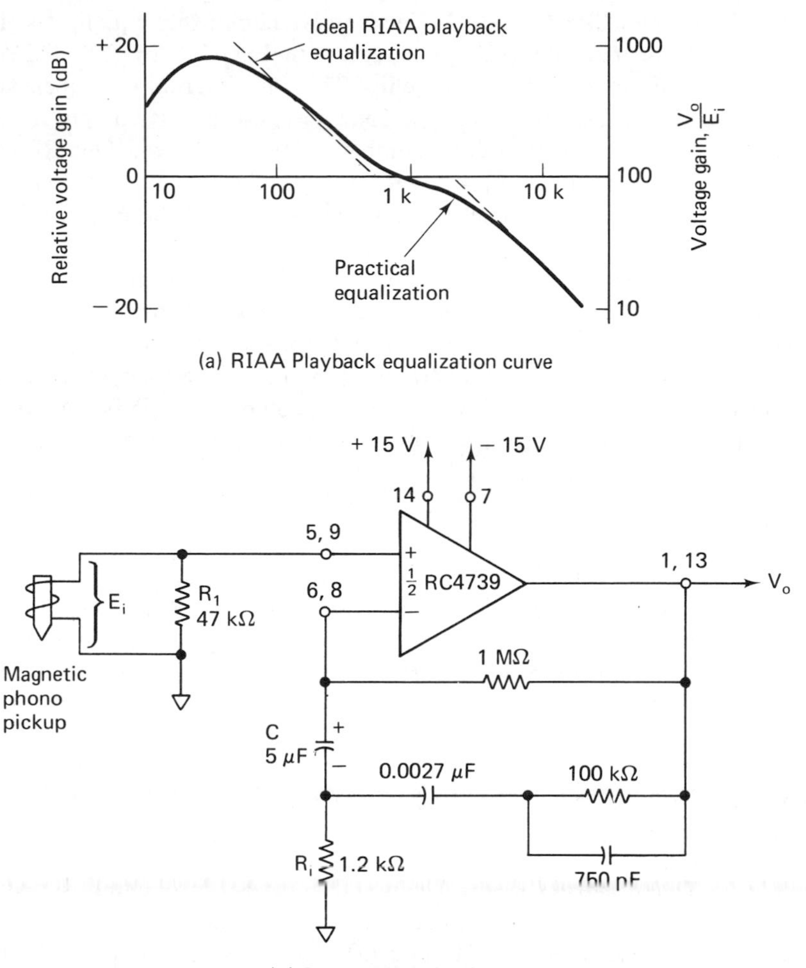

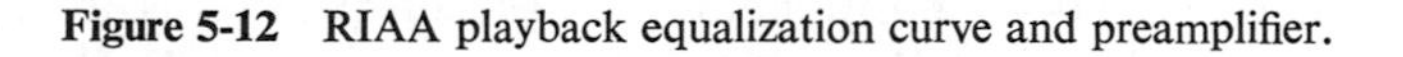

Figure 5-12 RIAA playback equalization curve and preamplifier.

Record Industry Association of America for long playing records.) The left-hand vertical axis of Fig. 5-12(a) shows how gain must be increased or decreased in decibels (dB) relative to gain at 1 kHz.

The preamplifier must also furnish voltage gain to raise the typically 5 mV signal from the pickup to at least 0.5 V for the input of an audio amplifier. The right-hand axis of Fig. 5-12(a) is a plot of the magnitude of the preamp's voltage gain. At 1 kHz, the voltage gain is 100, and if the input is 5 mV, the output is 5 mV $\times$ 100 = 0.5 V. The output of the preamp may go to a volume control before the input of the next stage.

5-10.3 Preamplifier Circuit. An inexpensive circuit that equalizes signal frequencies from magnetic pickups is shown in Fig. 5-12(b). The RC4739 has two low-noise op amps on one chip. They are internally compensated (see Chapter 10). (The μA 739 or MC 1303 may also be used as pin-for-pin replacements provided that external compensation is installed.) One RC4739 can equalize both channels of a stereo system. The first number on each terminal identifies the *A* channel op-amp and the second the *B* channel op-amp.

Resistor R_1 sets the preamp's input resistance to match the internal resistance of the pickup. R_i and the 100 kΩ feedback resistor set the gain at 1 kHz to be about 100. R_i and the 1-MΩ resistor set the gain at 50-Hz to be almost 1000. The 750-pF capacitor bypasses the 100-kΩ resistor above 2 kHz and rolls the gain off to 10 at 20 kHz. R_i and C roll off the gain below 30 Hz. The resulting practical playback equalization curve is shown by the solid line in Fig. 5-12(a).

5-11 Tone Control

5-11.1 Introduction. The preamplifier of Section 5-10 will deliver a flat frequency response at its output. In most high fidelity systems, the owner wants to have a tone-control feature that allows boosting or cutting the volume of bass or treble frequencies. A frequency-controlling network, made of resistors and capacitors, could be installed in series with the output of the preamplifier. However, this network would attenuate some of the frequencies by as much as 1/100 or 20 db. Much of the gain so carefully built into the preamplifier would be lost.

5-11.2 Tone-Control Circuit. The practical tone-control circuit shown in Fig. 5-13(a) (1) features boost or cut of bass frequencies below 500 Hz and of treble frequencies above 2 kHz and (2) eliminates attenuation. The top 50-kΩ audio taper potentiometer is the bass frequency control. With the wiper adjusted to full boost position, the voltage gain at 10 Hz is about $10R/R$ or 10. With the wiper at full bass cut, the voltage gain at 10 Hz is about $R/10R = 0.1$. In effect, the $10R$ pot is adjusted to be in series with R_i for cut or R_f for boost. The boost capacitors C_B begin to bypass the pot at frequencies between 50 Hz and 500 Hz, as shown in Fig. 5-13(b).

When adjusted to full boost, the treble control, $R/3$, and the capacitors C_T set the gain at 20 kHz to 10. At full cut, the gain is 0.1 as shown in Fig. 5-13(b). With both bass and treble control pots adjusted to the center of their rotation, the frequency response of the tone-control circuit will be flat. The input signal E_i delivered from the preamplifier should be about 0.2 V rms at 1 kHz. Therefore, the output of the tone control should be at about the same level.

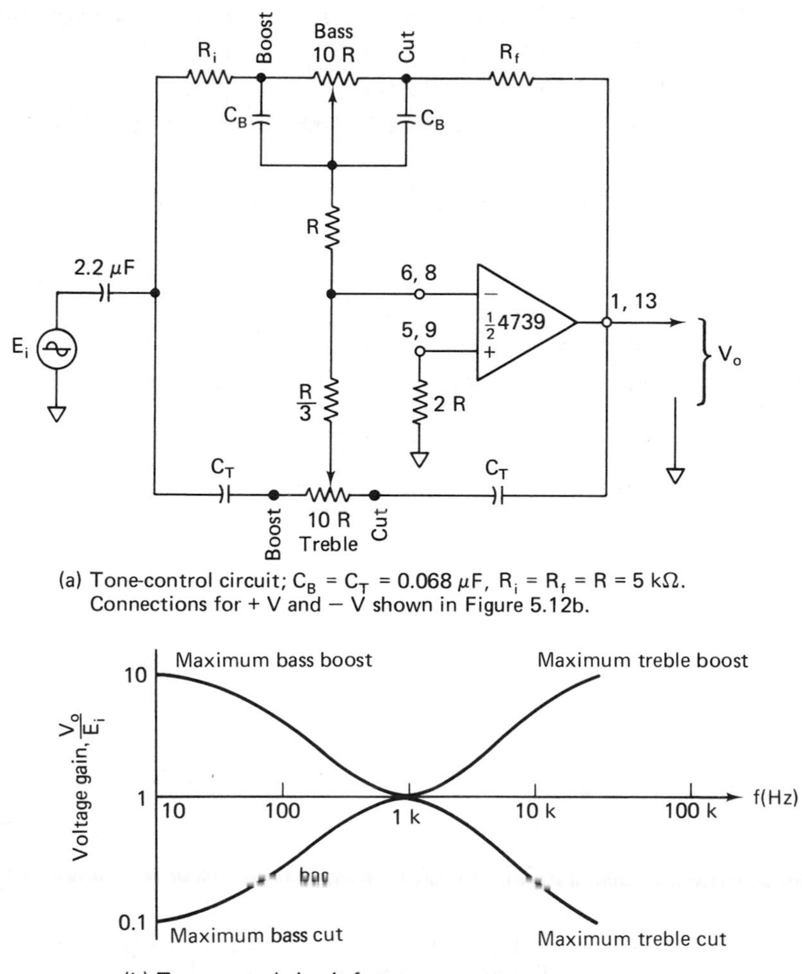

Figure 5-13 The tone-control circuit in (a) has the frequency response curves shown in (b).

More applications for the 741 and 301 are shown in Appendices 1 and 2, respectively.

Problems

5-1 In the circuit of Fig. 5-1, if $E_1 = 0.75$ V, what is the value of I_m?

5-2 Repeat Example 5-2 for $E_{FS} = 10$ V.

5-3 A microammeter with $I_{FS} = 100\ \mu$A is used in Fig. 5-1. If $R_i = 40$ kΩ, what is E_{FS}?

5-4 In Fig. 5-2, the full-scale meter current is 50 μA. Calculate R_i so that there is a full-scale deflection when $E_i = 10$ V rms.

5-5 Repeat Problem 5-4 for $E_i = 10$ V peak.

5-6 In Fig. 5-2, which diodes conduct for (a) positive half-cycles of E_i (b) negative half-cycles of E_i?

5-7 When the load is in the feedback loop, what type of circuit is this classified as?

5-8 V_o in Fig. 5-3(a) is measured to be 12 V; what is the zener voltage, V_Z?

5-9 If a germanium diode is put in the feedback loop of Fig. 5-3(b), determine (a) the approximate value of V_o and (b) I.

5-10 If $E_i = 5$ V in Fig. 5-3(b), determine (a) I and (b) V_o.

5-11 If β of the transistor in Fig. 5-4 is 50, find (a) the current through the LED and (b) I_o.

5-12 What name is used to classify circuits where the load current is independent of the load resistance?

5-13 For Fig. 5-5, let $E_1 = 5$ V, $R = 20$ kΩ, and $R_L = 4$ kΩ; find (a) I_L, (b) V_L, (c) V_o, (d) I_i, and (e) I_f.

5-14 If $V_o = 10$ V and $E_2 = 4$ V in Fig. 5-6, calculate (a) V_L and (b) I_L if $R_L = 6$ kΩ.

5-15 Using the values given in Problem 5-14, what is the total current, I_o, that would flow into the output terminal of the op amp? $R = 10$ kΩ.

5-16 In the circuit of Fig. 5-6, if the voltage at the (+) input, E_1, is 6 V and the voltage at the (−) input, E_2, is 10 V, determine (a) V_o, (b) I_L, and (c) V_L. $R = 10$ kΩ and $R_L = 500$ Ω.

5-17 Calculate (a) the load current, I_L, and (b) the current into pin 10 of Fig. 5-7 if $V_Z = 10$ V and $R_s = 40$ Ω. $\beta = 100$.

5-18 For the values given in Problem 5-17 what is the zener current?

5-19 If the resistance of the microammeter in Fig. 5-8(a) and (b) is 10 kΩ, what value of current will the meter measure?

5-20 (a) If $V_o = 12$ V and $R_f = 200$ kΩ in Fig. 5-8(c), determine I_{SC}. (b) What is the voltage across the 50-kΩ resistor?

5-21 In the circuit of Fig. 5-10, if $I_{SC} = 50$ μA, what is the value of m such that $I_L = 2$ mA?

5-22 If $mR = 49$ kΩ in Fig. 5-10, calculate I_L.

5-23 Calculate the phase shift of Fig. 5-11(b) if $R_i = 25$ kΩ. $f = 1$ kHz.

5-24 Determine the value of R_i in Fig. 5-11(b) for a phase shift of $-135°$. $f = 2$ kHz.

6

Signal Generators

6-0 Introduction

Up to now our main concern has been to use the op amp in circuits that process signals. In this chapter we concentrate on op amp circuits that generate signals. Four of the most common and useful signals are described by their shape when viewed on a cathode ray oscilloscope. They are the square wave, triangular wave, sawtooth wave and sine wave. Accordingly, the signal generator is classified by the shape of the wave it generates. Some circuits are so widely used that they have been assigned a special name. For example, the first circuit presented in Section 6-1 is a multivibrator that generates primarily square waves.

6-1 Free-Running Multivibrator

6-1.1 Multivibrator Action. A *free-running* or *astable multivibrator* is a square wave generator. The circuit of Fig. 6-1 is a multivibrator circuit and looks something like a comparator with hysteresis (Chapter 4), except that the input voltage is replaced by a capacitor. Resistors R_1 and R_2 form a voltage divider to feed back a fraction of the output to the (+) input. When V_o is at $+V_{sat}$, as shown in Fig. 6-1(a), the feedback voltage is called the upper threshold voltage, V_{UT}. V_{UT} is given in Eq. (4-1) and repeated here for convenience.

$$V_{UT} = \frac{R_2}{R_1 + R_2}(+V_{sat}) \qquad (6\text{-}1)$$

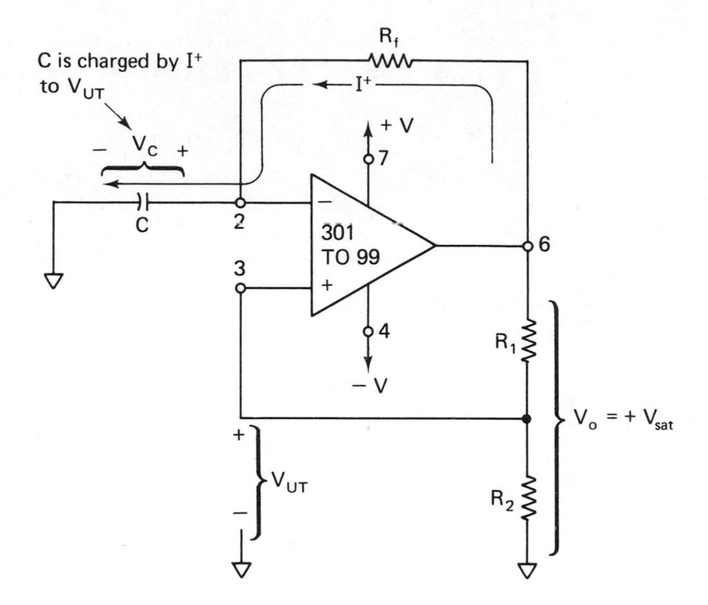

(a) When C is charged to V_{UT}, V_o switches to $-V_{sat}$

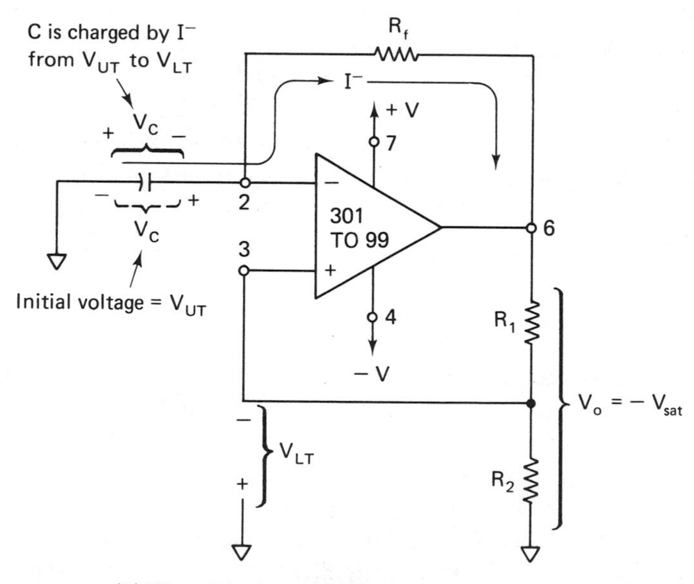

(b) When C is charged to V_{LT}, V_o switches to $+V_{sat}$

Figure 6-1 Free-running multivibrator. ($R_1 = 100\ \text{k}\Omega$, $R_2 = 86\ \text{k}\Omega$). Output voltage wave form shown in Fig. 6-2.

Resistor R_f provides a feedback path to the (−) input. When V_o is at $+V_{sat}$, current I^+ flows through R_f to charge capacitor C. As long as the capacitor voltage V_C is less than V_{UT}, the output voltage remains at $+V_{sat}$.

When V_C charges to a value slightly greater than V_{UT}, the (−) input goes positive with respect to the (+) input. This switches the output from $+V_{sat}$ to $-V_{sat}$. The (+) input is now held negative with respect to ground because the feedback voltage is negative and given by

$$V_{LT} = \frac{R_2}{R_1 + R_2}(-V_{sat}) \tag{6-2}$$

Equation (6-2) is the same as Eq. (4-2). Just after V_o switches to $-V_{sat}$, the capacitor has an initial voltage equal to V_{UT}. See Fig. 6-1(b). Now current I^- discharges C to 0 V and recharges C to V_{LT}. When V_C becomes slightly more negative than the feedback voltage V_{LT}, output voltage V_o switches back to $+V_{sat}$. The condition in Fig. 6-1(a) is re-established except that C now has an initial charge equal to V_{LT}. The capacitor will discharge from V_{LT} to 0 V and then recharge to V_{UT}, and the process is repeating. Free-running multivibrator action is summarized as follows:

1. When $V_o = +V_{sat}$, C charges from V_{LT} to V_{UT} and switches V_o to $-V_{sat}$.
2. When $V_o = -V_{sat}$, C charges from V_{UT} to V_{LT} and switches V_o to $+V_{sat}$.

The time needed for C to charge and discharge determines the frequency of the multivibrator.

6-1.2 Frequency of Oscillation. The capacitor and output voltage waveforms for the free-running multivibrator are shown in Fig. 6-2. Resistor R_2 is chosen to equal $0.86R_1$ to simplify calculation of capacitor charge time. Time intervals t_1 and t_2 show how V_C and V_o change with time for Figs. 6-1(a) and 6-1(b) respectively. Time intervals t_1 and t_2 are equal to the product of R_f and C.

The period of oscillation, T, is the time needed for one complete cycle. Since T is the sum of t_1 and t_2,

$$T = 2R_fC, \quad \text{for } R_2 = 0.86R_1 \tag{6-3a}$$

The frequency of oscillation, f, is the reciprocal of the period T and is expressed by

$$f = \frac{1}{T} = \frac{1}{2R_fC} \tag{6-3b}$$

where T is in seconds, f in hertz, R_f in ohms, and C in farads.

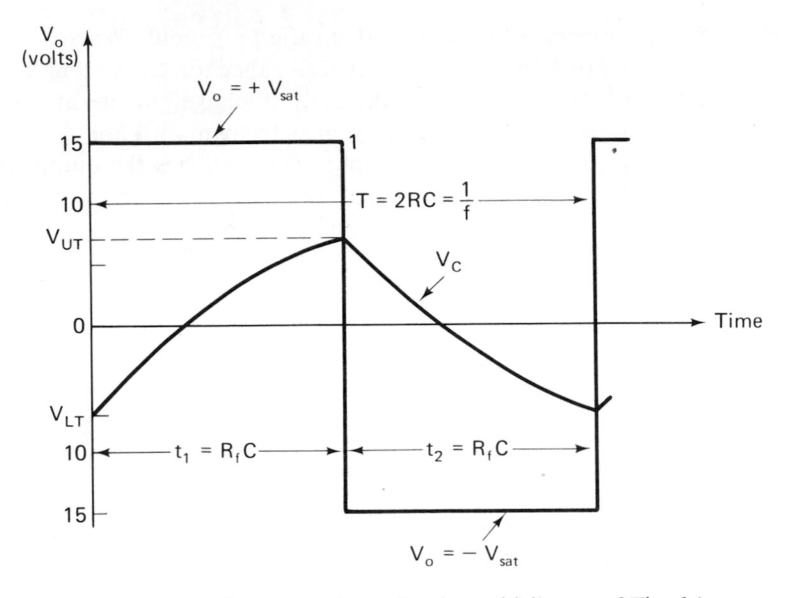

Figure 6-2 Voltage waveshapes for the multivibrator of Fig. 6-1.

Example 6-1: In Fig. 6-1, if $R_1 = 100\ \text{k}\Omega$, $R_2 = 86\ \text{k}\Omega$, $+V_{sat} = +15$ V, and $-V_{sat} = -15$ V, find (a) V_{UT} and (b) V_{LT}.
Solution: (a) By Eq. (6-1),

$$V_{UT} = \frac{86\ \text{k}\Omega}{186\ \text{k}\Omega} \times 15\ \text{V} \approx 7\ \text{V}.$$

(b) By Eq. (6-2),

$$V_{LT} = \frac{86\ \text{k}\Omega}{186\ \text{k}\Omega}(-15\text{V}) = -7\text{V}.$$

Example 6-2: Find the period of the multivibrator in Example 6-1 if $R_f = 100\ \text{k}\Omega$ and $C = 0.1\ \mu\text{F}$.
Solution: Using Eq. (6-3a), $T = (2)(100\ \text{k}\Omega)(0.1\ \mu\text{F}) = 0.020\ \text{s} = 20\ \text{ms}$.

Example 6-3: Find the frequency of oscillation for the multivibrator of Example 6-2.
Solution: From Eq. (6-3b), $f = \dfrac{1}{20 \times 10^{-3}\ \text{s}} = 50\ \text{Hz}$

6-2 One-Shot Multivibrator

6-2.1 Introduction. A *one-shot multivibrator* generates a single output pulse in response to an input signal. The length of the output pulse depends only on external components (resistors and capacitors) connected to the op amp. As shown in Fig. 6-3, the one-shot generates a single output pulse

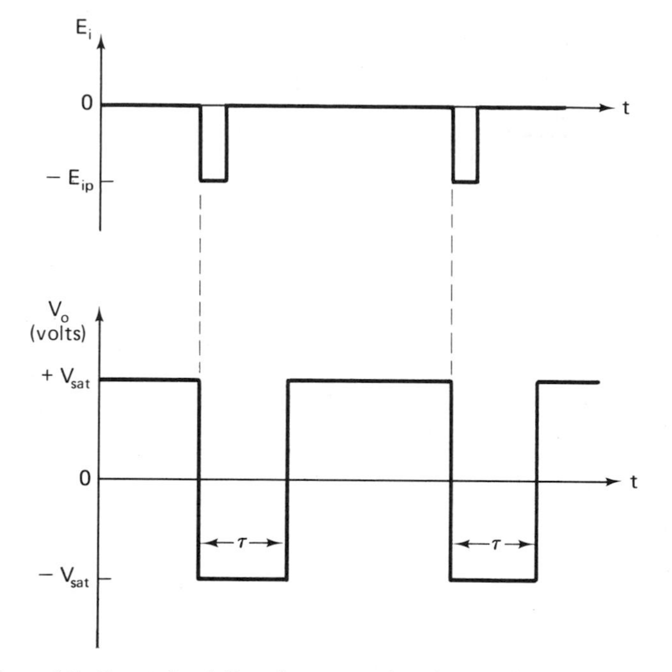

Figure 6-3 Input signal E_i and output pulse of a one-shot multivibrator.

on the negative-going edge of E_i. The duration of the input pulse can be longer or shorter than the expected output pulse. The duration of the output pulse is represented by τ in Fig. 6-3. Since τ can be changed only by changing resistors or capacitors, the one-shot can be considered a *pulse stretcher*. This is because the width of the output pulse is wider than the input pulse. Moreover, the one-shot introduces an idea of an adjustable delay, that is, the delay between the time when E_i goes negative and the time for V_o to go positive again. Operation of the one-shot will be studied in three parts: (1) the stable state, (2) transition to the timing state, and (3) the timing state.

6-2.2 Stable State. In Fig. 6-4(a), V_o is at $+V_{sat}$. Voltage divider R_1 and R_2 feeds back V_{UT} to the (+) input. V_{UT} is given by Eq. (6-1). The diode D_1 clamps the (−) input at approximately +0.5 V. The (+) input is positive with respect to the (−) input, and the high open-loop gain times the differential input voltage ($E_d = 2.1 - 0.5 = 1.6$ V) holds V_o at $+V_{sat}$.

6-2.3 Transition to the Timing State. If input signal E_i is at a steady dc potential as in Fig. 6-4(a), the (+) input remains positive with respect to (−) input and V_o stays at $+V_{sat}$. However, if E_i goes negative by a peak value E_{ip} approximately equal to twice V_{UT}, the voltage at the (+) input will be pulled below the voltage at the (−) input. Once the (+) input becomes negative with respect to the (−) input, V_o switches to $-V_{sat}$. With this change,

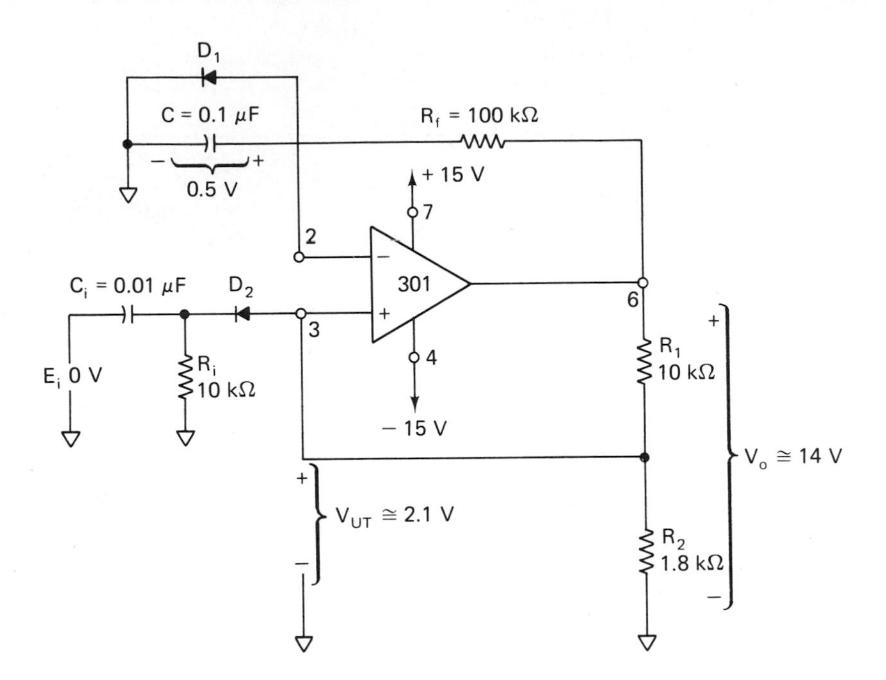

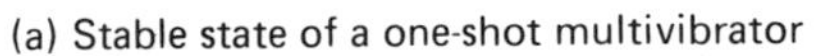
(a) Stable state of a one-shot multivibrator

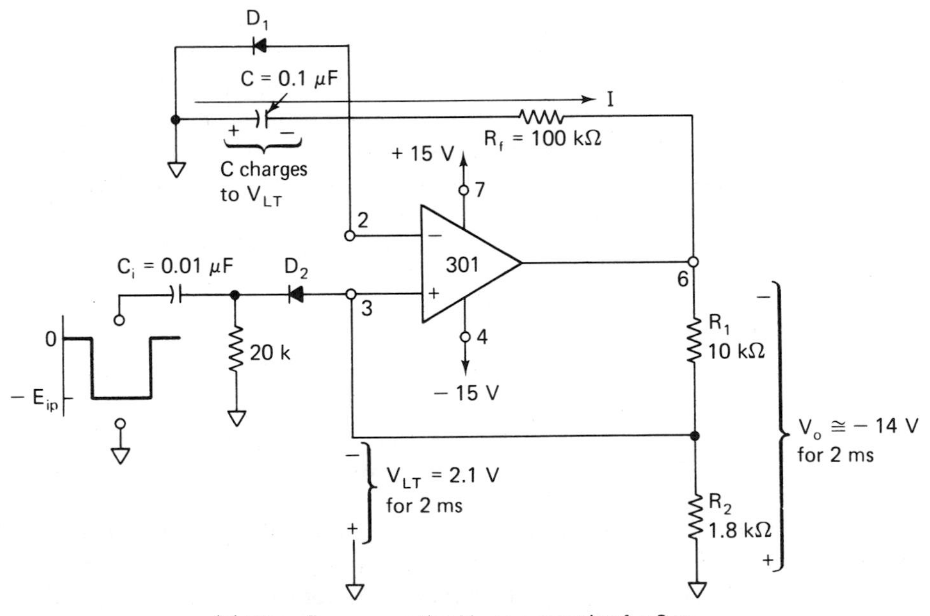

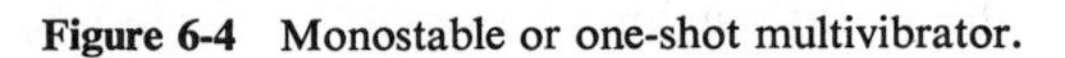
(b) When E_i goes negative, V_o goes negative for 2 ms

Figure 6-4 Monostable or one-shot multivibrator.

the one-shot is now in its timing state. For best results, C_i should be greater than 0.005 μF.

6-2.4 Timing State. The timing state is an unstable state; that is, the one-shot cannot remain very long in this state for the following reasons. Resistors R_1 and R_2 in Fig. 6-4(b) feed back a negative voltage ($V_{LT} = -2.1$ V) to the (+) input. The diode D_1 is now reversed-biased by $-V_{sat}$ and is essentially an open circuit. Capacitor C discharges to 0 and then recharges with a polarity opposite to that in Fig. 6-4(a). (See Fig. 6-4(b).) As C recharges, the (−) input becomes more and more negative with respect to ground. When the capacitor voltage is slightly more negative than V_{LT}, V_o switches to $+V_{sat}$. The one-shot has now completed its output pulse and is back to the stable state in Fig. 6-4(a). Since the one-shot has only one stable state, it is also called a *monostable multivibrator*.

6-2.5 Duration of Output Pulse. If R_2 is made about one-fifth of R_1, in Fig. 6-4, then the duration of output pulse is given by

$$\tau \approx \frac{R_f C}{5} \tag{6-4}$$

Example 6-4: Calculate τ for the one-shot of Fig. 6-4.
Solution: By Eq. (6-4),

$$\tau = \frac{(100\text{k}\Omega)(0.1\ \mu\text{F})}{5} = 2\text{ ms}.$$

For test purposes, E_i can be obtained from a square wave or pulse generator. Diode D_2 prevents the one-shot from coming out of the timing state on positive transitions of E_i. To build a one-shot that has a positive output pulse for a positive input signal, simply reverse the diodes.

6-3 Ramp Generator

6-3.1 Introduction. The op amp can be used to generate not only square waves but also ramp, triangular, sawtooth, and many other types of voltage waveforms. Circuits that generate either ramp, triangular, or sawtooth waves have one thing in common: a capacitor is charged by a constant current. For example in Fig. 6-5(a) the switch is closed at time $t = 0$ and constant current I charges capacitor C. To obtain an expression for voltage across the capacitor in terms of current, we begin with the relationship between current, charge, and time:

$$I = \frac{Q}{t}$$

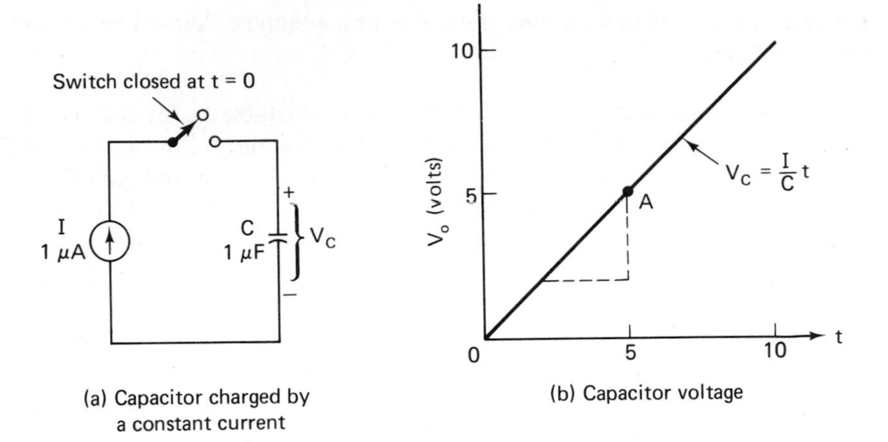

Figure 6-5 Voltage across a capacitor rises at a constant rate when the capacitor is charged by a constant current.

or

$$Q = It \tag{6-5a}$$

An equation relating voltage with charge and capacitance is

$$V_C = \frac{Q}{C} \tag{6-5b}$$

Substituting Eq. (6-5a) into Eq. (6-5b) to eliminate Q yields

$$V_C = \frac{I}{C} \times t \tag{6-5c}$$

where V_C is in volts, t in seconds, I in amperes, and C in farads. If we know I and C, the capacitor voltage V_C will be directly related to time elapsed after the switch is closed.

Example 6-5: Calculate the capacitor voltage V_C in Fig. 6-5(a) 5 seconds after the switch is closed.
Solution: By Eq. (6-5c), $V_C = \frac{1\ \mu\text{A}}{1\ \mu\text{F}}(5\ \text{s}) = 5\ \text{V}$.

Figure. 6-5(b) is a general plot of V_C versus time. The solution to Example 6-5 is point A on this diagram. Equation (6-5c) may be rearranged to show the *rate* at which V_o rises or falls in volts per second.

$$\frac{V_C}{t} = \frac{I}{C} = \frac{1\ \mu\text{A}}{1\ \mu\text{F}} = 1\ \frac{\text{V}}{\text{s}} \tag{6-5d}$$

V_C represents a continuous tally of how much voltage has been accumulated by the capacitor. For example, after 1 second, $V_C = 1$ V. For each following second the capacitor adds another volt. So V_C actually sums the voltage over a period of time and reads out the total voltage. In mathematics this type of summing process is called integration. Hence this circuit is called an *integrator*. The triangular shape of V_C is called a *ramp* and is the basis for generating many useful control signals. To make a basic single-ramp generator, we will use the op amp.

6-3.2 Ramp-Generator Circuit. The current source in Fig. 6-5(a) is replaced by input voltage E_i and R_i and an op amp as shown in Fig. 6-6. Current I is set by E_i and R_i at $I = E_i/R_i$. Substituting for I in Eq. (6-5c), we obtain V_o in terms of E_i and time t:

$$V_o = -E_i \times \frac{1}{R_i C} \times t \tag{6-6}$$

where R_i is in ohms, C in farads, t in seconds, V_o and E_i in volts. The minus sign in Eq. (6-6) shows that E_i is applied through R_i to the (−) input. Note that the capacitor voltage V_C equals V_o, so now the load current will be furnished from the op amp's output terminal and will not drain the capacitor.

There are two disadvantages to the circuit of Fig. 6-6. The first and most obvious is that V_o can go negative only to $-V_{sat}$. The second, not so obvious, is the fact that V_o will not remain at 0 V when $E_i = 0$ V. The reason for this is the inevitable presence of small bias currents that will charge the capacitor. (Bias currents will be studied in Chapter 9.) One method of stopping the capacitor from charging is to place a short circuit across it. V_C and V_o will then stay at 0 V. To start the ramp again, simply remove the short.

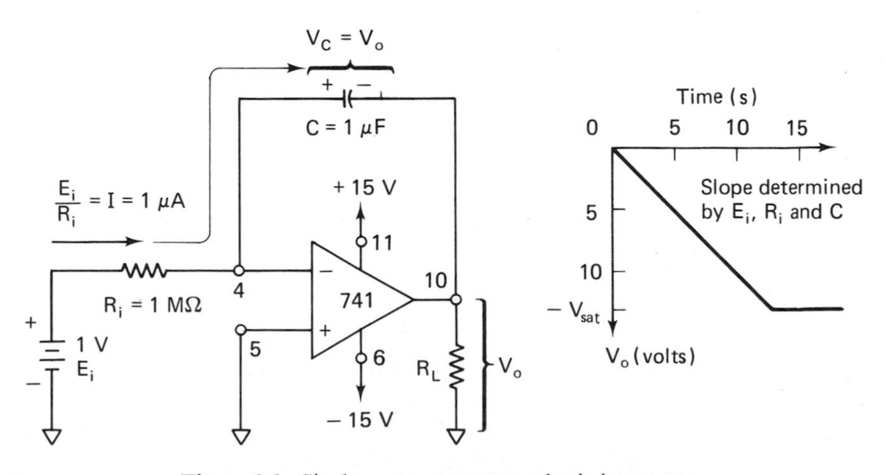

Figure 6-6 Single-ramp generator, a basic integrator.

If a positive-going ramp is needed, simply reverse E_i. Understanding how a single-ramp generator works is needed in designing adjustable timing circuits.

6-4 An Adjustable Timer

6-4.1 Circuit Description. The single-ramp voltage generator of Fig. 6-6 generates an output voltage that depends on time, in accordance with Eq. (6-6). If V_o is known, then the time that has elapsed since the capacitor began charging can be calculated. This principle is used in Fig. 6-7(a) to make an interval timer. Op amp A is a ramp generator that generates a negative-going ramp when the control switch is thrown to *start*. Op amp B is a comparator that monitors the ramp voltage with its negative input. An adjustable negative reference voltage is applied to the (+) input of op amp B. When the ramp voltage crosses the reference voltage, the output voltage of the comparator snaps to $+V_{sat}$. This action is shown by the wave shapes in Fig. 6-7(b). R_L can be a sensitive relay or other power-control device that would, for example, activate an alarm or turn off a lamp. The diode in series with R_L conducts current only when V_o goes positive.

6-4.2 Circuit Analysis. The circuit is analyzed beginning with an example.

Example 6-6: (a) Find the rate at which V_o drops in volts per second for the single ramp generator in Fig. 6-7(a). (b) Convert this *ramp-down* rate to volts per minute.
Solution: (a) Rewrite Eq. (6-6) to give volts per second:

$$\text{Volts per second} = \frac{V_{ramp}}{t} = -\frac{E_i}{R_iC}$$

Calculate $R_i \times C = 1\ \text{M}\Omega \times 60\ \mu\text{F} = 60\ \text{s}$. Substituting into Eq. (6-6) for E_i and R_iC,

$$\text{Volts per second} = \frac{V_{ramp}}{t} = -\frac{1\ \text{V}}{60\ \text{s}}$$

(b) Convert seconds to minutes:

$$\frac{V_{ramp}}{t} = -\frac{1\ \text{V}}{60\ \text{s}} \times \frac{60\ \text{s}}{1\ \text{min}} = -\frac{1\ \text{V}}{\text{min}}$$

We learn from Example 6-6 that E_i can control the rate at which V_{ramp} drops. For example if E_i is doubled to 2 V, the ramp will drop twice as fast, at a rate of 2 V/min.

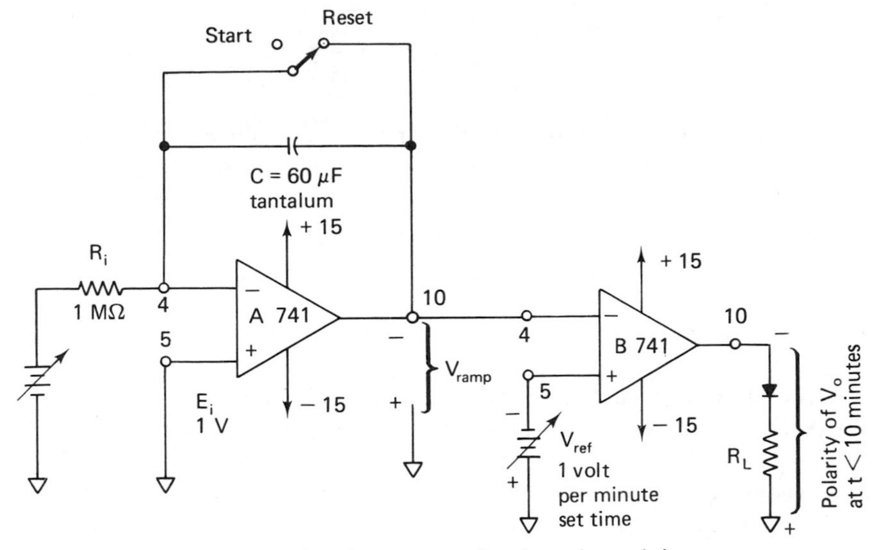

(a) Single-ramp generator A and comparator B make an interval timer

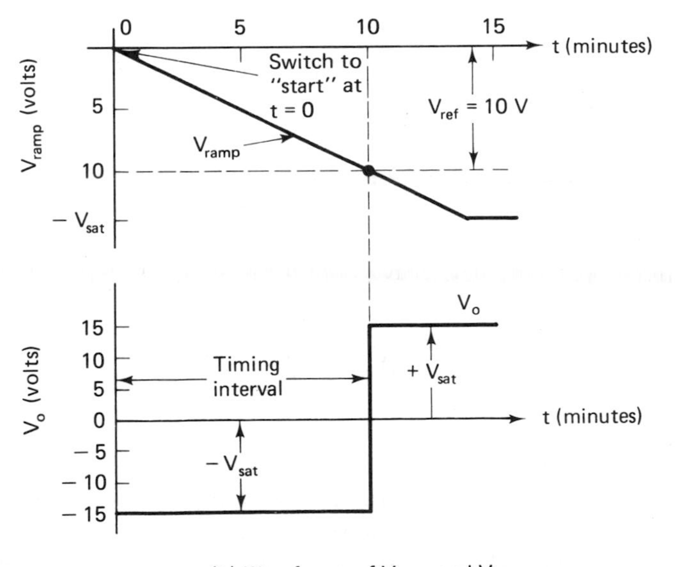

(b) Waveforms of V_{ramp} and V_O

Figure 6-7 Adjustable minute timer.

Example 6-7: If $V_{ref} = -10$ V in Fig. 6-7, how much time will elapse between the time the switch is opened (start) and V_o goes to $+V_{sat}$.

Solution: From Section 2-4, the comparator will switch when $V_{ramp} = V_{ref}$. The timing interval can be found from

$$\text{timing interval} = \frac{V_{\text{ref}}}{\dfrac{V_{\text{ramp}}}{t}} = \frac{10\text{ V}}{\dfrac{1\text{ V}}{\text{min}}} = 10\text{ min}$$

Now that a comparator has been connected to a single-ramp generator, it is natural to ask whether the output of the comparator can also control the input of the ramp generator. The answer is yes, and such an op amp circuit is used to build a triangular wave generator.

6-5 Triangular Wave Generator

6-5.1 Introduction. A triangular wave generator requires a minimum of two op amps. The resulting circuit operation is fairly complicated to analyze all at once. Understanding is simplified if we proceed in three logical steps. First, show how a basic triangular wave can be generated manually with one op amp, a resistor, a capacitor, and a switch. Second, select a comparator to replace the manual operation of the switch. Third, put the comparator and basic triangular wave generator together. We begin with applying the principles of the single-ramp generator studied in Section 6-3.

6-5.2 Basic Operation. A manually controlled triangular wave generator can be made by adding one switch and another dc control voltage to the single-ramp generator of Fig. 6-6. The resulting circuit is shown in Fig. 6-8(a). When the control switch is in the up position, E_i is -15 V and V_o ramps up. When the control switch is in the down position, E_i is $+15$ V and V_o ramps down. The rate of change of the ramp voltage is found from Eq. (6-6), for $E_i = +15$ V as

$$\frac{V_o}{t} = -\frac{E_i}{R_iC} = -\frac{15\text{ V}}{1\text{ M}\Omega \times 1\ \mu\text{F}} = -\frac{15\text{ V}}{\text{s}}$$

For $E_i = -15$ V, $V_o/t = 15$ V/s.

To see how the ramp voltages are converted to a triangular wave, refer to Fig. 6-8(b). At time $t = 0$, power is turned on with the switch in the "ramp down" position. E_i is positive, so V_o ramps down at -15 V/s. When V_o reaches a selected *lower-threshold voltage*, V_{LT}, the control switch is thrown to the "ramp up" position. E_i is now -15 V and V_o ramps up at $+15$ V/s. When V_o reaches a selected upper-threshold voltage, V_{UT}, the control switch is thrown back to the "ramp down" position. From then on the control switch position must be changed every time the ramp voltage V_o crosses one of the threshold voltages.

To make operation of the control switch automatic, we can replace it with a comparator. The type of comparator needed for the triangular wave

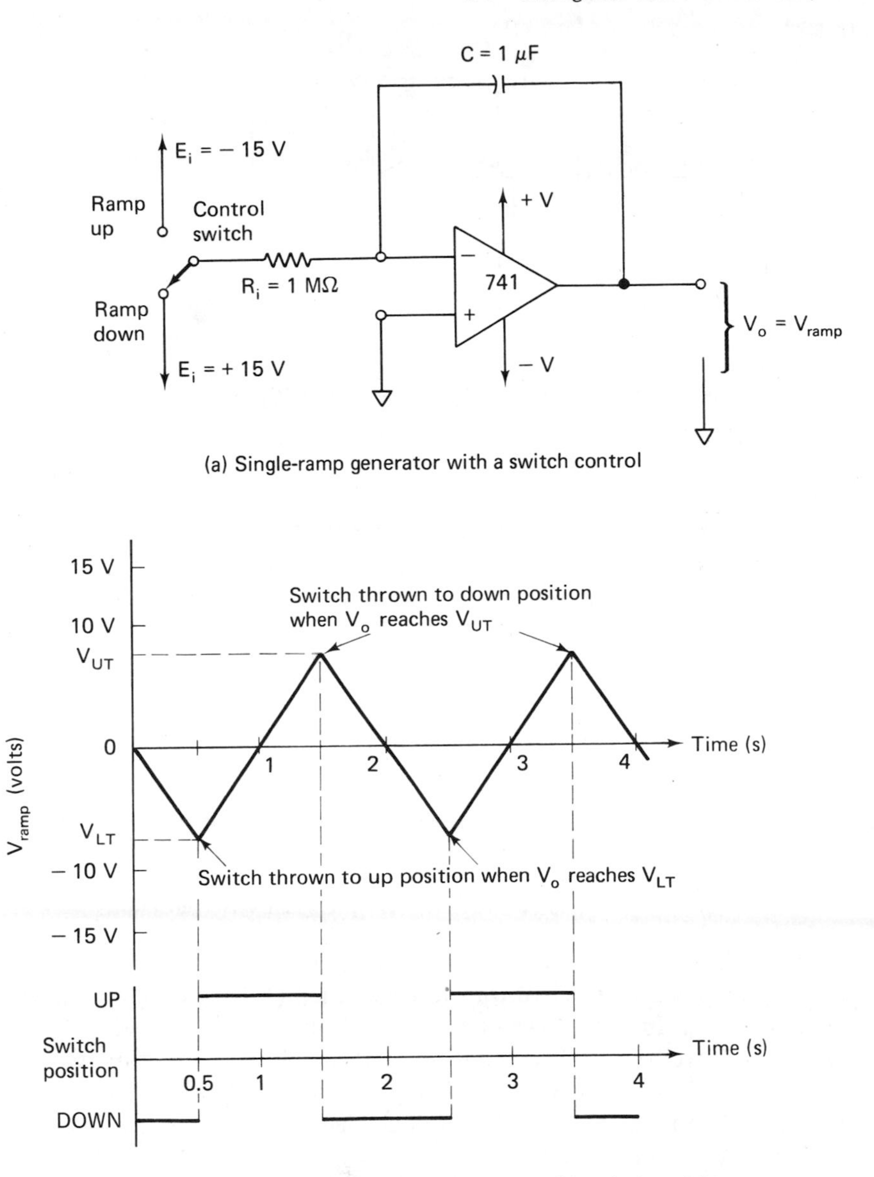

(a) Single-ramp generator with a switch control

(b) Timing diagram showing how V_o varies with switch position

Figure 6-8 Manually controlled triangular wave generator.

generator was not studied in Chapter 4. Therefore, we must introduce comparators again.

6-5.3 Comparator with Hysteresis and Voltage Sensing at the (+) Input. The output voltage V_{ramp} or Fig. 6-8 is applied to the input of the comparator in Fig. 6-9(a). To analyze comparator operation, assume that its output,

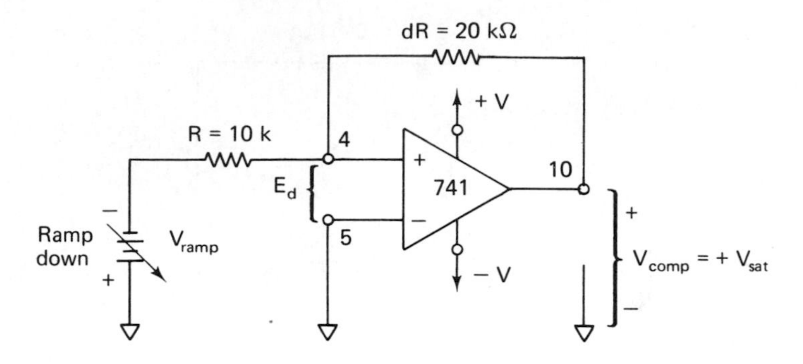

(a) When $V_{ramp} = V_{LT}$, V_{comp} switches from $+V_{sat}$ to $-V_{sat}$

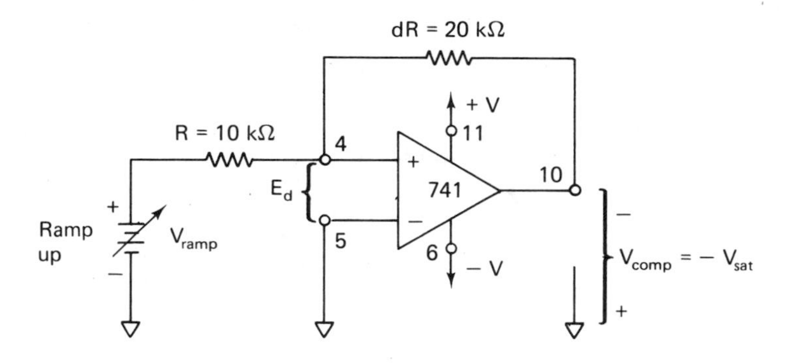

(b) When $V_{ramp} = V_{UT}$, V_{comp} switches from $-V_{sat}$ to $+V_{sat}$

Figure 6-9 Comparator operation for triangular wave generator.

V_{comp}, is locked at $+V_{sat}$. For any positive V_{ramp} in Fig. 6-9(a), the (+) input of the op amp will be positive with respect to the (−) input and V_{comp} will remain at $+V_{sat}$. In order to change V_{comp} to $-V_{sat}$, the input voltage V_{ramp} must go negative below 0 V to some lower-threshold voltage $-V_{LT}$. Only at the lower-threshold voltage is the op amp's (+) input equal to 0 V (ground potential). At this time $-V_{LT}$ is developed across R and $+V_{sat}$ is across dR. Expressed mathematically,

$$\frac{-V_{LT}}{R} = \frac{+V_{sat}}{dR}$$

or

$$V_{LT} = -\frac{+V_{sat}}{d} \quad \text{at } E_d \approx 0 \text{ V} \tag{6-7a}$$

For all negative input voltages greater than V_{LT}, the (+) input is now negative with respect to the (−) input and V_{comp} is locked at $-V_{sat}$. To switch

V_{comp} back to $+V_{sat}$, V_{ramp} must go positive (as shown in Fig. 6-9(b)) to an upper-threshold voltage, V_{UT}. When V_{ramp} reaches V_{UT}, all of V_{UT} will be developed across R and $-V_{sat}$ will be developed across dR, so $E_d \approx 0$ V. Thus

$$\frac{V_{UT}}{R} = -\frac{-V_{sat}}{dR}$$

or

$$V_{UT} = -\frac{-V_{sat}}{d} \quad \text{at } E_d \approx 0 \text{ V} \tag{6-7b}$$

The hysteresis voltage V_H (see Section 4-3) is expressed by the difference between threshold voltages, or

$$V_H = V_{UT} - V_{LT} \tag{6-7c}$$

Example 6-8: In Fig. 6-9, R = 10 kΩ and $dR = 20$ kΩ, so $d = 2$. Assume that $+V_{sat} = +15$ V and $-V_{sat} = -15$ V. Find (a) V_{UT}, (b) V_{LT}, and (c) V_H.
Solution: (a) By Eq. (6-7b),

$$V_{UT} = -\frac{(-15 \text{ V})}{2} = 7.5 \text{ V}$$

(b) By Eq. (6-7a)

$$V_{LT} = -\frac{15 \text{ V}}{2} = -7.5 \text{ V}$$

(c) By Eq. (6-7c),

$$V_H = 7.5 \text{ V} - (-7.5 \text{ V}) \times 15 \text{ V}$$

Wave shapes for this comparator will be shown in the triangle wave generator of Section 6-5.4.

6-5.4 Triangle Wave Generator. The output of the ramp generator in Fig. 6-8 is connected to the comparator input of Fig. 6-9, and the comparator's output is connected to the ramp's input. The resulting circuit is the triangle wave generator of Fig. 6-10(a).

To analyze this circuit, focus on the output voltage wave shapes in Fig. 6-10(b). From time 0 to time A, the comparator's output is at $+V_{sat}$. This corresponds to the condition shown in Fig. 6-8(a). During this time interval the comparator's positive output is forcing the ramp generator's output to ramp down (corresponding to a "ramp down" position in Fig. 6-8(a)). When the ramp voltage reaches V_{LT}, the comparator output switches to $-V_{sat}$ and forces the ramp generator's output to ramp back up. When the ramp voltage reaches V_{UT}, the comparator output switches back to $+V_{sat}$ and the ramp voltage starts ramping back down. The sequence repeats and we have a triangle wave generator.

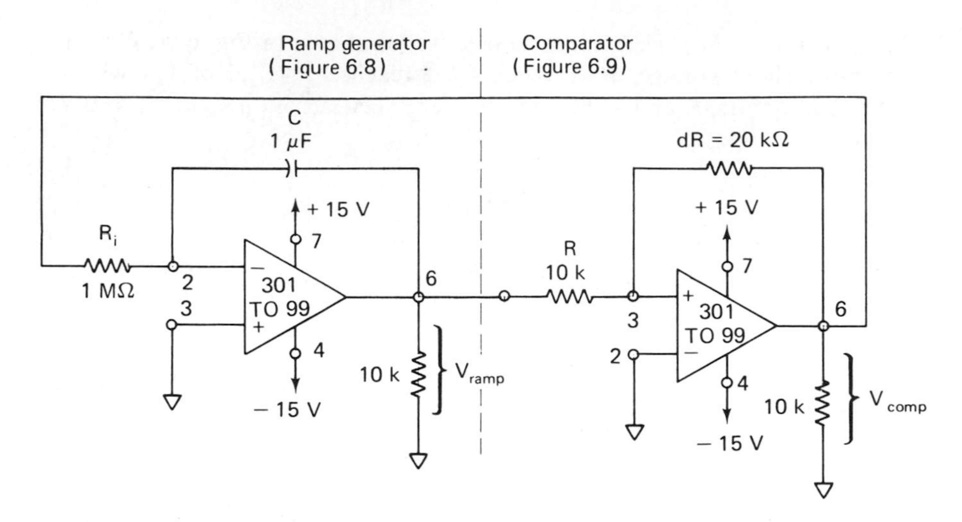

(a) Single-ramp generator plus comparator

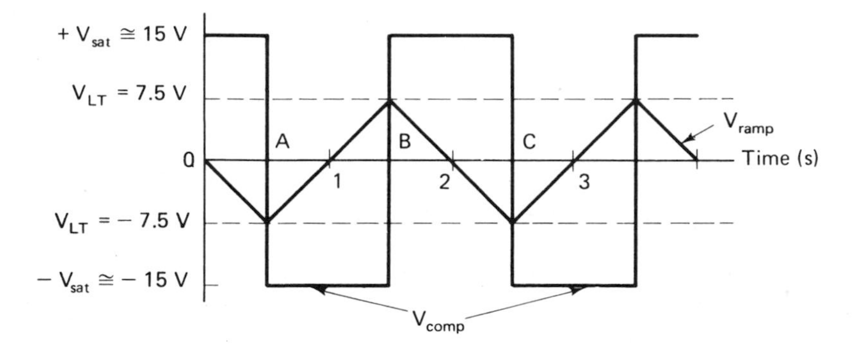

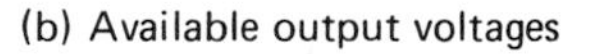
(b) Available output voltages

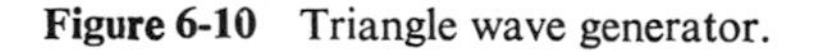
Figure 6-10 Triangle wave generator.

Example 6-9: What time interval is required for the triangle wave generator to complete one cycle? That is, what is the time interval from A to C in Fig. 6-10(b)?

Solution: Equation (6-6) is modified for Fig. 6-10. Calling the time interval from A to B *rise time*, t_r, and substituting the hysteresis voltage V_H for V_o and $-V_{sat}$ for E_i in Eq. (6-6), we obtain

$$t_r = -\frac{V_H}{-V_{sat}}(R_iC) = -\frac{15\text{ V}}{-15\text{ V}}(1\text{ M}\Omega \times 1\ \mu\text{F}) = 1\text{ s} \qquad \text{(6-8a)}$$

Calling the time interval from B to C *fall time*, t_f, and substituting $-V_H$ for V_o and $+V_{sat}$ for E_i into Eq. (6-6) yields

$$t_f = -\frac{-V_H}{+V_{sat}}(R_iC) = -\frac{-15\text{ V}}{15\text{ V}}(1\text{ M}\Omega \times 1\ \mu\text{F}) = 1\text{ s} \tag{6-8b}$$

The time interval from A to C is the period, T, of the wave:

$$T = t_R + t_F = 1\text{ s} + 1\text{ s} = 2\text{ s} \tag{6-8c}$$

The frequency of oscillation f is the reciprocal of the period T:

$$f = \frac{1}{T} = \frac{1}{2\text{ s}} = 0.5\text{ Hz} \tag{6-8d}$$

6-6 Sawtooth Wave Generator

6-6.1 Introduction. The single-ramp generator in Fig. 6-7 was used to make a minute timer. With a slight modification, we can make a sawtooth wave generator by continually resetting the timer. For example, in Fig. 6-11(a) suppose the switch is open. V_o will rise at a rate (see Example 6-6) of

$$\frac{V_o}{t} = -\frac{E_i}{R_iC} = -\frac{-1\text{ V}}{(100\text{ k}\Omega)(0.1\ \mu\text{F})} = \frac{1\text{ V}}{10\text{ ms}}$$

This expression shows that in every 10-ms time interval, V_o will go more positive by 1 V. If the switch is closed, the capacitor will quickly discharge through the short circuit, and V_o will drop to 0 V. If the switch is opened, the capacitor begins again to charge and V_o rises. By quickly closing and opening the switch every time V_o rises to a peak voltage V_p (for example 5 V), the sawtooth wave shape shown in Fig. 6-11(b) would be generated. The frequency f of the sawtooth wave generator is found from

$$f = \frac{E_i}{R_iC} \times \frac{1}{V_p} \tag{6-9a}$$

The period of oscillation is

$$T = \frac{1}{f} \tag{6-9b}$$

Example 6-10: In Fig. 6-11, what is (a) the frequency and (b) the period of oscillation?

Solution: (a) By Eq. (6-9a),

$$f = \frac{1\text{ V}}{10\text{ ms}} \times \frac{1}{5\text{ V}} = 20\text{ Hz}$$

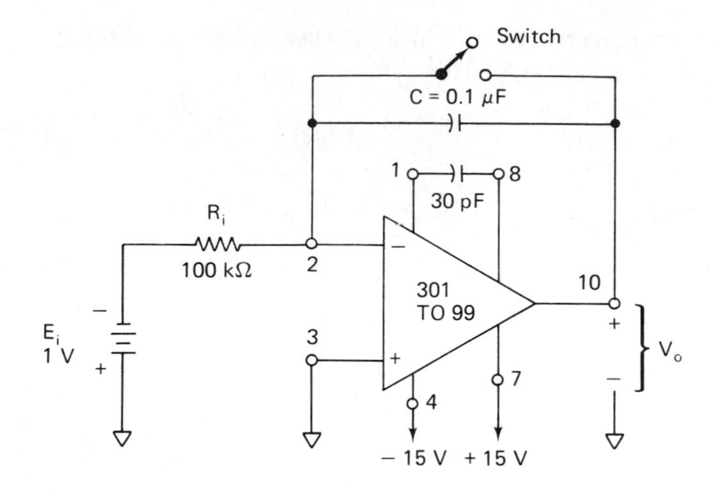

(a) Single-ramp generator with capacitor discharge switch

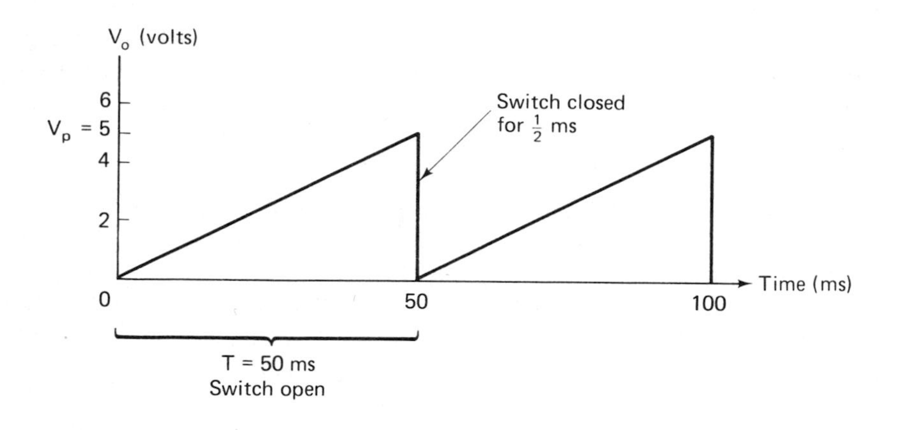

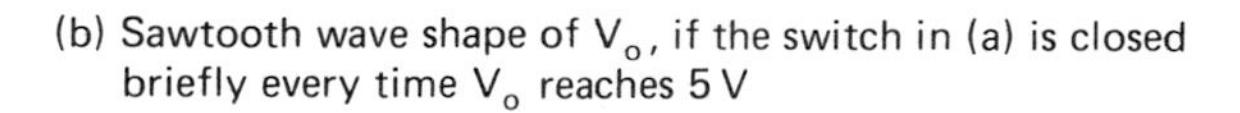
(b) Sawtooth wave shape of V_o, if the switch in (a) is closed briefly every time V_o reaches 5 V

Figure 6-11 Manually controlled sawtooth wave generator.

(b) By Eq. (6-9b),

$$T = \frac{1}{20\ \text{Hz}} = 50\ \text{ms}$$

To generate the sawtooth wave automatically, we need a device or circuit that will do four jobs in the following order:

1. Sense when the capacitor voltage reaches a desired peak value V_p.
2. Place a short circuit across the capacitor.

3. Finally, sense when the capacitor is almost completely discharged.
4. Remove the short circuit.

There is such a device, and it is inexpensive. It is the *programmable unijunction* transistor, PUT.

6-6.2 The Programmable Unijunction Transistor, PUT. The PUT is a three-terminal device that acts as a voltage-sensitive switch. Its schematic diagram is shown in Fig. 6-12(a). The switch terminals are labeled anode *A* and cathode *K*. Current flows only from anode to cathode (as indicated by the arrow).

Normally, the switch terminals act as an open circuit. To make the switch terminals act as a short circuit whenever the desired voltage level V_p is reached, the PUT must compare this changing voltage level against a fixed reference voltage. Therefore, a third terminal is needed where the reference voltage is applied. This third terminal is called the *gate*, *G*. Apply the voltage you want to sense to the anode terminal. When the anode voltage becomes more positive than the gate voltage (by a few tenths of a volt), the PUT's anode and cathode suddenly act as a short circuit. They remain a short circuit independently of the gate terminal until the anode-to-cathode current drops below the PUT's *holding current*, I_H. (I_H is typically a few milliamperes.) Then the anode and cathode terminals abruptly act as an open circuit. The cathode *K* of a PUT should be connected to ground, as it is by the virtual ground at pin 2 in Fig. 6-12(a). How the PUT's characteristics are used in the sawtooth wave generator is shown in Section 6-6.3.

6-6.3 Sawtooth Wave Generator Operation. The simplest sawtooth wave generator is shown in Fig. 6-12(a) The desired peak voltage of the sawtooth wave, V_p, is applied to the gate terminal of the PUT. E_i drives V_o positive at 1 V/10 ms until V_o exceeds V_p by a few tenths of a volt. The PUT's anode and cathode terminals then abruptly act as a short circuit and discharge capacitor C to about 1 V. When capacitor discharge current becomes less than the PUT's holding current, the PUT's anode and cathode terminals abruptly act as an open circuit. Unfortunately, the PUT does not act as an ideal short circuit because its A–K terminal voltage drops only to a *forward voltage* V_F equal to about 1 V, not 0 V. But by increasing V_p by about 1 V to compensate for V_F, we can generate a sawtooth wave with a frequency of 20 Hz as in Fig. 6-12(b). Note that V_o now varies between $V_F \approx 1$ V and $V_p \approx 6$ V. The frequency of oscillation is expressed by

$$f = \frac{E_i}{R_i C} \times \frac{1}{(V_p - 1)} \tag{6-10}$$

Equation 6-10 will be studied in more detail in Section 6-7.

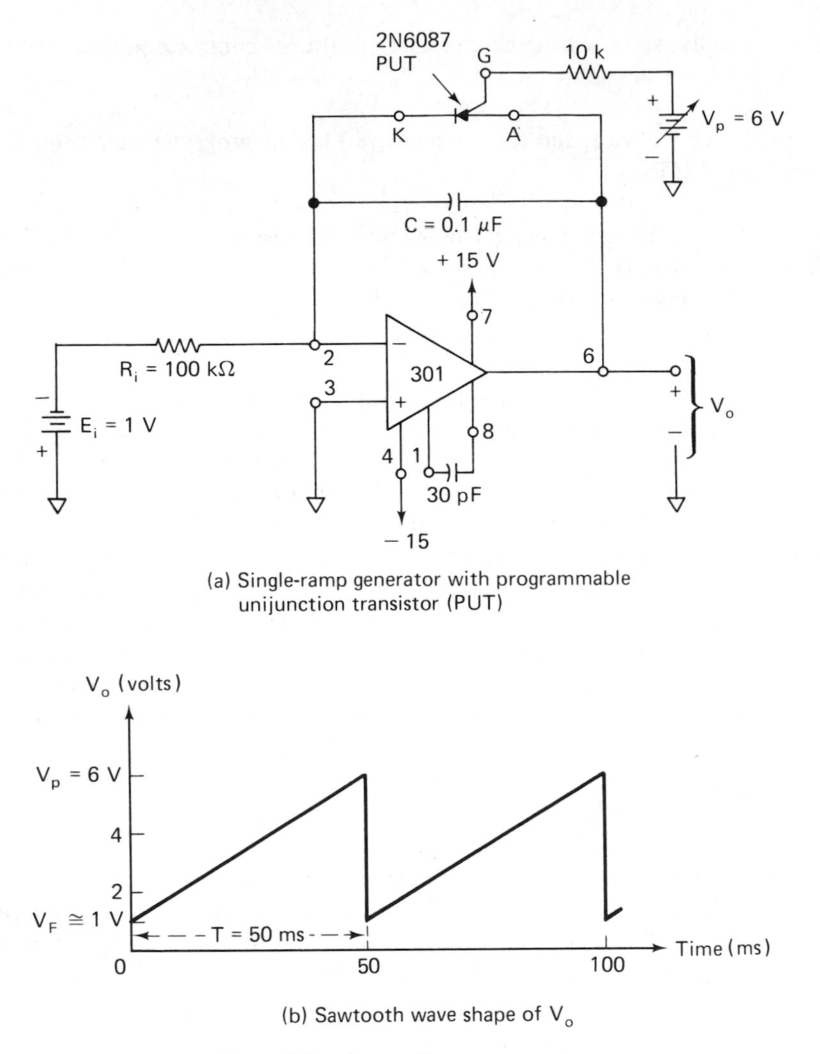

(a) Single-ramp generator with programmable unijunction transistor (PUT)

(b) Sawtooth wave shape of V_o

Figure 6-12 Sawtooth wave generator.

6-7 Voltage-to-Frequency Converters

6-7.1 Voltage-Controlled Oscillator. Equation (6-10) shows that the frequency of a sawtooth wave generator depends on two factors:

1. how fast V_o rises in volts per second, E_i/R_iC
2. the value to which V_o can rise, $(V_p - 1)$ V

Therefore, frequency can be controlled by either voltage E_i or V_p. This is shown in the next two examples.

Example 6-11: If E_i is doubled to 2 V in Fig. 6-12(a), what is the new frequency of oscillation?
Solution: From Eq. (6-10),

$$f = \frac{2\text{ V}}{(100\text{ k}\Omega)(0.1\ \mu\text{F})} \times \frac{1}{(6-1)\text{ V}} = 40\text{ Hz}$$

The frequency is doubled because V_o rises twice as fast.

Example (6-12): If V_p is approximately halved to 3.5 V in Fig. 6-12(a), what is the new frequency of oscillation? $E_i = 1$ V.
Solution: From Eq. (6-10),

$$f = \frac{1\text{ V}}{(100\text{ k}\Omega)(0.1\ \mu\text{F})} \times \frac{1}{(3.5-1)\text{ V}} = \frac{1\text{ V}}{0.010\text{ s}} \times \frac{1}{2.5\text{ V}} = 40\text{ Hz}$$

The frequency is doubled because V_o now has to rise only to 2.5 V rather than to 5 V.

Both examples show that our sawtooth wave generator can convert a voltage E_i or V_p to a frequency; therefore, it is a voltage-to-frequency converter.

6-7.2 Frequency Modulation and Frequency Shift Keying. Examples 6-11 and 6-12 indicate one way of achieving *frequency modulation*, FM. Thus, if the amplitude of E_i varies, the frequency of the sawtooth oscillator will be changed or modulated. If E_i is keyed between two voltage levels, the sawtooth oscillator changes frequencies. This type of application is called *frequency shift keying*, FSK, and is used for data transmission. These two preset frequencies correspond to "0" and "1" states (commonly called *space* and *mark*) in binary.

6-8 Sine Wave Oscillator

6-8.1 Oscillator Theory. A basic sine wave oscillator should generate a sine wave at only one frequency. In order to do this, assume that different frequencies are present at some starting point in a circuit. The frequencies are passed through a frequency-selection network where they are reduced in amplitude and exit with different phase angles. Next they are passed through an amplifier to replace the amplitude lost in the selection circuit. We hope that only *one* of the frequencies exiting from the amplifier will be an exact copy of one of the original frequencies both in magnitude and phase angle. Then, by connecting the amplifier output back to the starting point, we have an oscillator that will oscillate at only this one frequency.

To explore this principle in detail, refer to Fig. 6-13(a), in which three frequencies are considered. E_o is the desired oscillating frequency. E_H is a

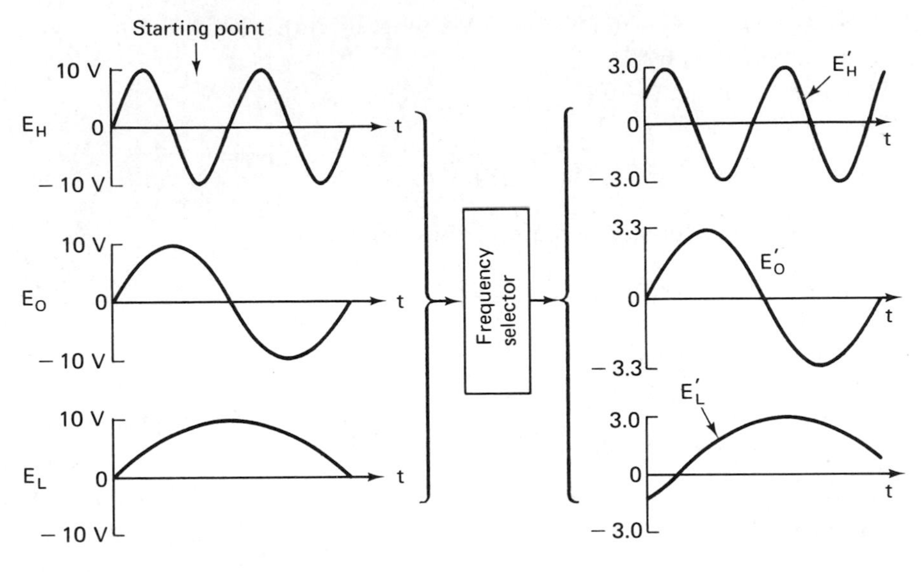

(a) The frequency selector transmits only one frequency, E_o, with no change in phase angle

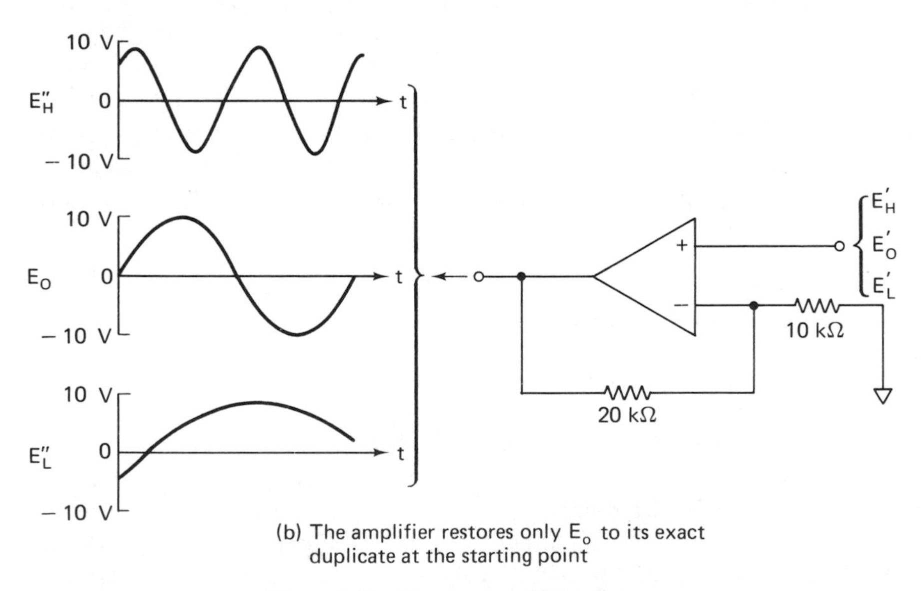

(b) The amplifier restores only E_o to its exact duplicate at the starting point

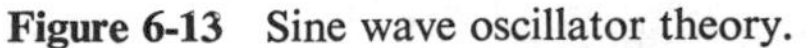
Figure 6-13 Sine wave oscillator theory.

higher frequency, and E_L is a lower frequency. At the output of the frequency selector, E_o' has the same phase angle as input E_o, although its peak value is reduced by one-third. (Note the different voltage scales between input and output.) Output E'_H is not in phase with E_H, nor is E'_L in phase with E_L.

Visualize all three outputs of the frequency selector applied to a noninverting amplifier with a gain of 3. As shown in Fig. 6-13(b), only one frequency, E_o, will emerge from the amplifier with exactly the same amplitude and, more important, exactly the same phase as the original E_o.

By connecting the amplifier output back to the input of the frequency selector, we close a loop. Now a signal of only one frequency can be transmitted completely around the loop without changing its magnitude or phase angle. To put it another way, only one frequency has a loop gain of 1. All other signals will be attenuated after each pass around the loop and experience a phase shift that will act to dampen them out.

One final point: Where did E_o come from in the first place? The answer is that either a transient voltage generated by turning the power on or the inevitable presence of noise introduces a number of frequencies within the loop. Among them will be the one oscillating frequency E_o that will see a loop gain of 1. Actually, the loop gain should be slightly greater than 1 so that E_o can build up. As will be shown in Section 6-8.3, we must make provision to prevent E_o from building up indefinitely and forcing the amplifier into saturation.

6-8.2 Setting up an Oscillator. Figure 6-14 presents a practical circuit to illustrate the ideas presented in Section 6-8.1. E_i is a voltage from an audio oscillator. The *RC* network is a Wein Bridge type of frequency selector. The output of the Wein Bridge, E', is applied to the noninverting amplifier. The gain of the amplifier is adjusted with the 50 kΩ pot. If the frequency of E_i is varied, only one frequency will be developed at E' that has phase shift

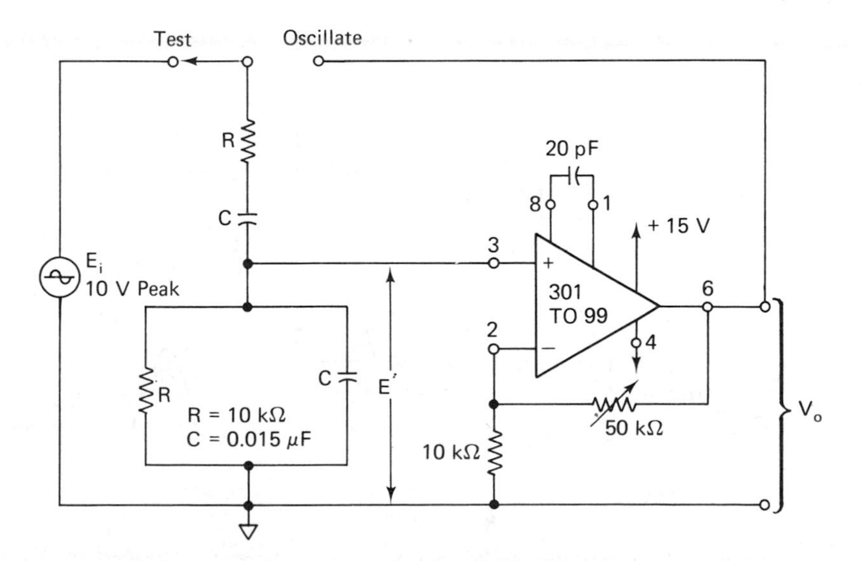

Figure 6-14 Test circuit to find the frequency of oscillation $f_o = 1/(2\pi RC)$.

of 0°. This frequency, f_o, is calculated from

$$f_o = \frac{1}{2\pi RC} \tag{6-11}$$

Attenuation of the frequency selector at f_o is one-third, or

$$E' = \frac{1}{3}E_i \quad \text{at } f_o$$

By adjusting the amplifier's gain (50 kΩ pot), we can amplify E' by a gain of 3 so that the amplifier output V_o is precisely equal to E_i for amplifier gain = 3, and frequency = f_o.

Next, monitor V_o with a CRO and flip the switch from "test" to "oscillate." This action closes the loop, which has a net gain of 1 ($\frac{1}{3} \times 3 = 1$). One of two events will then be seen on the CRO at V_o. Either V_o will switch from $+V_{sat}$ to $-V_{sat}$ at the frequency f_o, or V_o will not show any oscillation and lock at $+V_{sat}$ or $-V_{sat}$. By adjusting the 50-kΩ potentiometer, you will see briefly a nice sine wave, whose amplitude will either decay or build up. The oscillator is almost working but needs some amplitude control. One technique for introducing amplitude control is given in section 6-8.3.

6-8.3 Wein Bridge Oscillator. Experience with the test circuit of Fig. 6-14 showed that the output voltage V_o could increase without limit once

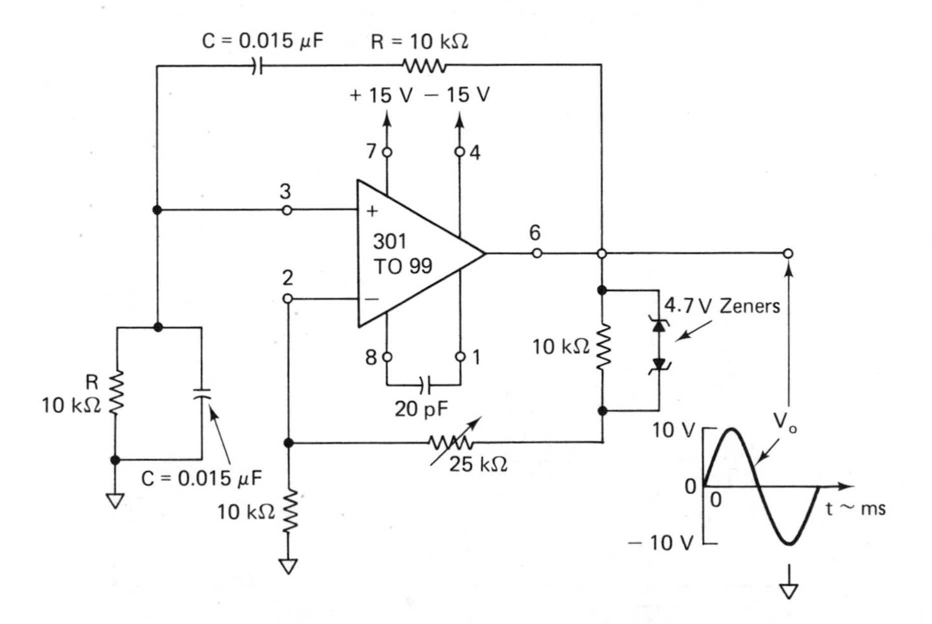

Figure 6-15 Practical Wein Bridge sine wave oscillator, $f_o \cong 1$ kHz.

oscillation began. What is needed is a circuit that will sense the amplitude of the output voltage and reduce amplifier gain when the voltage exceeds a specified level. We add a back-to-back zener diode and one resistor to do this job in the practical Wein Bridge oscillator of Fig. 6-15. When the oscillator's output voltage increases above the zener voltage, one zener or the other (depending on the polarity of V_o) breaks down. The zener then shunts the 10-kΩ resistor to reduce the gain of the amplifier and prevent V_o from being driven to $\pm V_{sat}$. The 25-kΩ resistor allows adjustment of V_o from peak values of about $1.5\ V_Z \approx 8$ V to $\pm V_{sat}$. The resulting sine wave output has very little distortion. For best results, the oscillator's output should be connected to a voltage follower to avoid undue loading.

Problems

6-1 In Example 6-1, saturation voltages are ± 10 V. Find V_{UT} and V_{LT}.

6-2 Find the period of the multivibrator in Fig. 6-1 if $R_f = 10$ kΩ and $C = 0.01\ \mu$F.

6-3 Find the frequency of oscillation in Problem 6-2 above.

6-4 E_i must go negative by what value in Fig. 6-4(a) to trigger the one-shot into its timing state?

6-5 What is another name for a one-shot?

6-6 In Fig. 6-4, $R_f = 10$ kΩ and $C = 0.1\ \mu$F. Calculate the duration of the output pulse.

6-7 If D_1 is reversed in Fig. 6-4, what is the effect on circuit operation?

6-8 Calculate the capacitor voltage 10 seconds after the switch is closed in Fig. 6-5(a).

6-9 R_i is changed to 1 MΩ and C to 10 μF in Fig. 6-7(a). At what rate will V_o drop in volts per second?

6-10 V_{ref} is changed to -5 V in Example 6-7. How long does it take for the op amp to saturate?

6-11 The saturation voltages are ± 10 V in Example 6-8. Find the new values of V_{UT}, V_{LT}, and V_H.

6-12 The saturation voltages in Example 6-9 are changed to ± 10 V. What is the new frequency of oscillation?

6-13 In Fig. 6-11(a), $R_i = 1$ MΩ and $C = 1\ \mu$F; what is the frequency of oscillation, if the switch is momentarily closed when $V_C = 5$ V?

6-14 If E_i is halved to 0.5 V in Fig. 6-12(a), what is the new frequency of oscillation?

7

Power Supplies and Power Amplifiers

7-0 Introduction

Most electronic devices require dc voltages to operate. Batteries are useful in low-power or portable devices, but operating time is limited unless the batteries are recharged or replaced. The most readily available source of power is the 60-Hz, 110-V ac wall outlet. The circuit that converts this ac voltage to a dc voltage is called a *dc power supply*.

The most economical dc power supply is some type of rectifier circuit. Unfortunately, some ac ripple voltage rides on the dc voltage, so the rectifier circuit does not deliver pure dc. An equally undesirable characteristic is a reduction in dc voltage as more load current is drawn from the supply. Since dc voltage is *not* regulated (that is, constant with changing load current), this type of power supply is classified as *unregulated*. Unregulated power supplies are introduced in Sections 7-1 and 7-2. It is necessary to know their limitations before such limitations can be minimized or overcome by adding regulation.

A circuit that essentially eliminates variation of power supply voltage with changing load currents is called a *voltage regulator*. Normally, all of the ac ripple voltage is also eliminated by the regulator circuit. When a voltage regulator is connected to a rectifier circuit, the result is a *regulated power supply*.

Voltage regulators with outstanding performance can be made quickly and easily by using the op amp. An even greater benefit provided by the op amp is the ease of adding options such as short-circuit protection of the regulated power supply. With a few minor changes and at a very nominal cost, the regulator can be converted to a power amplifier that rivals high fidelity amplifiers in performance.

7-1 Introduction to the Unregulated Power Supply

7-1.1 Power Transformer. A transformer is required for reducing the 110-V ac wall current to the lower ac value required by transistor, ICs, and other electronic devices. Transformer voltages are given in terms of rms values. In Fig. 7-1, the transformer is rated as 110 V to 24 V center tap. With 110 V rms connected to the primary, 24 V rms is developed between secondary terminals 1 and 2. A third lead, brought out from the center of the secondary, is called a *center tap*, CT. Between terminals CT and 1 or CT and 2, the rms voltage is 12 V.

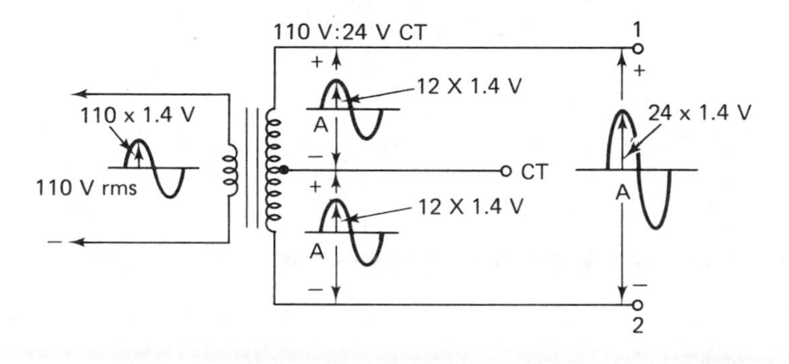

(a) Peak voltages for the positive half-cycle

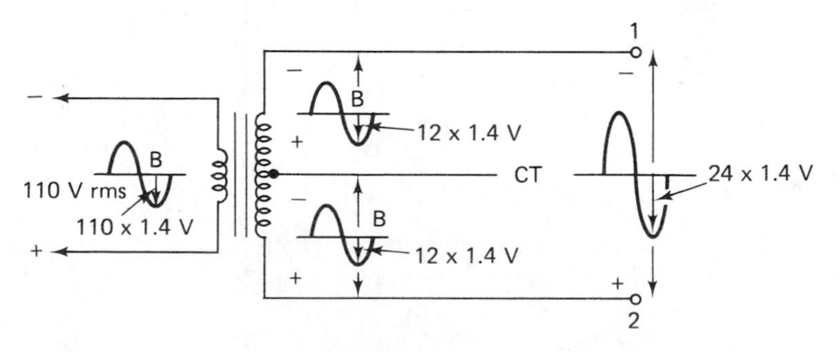

(b) Peak voltages for the negative half-cycle

Figure 7-1 Power transformer.

An oscilloscope would give the sinusoidal voltages shown in Fig. 7-1. The maximum instantaneous voltage, E_m, is related to the rms value E_{rms} by

$$E_m = 1.4(E_{rms}) \tag{7-1}$$

In Fig. 7-1(a), voltage polarities are shown for the positive half-cycle; those for the negative half-cycle are shown in Fig. 7-1(b).

Example 7-1: Find E_m in Fig. 7-1 between terminals 1 and 2.
Solution: By Eq. (7-1), $E_m = 1.4(24 \text{ V}) = 34$ V.

7-1.2 Rectifier Diodes. In Fig. 7-2(a), four diodes are arranged in a diamond configuration called a *full-wave bridge rectifier*. They are connected to terminals 1 and 2 in the transformer of Fig. 7-1. When terminal 1 is positive with respect to terminal 2, diodes D_1 and D_2 conduct. When terminal

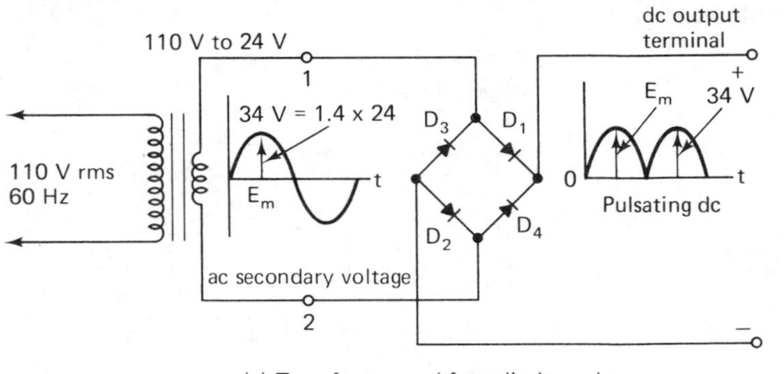

(a) Transformer and four diodes reduce from the power supply

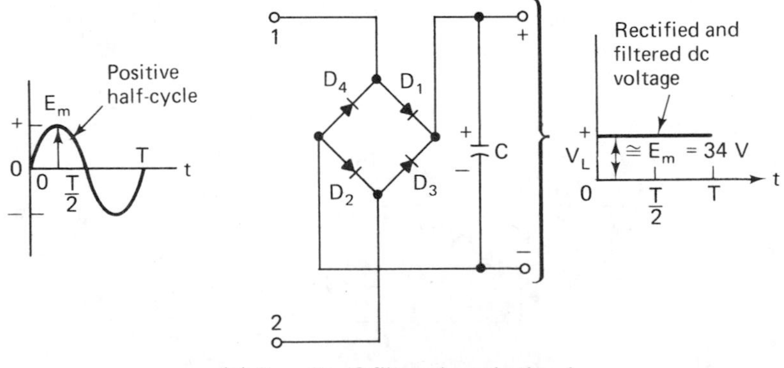

(b) Capacitor C filters the pulsating dc in (a) to give a dc load voltage

Figure 7-2 Transformer plus rectifier diodes plus filter capacitor equals unregulated power supply.

2 is positive with respect to terminal 1, diodes D_3 and D_4 conduct. The result is a pulsating dc voltage between the output terminals.

7-1.3 Filter Capacitor. The pulsating dc voltage in Fig. 7-2(a) is not pure dc, so a filter capacitor is placed across the dc output terminals, as shown in Fig. 7-2(b). This capacitor smooths out the pulsations and gives an almost pure dc output voltage, V_L. V_L is the unregulated voltage that supplies power to the load. The filter capacitor is typically a large electrolytic capacitor, 500 μF or more.

7-1.4 The Load. In Fig. 7-2(b), nothing other than the filter capacitor is connected across the dc output terminals. The unregulated power supply is said to have no load. This means that the *no-load current*, or 0 load current, I_L, is drawn from the output terminals. Usually, the maximum expected load current, or full-load current, to be furnished by the supply is known. The load is modeled by resistor R_L as shown in Fig. 7-3(a). As stated in Section

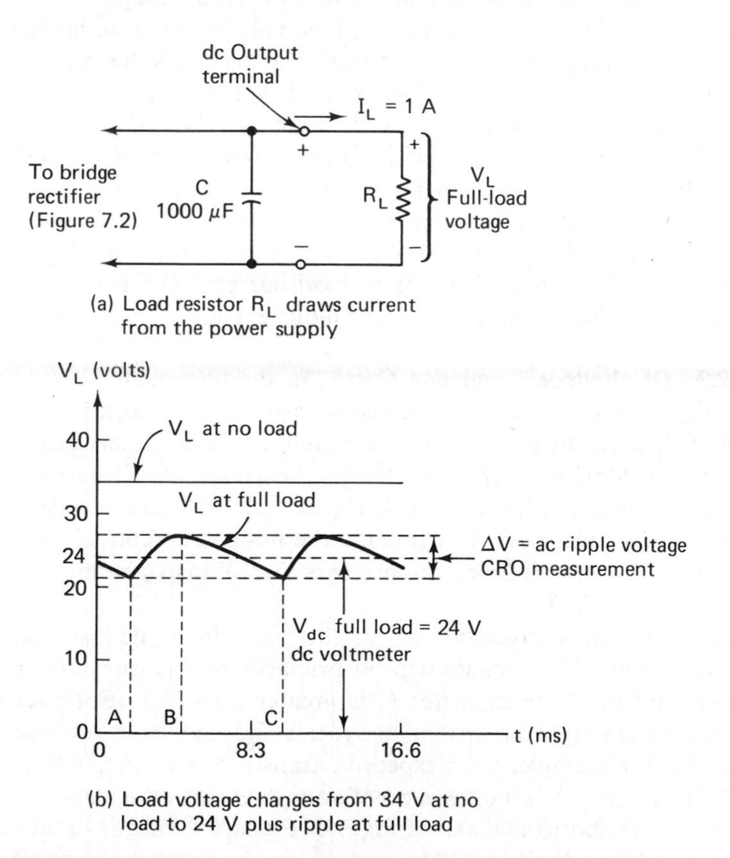

Figure 7-3 Variation of dc load voltage and ac ripple voltage from no-load current to full-load current.

7-0, the load voltage changes as the load current changes in an unregulated power supply. The manner in which this occurs is examined next.

7-2 Predicting DC Voltage Regulation and AC Ripple Voltage

7-2.1 Load Voltage Variations with Load Current. A dc voltmeter connected across the output terminals in Fig. 7-2(b) measures the no-load voltage, or

$$V_{\text{dc no load}} = E_m$$

From Example 7-1, $V_{\text{dc no load}}$ is 34 V. An oscilloscope would also show the same value with no ac ripple voltage, as in Fig. 7-3(b). Now suppose that a load R_L was connected to draw a full-load dc current of $I_L = 1$ A, as in Fig. 7-3(a). An oscilloscope now shows that the load voltage V_L has a lower *average*, or dc value, V_{dc}. Moreover, the load voltage has an ac ripple component, ΔV_o, superimposed on the dc value. The average value measured by a dc voltmeter is 24 V. The peak-to-peak ripple voltage is $\Delta V_o = 5$ V.

There are two conclusions to be drawn from Fig. 7-3(b). First, the dc load voltage goes down as dc load current goes up; how much the load voltage drops can be estimated by a simple technique explained in Section 7-2.2. Second, the ac ripple voltage increases from 0 V at no-load current to a large value at full-load current. As a matter of fact, the ac ripple voltage increases directly with an increase in load current. The amount of ripple voltage can also be estimated, by a technique explained in Section 7-2.3.

7-2.2 DC Voltage Regulation Curve. In the unregulated power supply circuit Fig. 7-4(a), the load R_L is varied so that we can record corresponding values of dc load current and dc load voltage. The dc meters respond only to the average (dc) load current or voltage. If corresponding values of current and voltage are plotted, the result is the *dc voltage regulation curve* of Fig. 7-4(b). For example, point 0 represents the no-load condition, $I_L = 0$ and $V_{\text{dc no load}} = E_m = 34$ V. Point A represents the full-load condition, $I_L = 1$ A and $V_{\text{dc full load}} = 24$ V.

There is a general procedure to estimate the value of dc load voltage for any load current. These values depend primarily on the transformer rating, provided that the filter capacitor C is greater than 200 μF. Power-supply transformers are rated by rms voltage ratios and maximum secondary load current I_T. For example, the inexpensive transformer of Fig. 7-4(a) is rated as 110 V: 24 V at 1 A. Find the ratio of your load current I_L to I_T and locate this point on the horizontal axis of Fig. 7-5. Proceed vertically to the estimating curve and then horizontally to read V_{dc} on the vertical axis as a fraction of E_m. The procedure is illustrated by an example.

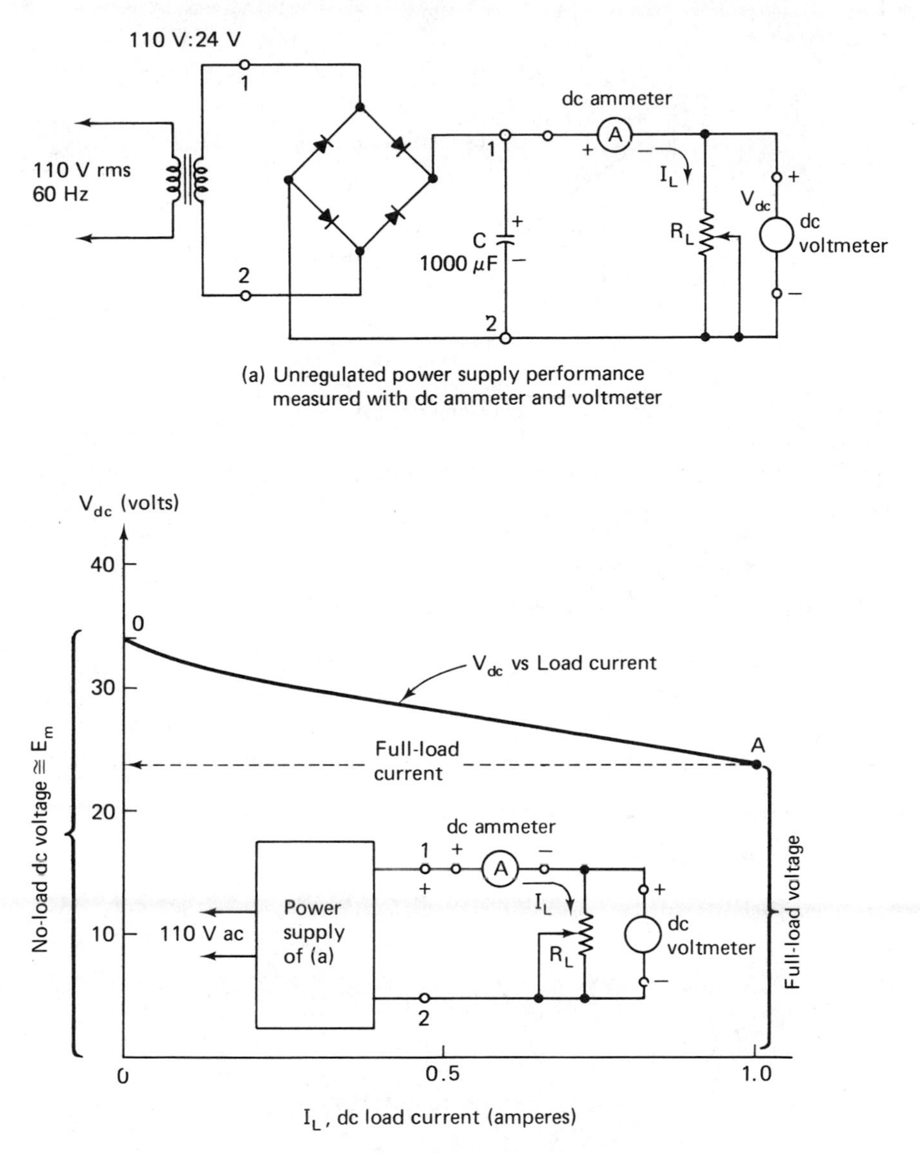

Figure 7-4 Dc load voltage varies with load current in (a) as shown by the voltage regulation curve in (b).

Example 7-2: In Fig. 7-4, $E_m = 24\text{ V} \times 1.4 = 34\text{ V}$. $I_T = 1$ A. Find the dc load voltage V_{dc} at (a) $I_L = 0.5$ A and (b) $I_L = 1.0$ A.

Solution: (a) $I_L/I_T = 0.5\text{ A}/1.0\text{ A} = 0.5$. Locate point *M* in Fig. 7-5 and read

$$V_{dc} = 0.85E_m = 0.85 \times 34 = 29\ V$$

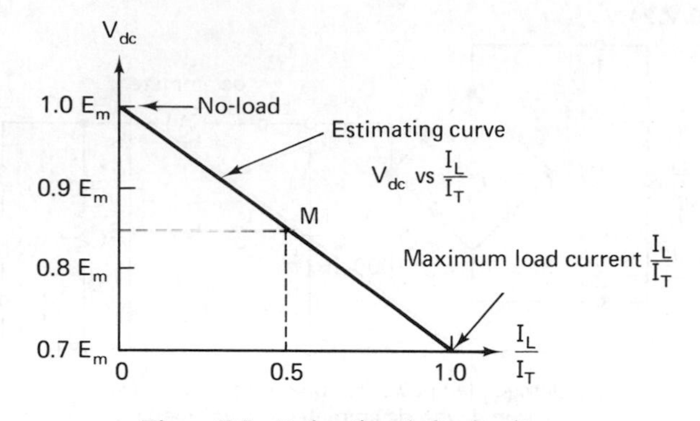

Figure 7-5 Estimating dc load voltage.

(b) $I_L/I_T = 1.0\text{ A}/1.0\text{ A} = 1.0$. Locate the maximum load point where $V_{dc} = 0.7E_m = 0.7 \times 34 = 24$ V.

If a transformer were rated for 110 V: 24 V and $I_T = 2$ A, V_{dc} would be approximately 31 V and 29 V for parts (a) and (b) in Example 7-2.

7-2.3 Estimating and Reducing AC Ripple Voltage. It was concluded in Section 7-2.1 and Fig. 7-3(b) that ac ripple increases as dc load current increases. The worst case of ac ripple occurs at maximum load current. The peak-to-peak ac ripple voltage ΔV can be estimated from Fig. 7-6 as shown by the following example.

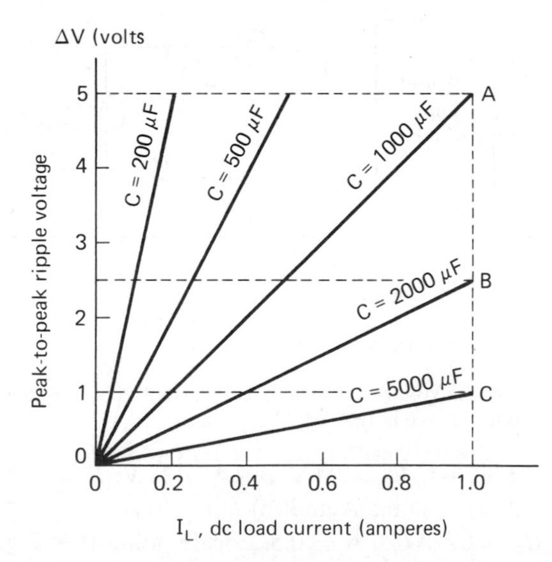

Figure 7-6 Estimating peak-to-peak ac ripple voltage ΔV.

Example 7-3: What is ΔV for the power supply of Fig. 7-4 if the load current is 1 A and (a) $C = 1000\ \mu F$, (b) $C = 2000\ \mu F$, and (c) $C = 5000\ \mu F$?
Solution: In Fig. 7-6, enter the horizontal axis at $I_L = 1.0$ A. Proceed vertically to intersections A, B, and C with curves respectively labeled $C = 1000\ \mu F$, $C = 2000\ \mu F$, and $C = 5000\ \mu F$. Read ΔV from the vertical axis as shown to get (a) $\Delta V = 5$ V, (b) $\Delta V = 2.5$ V, and (c) $\Delta V = 1$ V.

Conclusion: From Fig. 7-6, ac ripple is shown to be reduced by increasing the value of filter capacitor C. Doubling C halves the ripple. If an oscilloscope is not available, an ac voltmeter can be connected across the load R_L to measure the approximate rms ripple voltage V_r. V_r is related to ΔV by

$$V_r = \frac{\Delta V}{3} \qquad (7\text{-}2)$$

Example 7-4: Find the ac voltmeter readings for the peak-to-peak ripple voltages in Example 7-3.
Solution: (a) $V_r = 5\ \text{V}/3 = 1.7$ V; (b) $V_r = 2.5\ \text{V}/3 = 0.8$ V; (c) $V_r = 1\ \text{V}/3 = 0.3$ V.

7-2.4 Minimum Instantaneous Load Voltage. Refer to Fig. 7-3(b) and note that ΔV is centered on the dc load voltage V_{dc}. The minimum *instantaneous* value of load voltage V_L will be

$$\text{minimum } V_L = V_{dc} - \frac{\Delta V}{2} \qquad (7\text{-}3)$$

Since V_{dc} goes down and ΔV goes up with increasing load current, the lowest minimum value of V_L occurs at full load current. This value of V_L places a limit on any voltage regulator that will be attached to the unregulated power supply. But before we proceed to the voltage regulator, another type of unregulated power supply will be studied so that it can be used with the power amplifier.

7-3 Bipolar and Two-Value Unregulated Power Supplies

7-3.1 Bipolar or Positive and Negative Power Supplies. Many electronic devices need both positive (+) and negative (−) voltages. These voltages are measured with respect to a third common (or grounded) terminal. To obtain a positive and negative voltage, either two secondary transformer windings or one center-tapped secondary winding is needed.

A transformer rated at 110 V : 24 V CT is shown in Fig. 7-7. Diodes D_1 and D_2 make terminal 1 positive with respect to center tap CT. Diodes

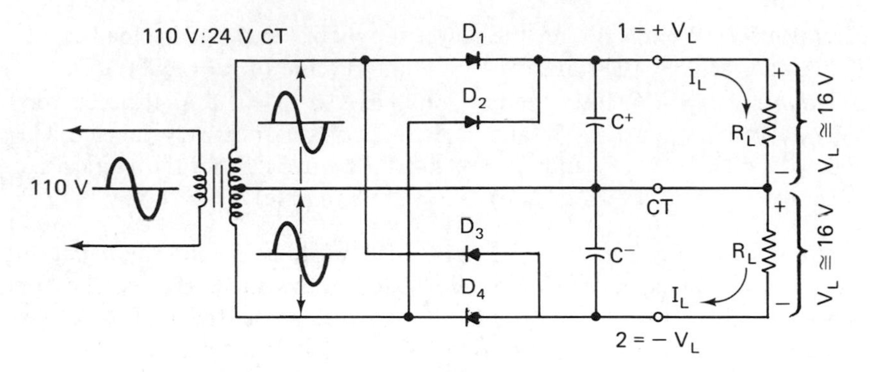

Figure 7-7 Bipolar power supply.

D_3 and D_4 make terminal 2 negative with respect to the center tap. From Eq. 7-1 and Section 7-2.1, both no-load dc voltages are 1.4×12 V rms = 16 V. Capacitors $C+$ and $C-$ respectively filter the positive and negative supply voltages. As outlined in Figs. 7-5 and 7-6, the ac ripple voltage and dc voltage regulation may be predicted for both load voltages.

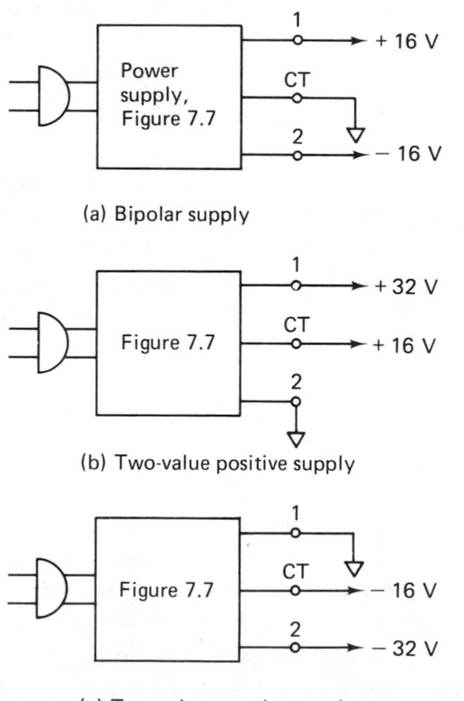

Figure 7-8 Bipolar and two-value power supplies.

7-3.2 Two-Value Power Supplies. If the center tap of the power supply of Fig. 7-7 is grounded, we have a *bipolar* power supply. It is shown schematically in Fig. 7-8(a). If terminal 2 is grounded as in Fig. 7-8(b), we have a two-value positive supply. Finally, by grounding terminal 1 in Fig. 7-8(c), we get a two-value negative power supply. This indicates the versatility of the center tapped transformer. The bipolar power supply is used with an audio amplifier in Section 7-11.

7-4 Need for Voltage Regulation

Previous sections have shown that the unregulated power supply has two undesirable characteristics: the dc voltage decreases and the ac ripple voltage increases as load current increases. Both disadvantages can be minimized by adding a voltage-regulator section to the unregulated supply as in Fig. 7-9. The resulting power supply is classified as a *voltage-regulated supply.* Before turning to the op amp for an almost ideal voltage regulator, we must first learn something about an element that gives a stable reference voltage, the zener diode. The zener diode and its applications as a basic voltage regulator are studied in Section 7-5.

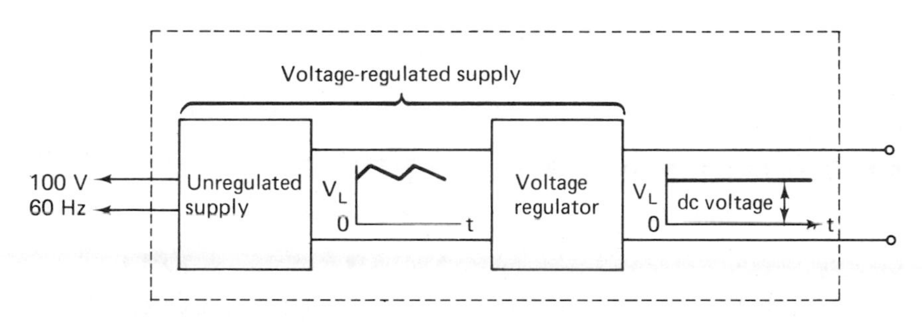

Figure 7-9 Unregulated supply plus a voltage regulator gives a voltage-regulated power supply.

7-5 Zener Diode Regulator

7-5.1 Characteristics of the Zener Diode. The symbol for a zener diode is shown in Fig. 7-10. When forward-biased (with the + terminal of E_i connected to the arrow-head), the zener behaves as any silicon diode does. Its terminal voltage stays at about 0.6 V to 0.7 V no matter what current it conducts. However, let a reverse-biased voltage (with the − terminal of E_i connected to the arrow-head) be connected across the zener. As the reverse-biased voltage is increased, the zener voltage will increase until a zener *breakdown voltage*, V_Z, is reached. For all values of E_i greater than V_Z, the zener's

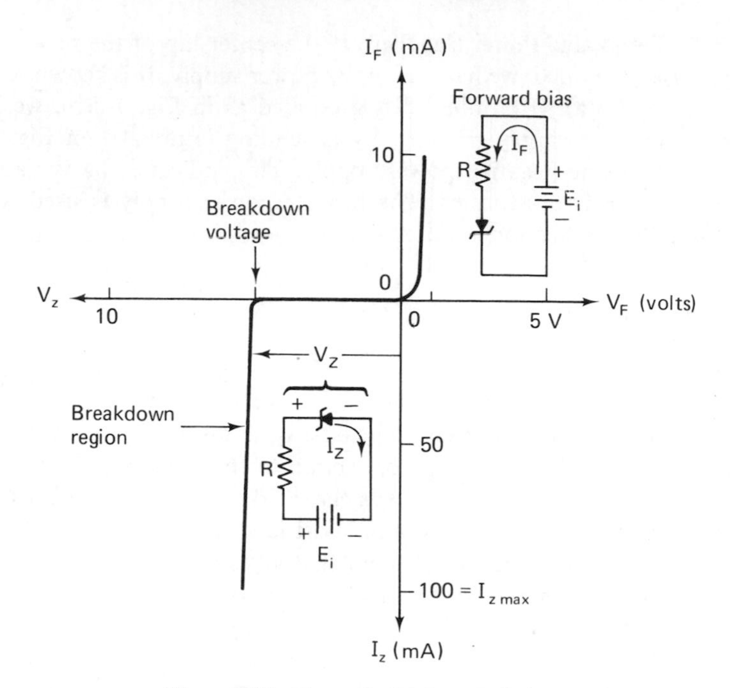

Figure 7-10 Zener diode characteristics.

terminal voltage stays constant no matter what current flows through the zener. As shown in Fig. 7-10, this characteristic of the zener is called the *breakdown region.* Zener diodes are designed to operate in the breakdown region. They are available with zener voltage ratings from a few volts to a few hundred volts. In Fig. 7-10, $V_Z = 5$ V.

7-5.2 Zener Voltage Regulator. If a load is connected across a zener as in Fig. 7-11(a), the load voltage equals the zener voltage. As the load current varies in Fig. 7-11(b), the load voltage stays at V_Z. What happens is that the zener absorbs all current not drawn by the load. We shall not go into the analysis and design of zener regulators. Our objective is to reach the following conclusions.

1. Once the zener is reverse-biased, its terminal voltage stays constant at V_Z. V_Z is set by the zener one buys.
2. The zener needs to be reverse-biased into breakdown to set up our reference voltage. To insure breakdown, the zener must conduct a few mA, I_{zon}, typically 5 mA. This current is set by resistor R_D, V_Z, and the input voltage E_i according to

$$I_{zon} = \frac{E_i - V_Z}{R_D} \cong 5 \text{ mA} \tag{7-4}$$

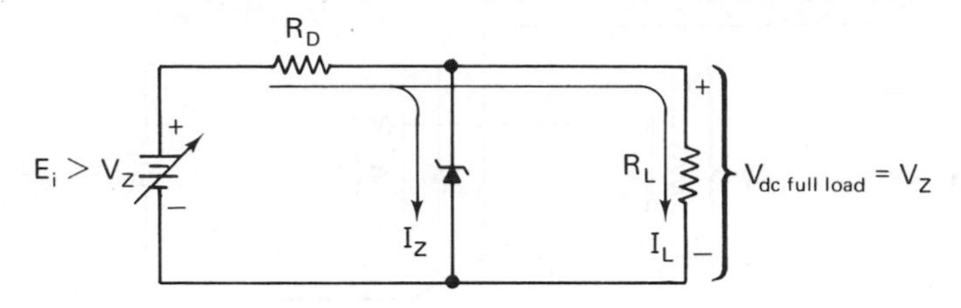

(a) Circuit currents at full-load

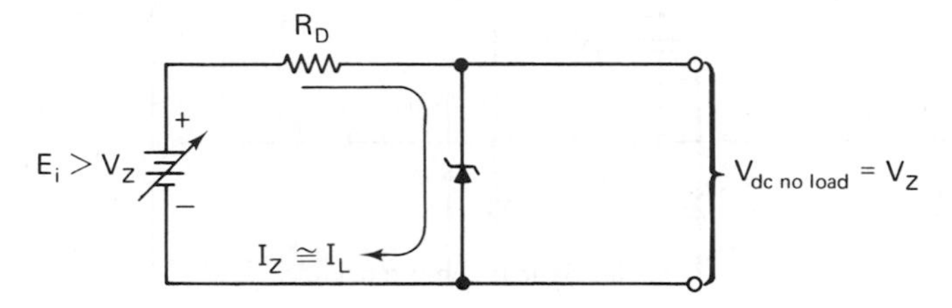

(b) Circuit currents at no-load

Figure 7-11 Zener voltage regulator.

7-6 Basic Op Amp Voltage Regulator

7-6.1 Op Amp Regulator. Close examination of the op amp regulator in Fig. 7-12(a) shows that it is the noninverting amplifier of Section 3-6. The load voltage V_o is set by the battery reference voltage V_{ref} and feedback resistors R_f and R_i according to

$$V_o = \frac{R_f + R_i}{R_i} V_{ref} \tag{7-5}$$

As load current I_L varies, V_o is held constant according to Eq. (7-5). The ripple voltage and changing dc voltage of the unregulated supply are absorbed by the op amp. Voltage between pins 11 and 6 should not exceed 36 V. Output pin 10 and input pin 5 must always be at least 2 V below pin 11; otherwise the op amp will go into negative or positive saturation, respectively.

Either resistor R_i or R_f or the battery V_{ref} may be located away from the op amp and varied to change the dc output voltage V_o. Such an arrangement is called *remote programming*. The op amp circuit now regulates the load voltage to be constant and independent of the load current.

7-6.2 Op Amp Regulator with Zener Reference. In Fig. 7-12(b), the battery is replaced by a zener diode and bias resistor R_D. R_D is determined by rearranging Eq. 7-4. Thus the voltage at pin 10 must always be a few volts more positive than pin 5 to keep the zener diode operating in its breakdown region.

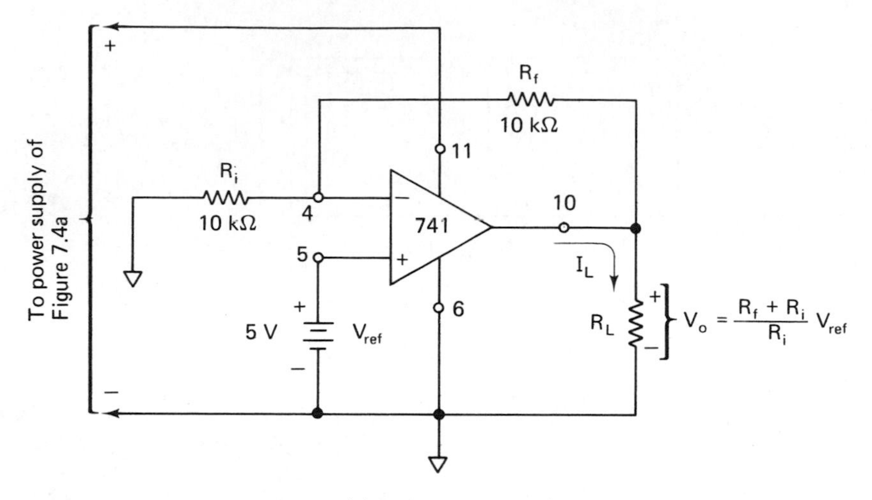

(a) Basic op amp regulator

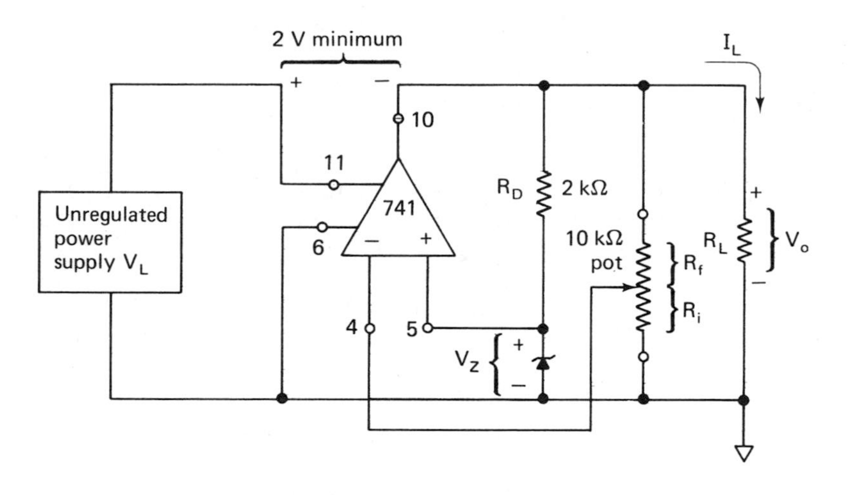

(b) Practical op amp voltage-regulated power supply

Figure 7-12 Basic op amp regulators.

Pin 11 must always be about 2 V more positive than pin 10 to avoid op amp saturation. The 2 V difference between minimum unregulated supply voltage V_L and regulated supply voltage V_o is so important that it is specified for all IC regulated power supplies as *minimum input-output differential voltage*. These points are illustrated by the following design procedure.

To design Fig. 7-12(b) as a 15 V voltage regulator,

1. Select a zener about 1/3 to 1/2 of V_o or $V_z = 5$ V.

2. Calculate R_D from Eq. (7-4),

$$R_D = \frac{15\text{ V} - 5\text{ V}}{5\text{ mA}} = 2\text{ k}\Omega$$

3. Pick $R_f + R_i$ to draw about 1 mA from the regulated output $V_o =$ 15 V or $R_f + R_i = 15\text{ V}/1\text{ mA} \cong 15\text{ k}\Omega$; select a 10-kΩ potentiometer.
4. From Eq. 7-5, solve for R_i:

$$R_i = (R_f + R_i)\frac{V_{ref}}{V_o} = 10\text{ k}\Omega\left(\frac{5\text{ V}}{15\text{ V}}\right) = 3.33\text{ k}\Omega$$

Set the pot for $R_i = 3.33\text{ k}\Omega$ or install fixed resistors of $R_i = 3.33\text{ k}\Omega$ and $R_f = 10\text{ k}\Omega - 3.33\text{ k}\Omega = 6.67\text{ k}\Omega$.

5. Minimum unregulated supply voltage (voltage at pin 11) must be $V_o + 2\text{ V} = 17\text{ V}$.

7-7 Current-Boosting the Op Amp Regulator

7-7.1 Pass Transistor. The op amp regulator of Fig. 7-12 can only furnish load currents of up to about 5 mA. By adding a pass transistor as in Fig. 7-13(a), the maximum load current capability of the voltage regulator is extended by a factor of approximately 100 to 0.5 A. To see how this is accomplished, assume that V_o is set at 15 V according to Eq. (7-5) or the design procedure in Section 7-6.2. Then let R_L be 30 Ω so that the load current is $I_L = V_o/R_L = 15\text{ V}/30\ \Omega = 0.5\text{ A}$. I_L is furnished by *emitter* terminal *E* of the pass transistor. If β (the current gain) of the transistor is 100, the transistor requires a base current drive of $I_L/\beta = 500\text{ mA}/100 = 5\text{ mA}$. The op amp can furnish 5 mA from its output terminal (pin 10) to the base terminal *B*.

Changes in unregulated supply voltage V_L due to either ac ripple or poor dc voltage regulation are absorbed by the collector of the pass transistor. The pass transistor should be heat-sinked to have a power rating P_D of

$$P_D = (V_L - V_o)I_{L\text{ full load}} \tag{7-6}$$

Example 7-5: If $V_L = 24\text{ V}$, $I_{L\text{ iull load}} = 0.5\text{ A}$, and $V_o = 15\text{ V}$ in Fig. 7-13(a), find the power dissipation of the pass transistor.
Solution: Eq. (7-6) $P_D = (24 - 15)\text{ V} \times 0.5\text{ A} = 4.5\text{ W}$

7-7.2 Current-Boost Transistor. The pass transistor of Fig. 7-13(a) can be replaced by two transistors for further increase in load current-capability. As shown in Fig. 7-13(b), the op amp can furnish up to 5 mA into the base of the pass transistor. The current gain of the pass transistor results in

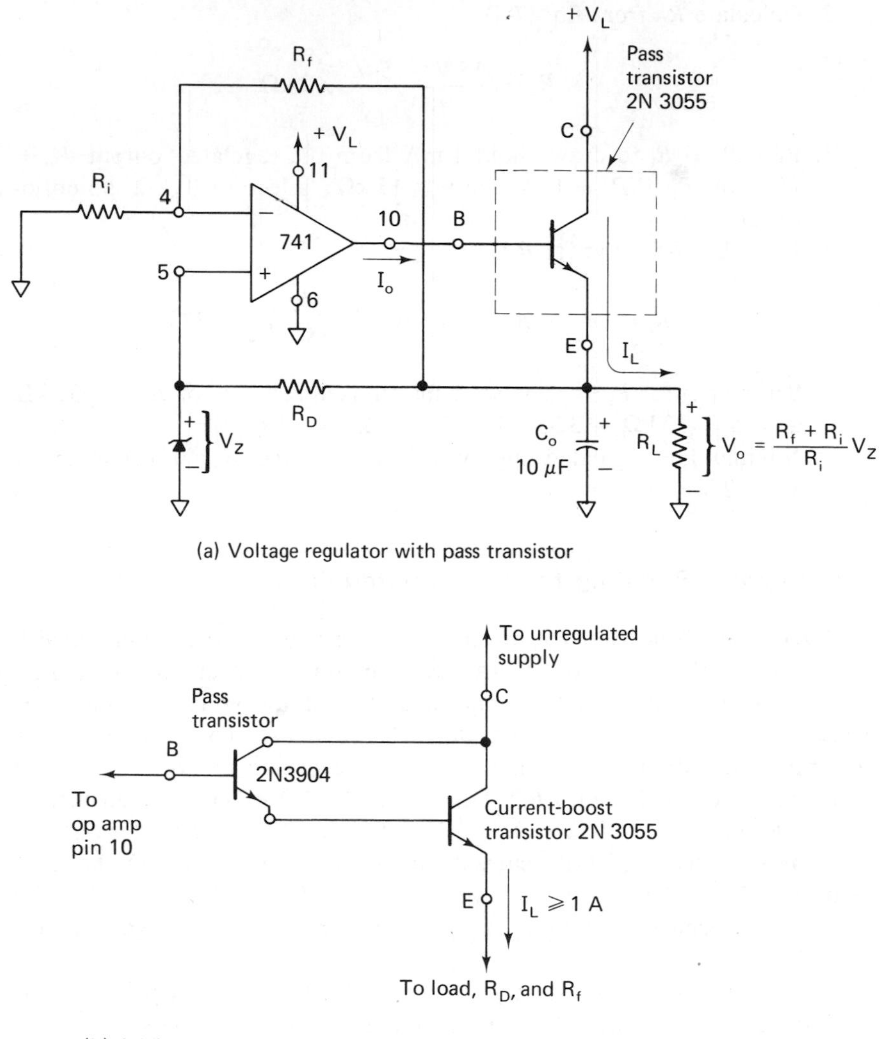

(a) Voltage regulator with pass transistor

(b) Adding a current-boost transistor for large load currents

Figure 7-13 Increasing the load-current capability of a basic op amp regulator.

100 × 5 mA or 0.5 A as the base current drive into the base of the current-boost transistor. The current gain of the current-boost transistor allows an emitter or load current that exceeds 1 A. In practice, the power dissipation limitation of the boost transistor limits load currents to a few amperes. For more current output, more current-boost transistors must be added in parallel with the original boost transistor. (See *Solid State Devices and Applications* by Driscoll and Coughlin, page 244, Prentice-Hall, Englewood Cliffs, NJ, 1975.)

7-8 Short-Circuit or Overload Protection

Short-circuit protection is easily incorporated into the voltage regulator by adding a current-limiting transistor Q_S and a current-sensing resistor R_S as shown in Fig. 7-14. The load current I_L flows through R_S. Usually, the regulator is designed to furnish up to a specified maximum load current, $I_{L\,max}$. $I_{L\,max}$ is large with respect to the currents through R_D and R_f.

Resistor R_S monitors the load current I_L to prevent the regulator from drawing too much load current. When I_L reaches $I_{L\,max}$, the voltage drop across R_S rises to about 0.6 V. Transistor Q_S then turns on, and its collector conducts the op amp's output current away from the base of the 2N 3055 pass transistor. Once Q_S turns on, the emitter current of the pass transistor, and consequently the load current, is held constant at $I_{L\,max}$. R_S sets the value of $I_{L\,max}$ according to

$$I_{L\,max} = \frac{0.6\text{ V}}{R_S} \tag{7-7}$$

Example 7-6: Choose R_S for a maximum load current of 0.5 A for the regulator of Fig. 7-14.

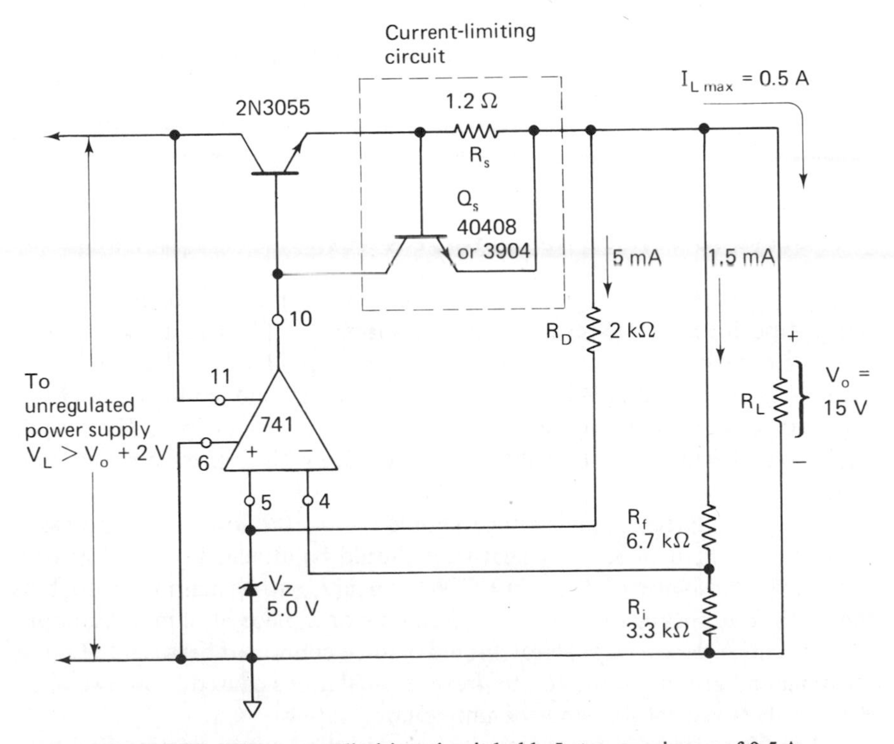

Figure 7-14 The current-limiting circuit holds I_L to a maximum of 0.5 A even if R_L is short-circuited.

Solution: Rearranging Eq. (7-7),

$$R_S = \frac{0.6\ \text{V}}{I_{L\ max}} = \frac{0.6\ \text{V}}{0.5\ \text{A}} = 1.2\ \Omega$$

The worst overload occurs when the output terminals are short-circuited. In this case, the pass transistor must be heat-sinked to be able to dissipate power equal to

$$P_D = V_L I_{L\ max} \tag{7-8}$$

For example, if $V_L = 20$ V and $I_{L\ max} = 0.5$ A, the transistor must dissipate 10 W.

The features of voltage programming, current boost, and overload protection are incorporated in *integrated circuit regulators*. IC voltage regulators are versatile, convenient to use, and relatively inexpensive. They are available in four classifications:

1. Positive voltage
2. Negative voltage
3. Dual voltage
4. Special-purpose voltage regulators

Data sheets and application notes are available from manufacturers (at no cost) and show how to use the devices. Two IC regulators are selected from the vast number available to illustrate how to design and build your own regulator.

7-9 A 5-V IC Voltage Regulator, the 109

The popular digital-logic family called transistor–transistor logic (TTL) needs a positive voltage-regulated power supply of 5 V. The design of such a regulator is very simple: buy a LM109 voltage regulator and hook up three leads. As shown in Fig. 7-15, the input lead and ground lead go to any unregulated power supply whose voltage lies between 7 V and 35 V. The output lead gives a regulated voltage that is 5 V positive with respect to the ground lead.

The unregulated supply voltage should be on the low side if possible, closer to 7 V than to 35 V. A heat-sink should be attached to the 109 with a thermal resistance of $\theta_{SA} = 10°\text{C/W}$. An alternative method is to bolt the 109 and a mica washer to a metal chassis or a piece of aluminum about 5 in × 5 in. A 0.22-μF capacitor should also be connected between the input terminal and ground of the 109 to prevent oscillations caused by an excessive distance between the 109 and the unregulated supply.

The 109 is designed to provide its own short-circuit protection and provide a regulated voltage for load currents of over 1 A. They also have internal

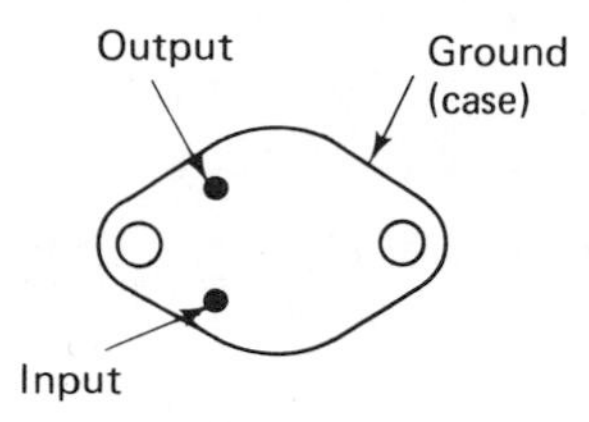

(a) 109 connection diagram TO-3 package, bottom view

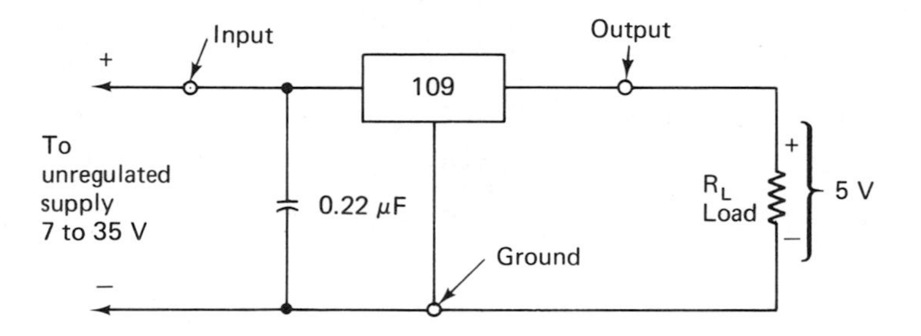

(b) 109 as a 5-V regulator

Figure 7-15 The LM109 or RC109 5-V voltage regulator.

thermal shutdown circuitry. Such a circuit uses a transistor junction to sense when the 109's pass transistor gets to 175°C and turns the regulator off. When the excessive load current that caused the heat is removed, the 109 cools down and turns itself back on. It is essentially burn-out proof.

7-10 Bipolar Regulated Power Supply

In the op amp circuits studied this far, a $\pm$15-V bipolar regulated voltage power supply was necessary. One excellent IC voltage regulator made exactly for this application is the RC 4195 (available from Raytheon Semi-Conductor Division, 350 Ellis Street, Mountain View, California, 94040).

To use the 4195, simply connect its input terminals $+V_{in}$ and $-V_{in}$ and ground (gnd) to an unregulated supply of $\pm$18 V to $\pm$30 V. For example, let the transformer of Fig. 7-7 be a 110 V: 30 V CT, which gives respective unregulated dc voltages of +21 V and −21 V at terminals 1 and 2. Make the following connections:

Fig. 7-7	*Fig. 7-16*
from terminal 1	to $+V_{in}$
from terminal CT	to gnd
from terminal 2	to $-V_{in}$

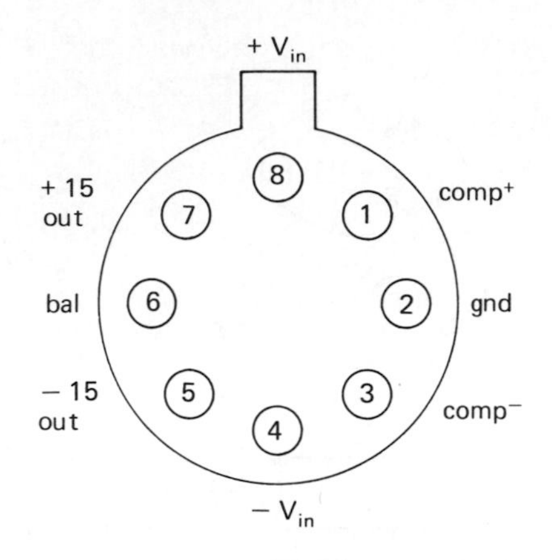

(a) RC4195 Dual regulator TO-99 package, bottom view

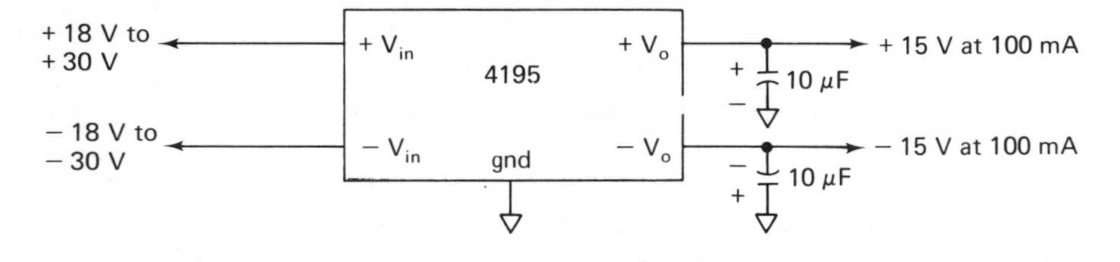

(b) Connections for a ± 15-V voltage regulator

Figure 7-16 The RC 4195, a ±15-V dual tracking voltage regulator.

Terminals $+V_o$ and $-V_o$ of the 4195 give +15 V and −15 V with respect to ground. The manufacturer recommends installation of two 10-μF bypass capacitors at the output terminals as shown in Fig. 7-16.

The 4195 features thermal shutdown and internal short-circuit protection. A clip-on heat sink should be installed to achieve high load-current capability. A single 4195 IC regulator can supply power to more than ten op amps. This same power supply is used in the breadboards of Chapter 1.

7-11 Power Amplifier

The ideas of current boost, the basic op amp regulator, and an inverting amplifier can be put together to form an inexpensive but excellent power amplifier, as shown in Fig. 7-17. The 301 op amp has feed-forward compensation (see Chapter 10) to improve high-frequency response. R_f and R_i set

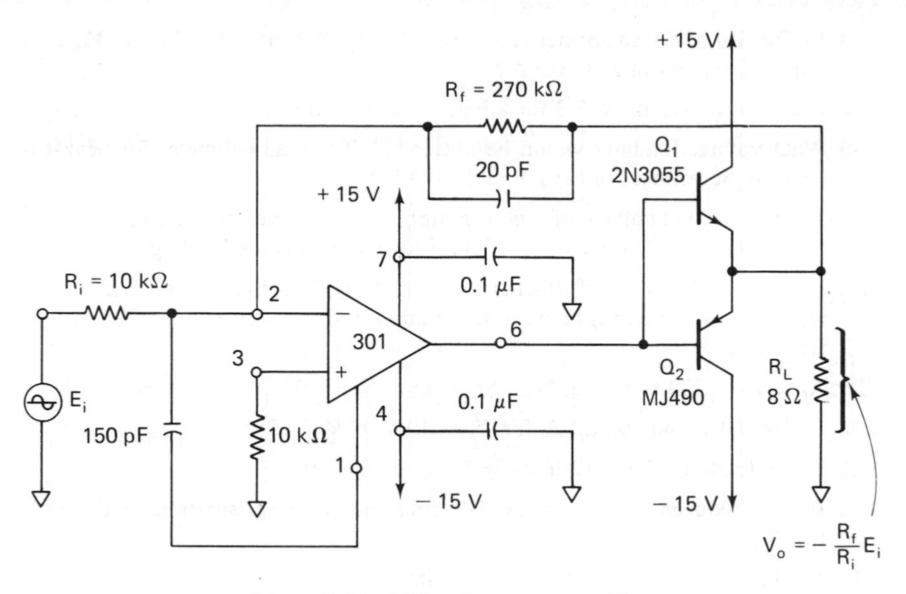

Figure 7-17 5-W audio power amplifier.

the amplifier's voltage gain at 27. When E_i is positive, Q_1 conducts to furnish a positive output voltage. If the current gain of Q_1 is high ($\beta \geqq 100$), then Q_1 boosts the maximum output current capability of the op amp (approximately 10 mA) to over 1 A. When E_i is negative, Q_2 conducts to furnish the output voltage to load R_L.

The frequency response of the amplifier is essentially flat from dc to over 20 kHz. The power output is a maximum of about 5 W to an 8-Ω or 4-Ω speaker.

The amplifier circuit may also be used for a positive or negative voltage regulator for any load voltage between −12 V and +12 V. This is done by replacing E_i with a dc voltage. The circuit of Fig. 7-17 is connected as an inverting amplifier; therefore E_i must be opposite in polarity to the desired regulated load voltage. Choose R_f and R_i to give the correct value of V_o in accordance with $V_o = -E_i R_f / R_i$.

Problems

7-1 A transformer is rated at 115 V to 28 V rms at 1 A. What is the peak secondary voltage?

7-2 A 115 V to 28 V transformer is used in Fig. 7-4(a). Find $V_{\text{dc no load}}$.

7-3 As dc load current decreases, what happens to (a) dc load voltage and (b) ac ripple voltage?

7-4 In Fig. 7-4, the transformer rating is 115 V to 28 V at 1 A. What is V_{dc} at a full load current of $I_L = 0.5$ A?

7-5 Recalculate ΔV in Ex. 7-3 for a load current of 0.5 A.

7-6 What voltage readings would be obtained with an ac voltmeter for peak-to-peak ripple voltages of (a) 1 V and (b) 3 V?

7-7 The dc full-load voltage of a power supply is 28 V, and the peak-to-peak ripple voltage is 6 V. Find the minimum instantaneous load voltage.

7-8 A 110 V: 28 V CT transformer is installed in Fig. 7-7. What no-load dc voltage would be measured (a) between terminals 1 and 2, (b) from 1 to CT, and (c) from 2 to CT?

7-9 $R_f = R_i = 10$ kΩ in Fig. 7-12. Show that $V_o = 10$ V.

7-10 In Fig. 7-12, find R_f and R_i for $V_o = 10$ V, if $V_Z = 5$ V.

7-11 Evaluate R_D in Fig. 7-12 if $V_Z = 5$ V and $V_o = 10$ V.

7-12 If V_o is reduced to 10 V in Ex. 7-5, find the power dissipation in the pass transistor.

7-13 In Fig. 7-14, $R_S = 1$ Ω. Find the maximum load current.

7-14 What is the amplifier gain in Fig. 7-17 if R_f is changed to 220 kΩ?

8

Differential, Instrumentation, and Bridge Amplifiers

8-0 Introduction

The most useful amplifier for measurement, instrumentation, or control is the *instrumentation amplifier*. It is designed with several op amps and precision resistors, which make the circuit extremely stable and useful where accuracy is important. There are now many integrated circuits and modular versions available in single packages. Unfortunately, these packages are relatively expensive (from $10 to over $100). But when performance and precision are required, the instrumentation amplifier is well worth the price, because its performance cannot be matched by the average op amp.

An inexpensive first cousin to the instrumentation amplifier is the basic *differential amplifier*. This chapter begins with the differential amplifier to show in which applications it can be superior to the ordinary inverting or noninverting amplifier. The differential amplifier, with some additions, leads into the instrumentation amplifier, which is discussed in the second part of this chapter. The final sections consider *bridge amplifiers*, which involve both instrumentation and basic differential amplifiers.

8-1 Basic Differential Amplifier

8-1.1 Introduction. The differential amplifier can measure as well as amplify small signals that are buried in much larger signals. How the differential amplifier accomplishes this task will be studied in Section 8-2. But first, let us build and analyze the circuit performance of the basic differential amplifier.

Four precision (1%) resistors and an op amp make up a differential amplifier, as shown in Fig. 8-1. There are two input terminals, labeled (—) input, and (+) input, corresponding to the closest op amp terminal. If E_1 is replaced by a short circuit, E_2 sees an inverting amplifier with a gain of $-m$. Therefore, the output voltage due to E_2 is $-mE_2$. Now let E_2 be short-circuited; E_1 divides between R and mR to apply a voltage of $E_1m/(1 + m)$ at the op amp's (+) input. This divided voltage sees a noninverting amplifier with a gain of $(m + 1)$. The output voltage due to E_1 is the divided voltage, $E_1m/(1 + m)$, times the noninverting amplifier gain, $(1 + m)$, which yields mE_i. Therefore, E_1 is amplified at the output by the multiplier m to mE_1. When both E_1 and E_2 are present at the (+) and (—) inputs respectively, V_o is $mE_1 - mE_2$, or

$$V_o = mE_1 - mE_2 = m(E_1 - E_2) \tag{8-1}$$

Equation (8-1) shows that the output voltage of the differential amplifier, V_o, is proportional to the *difference* in voltage applied to the (+) and (—) inputs. Multiplier m is called the *differential gain* and is set by the resistor ratios.

Example 8-1: In Fig. 8-1, the differential gain is found from

$$m = \frac{mR}{R} = \frac{100\ \text{k}\Omega}{1\ \text{k}} = 100$$

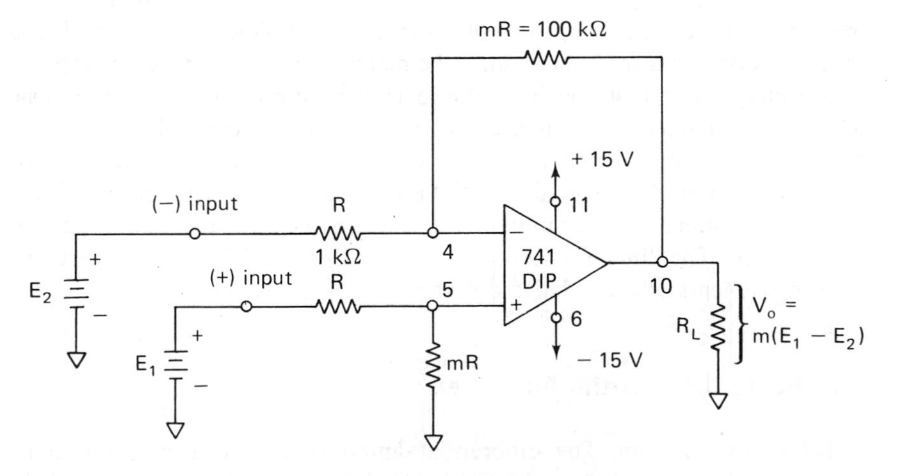

Figure 8-1 Basic differential amplifier.

Find V_o for $E_1 = 10$ mV and (a) $E_2 = 10$ mV, (b) $E_2 = 0$ mV, and (c) $E_2 = -20$ mV.

Solution: By Eq. (8-1), (a) $V_o = 100(10 - 10)\text{ mV} = 0$; (b) $V_o = 100(10 - 0)\text{ mV} = 1.0\text{ V}$; (c) $V_o = 100[10 - (-20)]\text{ mV} = 100(30\text{ mV}) = 3\text{ V}$.

As expected from Eq. (8-1) and shown from part (a) of Example 8-1, when $E_1 = E_2$ the output voltage is 0. To put it another way, when a common (same) voltage is applied to the input terminals, $V_o = 0$. Section 8-1.2 examines this idea of a common voltage in more detail.

8-1.2 Common-Mode Voltage. The output of the differential amplifier should be 0 when $E_1 = E_2$. The simplest way to apply equal voltages is to wire both inputs together and connect them to the voltage source (see Fig. 8-2). For such a connection, the input voltage is called the *common-mode input voltage*, E_{CM}. Now V_o will be 0 if the resistor ratios are equal (mR to R for the inverting amplifier gain equals mR to R of the voltage-divider network). Practically, the resistor ratios are equalized by installing a potentiometer in series with one resistor, as shown in Fig. 8-2. The potentiometer is trimmed until V_o is reduced to a negligible value. This causes the *common-mode voltage gain*, V_o/E_{CM}, to approach 0. It is this characteristic of a differential amplifier that allows a small signal to be picked out of a larger signal. It may be possible to arrange the circuit so that the larger undesired signal is the common-mode input voltage and the small signal is the differential

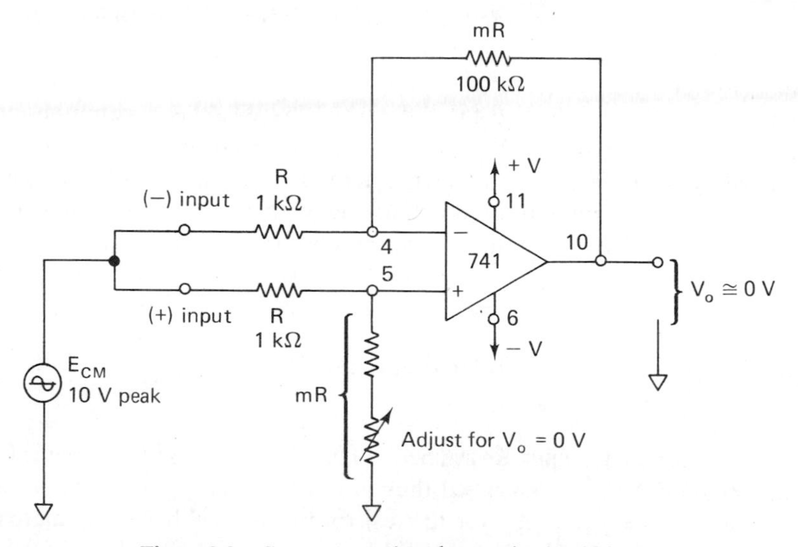

Figure 8-2 Common-mode voltage gain should be 0.

input voltage. Then the differential amplifier's output voltage will contain only an amplified version of the differential input voltage. This possibility is investigated in Section 8-2.

8-2 Measuring Remote Voltage

8-2.1 Measurement with a Single Input Amplifier. A natural way to try amplifying signals from a remote source E_i is shown in Fig. 8-3(a). Two wires with resistances R_{W1} and R_{W2} are used to connect the signal generator to the amplifier. The third terminal of the signal generator is wired to ground. Inevitably, the ground return wire, whose resistance is R_G, also carries return currents from other apparatus on the same power feeder. These return currents may be on the order of amperes and develop a voltage drop along the ground wire modeled by V_G. It is easier to see how V_G affects circuit operation from the schematic drawings in Fig. 8-3(b) and (c). Since V_G is also measured across R_{W2}, R_{W2} may be neglected and V_G is in series with E_i, as shown in Fig. 8-3(c). Now it is obvious that the voltage from the 60-Hz power frequency or noise signals on the ground wire are amplified by the same gain as the desired signal E_i. As shown in Fig. 8-3(c), the output voltage due to V_G can be much larger (10 V) than the output component due to E_i (1 V). We will now use the differential amplifier to measure E_i and show that only E_i will be amplified, while V_G will be balanced out.

8-2.2 Measurement with a Differential Amplifier. In Fig. 8-4(a), the signal generator is connected to the inputs of a differential amplifier. Currents in the ground wire still cause a voltage V_G to be developed across the ground wire resistance R_G. However, as shown in Fig. 8-4(b), the ground noise voltage V_G is now applied to the differential amplifier as a common-mode voltage. As was shown in Section 8-1.2, the common-mode voltage is *not* amplified by the differential amplifier. Therefore, the output voltage V_o is an amplified version of only the differential input voltage E_i. As was stated in Section 8-1.1, the differential amplifier can amplify a small signal ($E_i = 10$ mV) that normally would be buried in a much larger signal ($V_G = 0.1$ V).

8-3 Improving the Basic Differential Amplififer

8-3.1 Increasing Input Resistance. There are two disadvantages to the basic differential amplifier studied thus far: It has low input resistance, and changing gain is difficult, because the resistor ratios must be closely matched. The first disadvantage is eliminated by *buffering* or isolating the inputs with voltage followers. This is accomplished with two op amps connected as

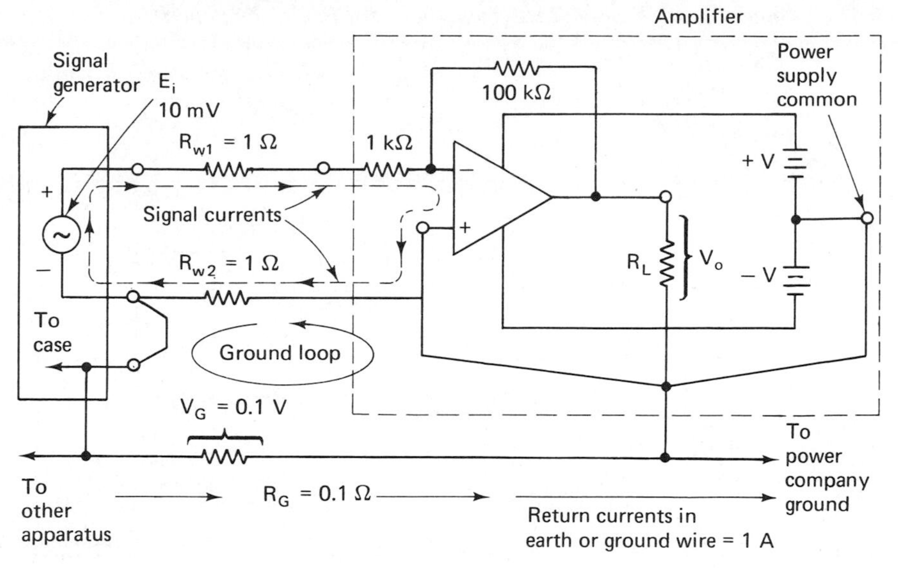

(a) Using an inverting amplifier to measure signals at a distance

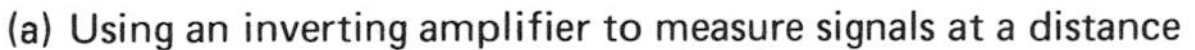

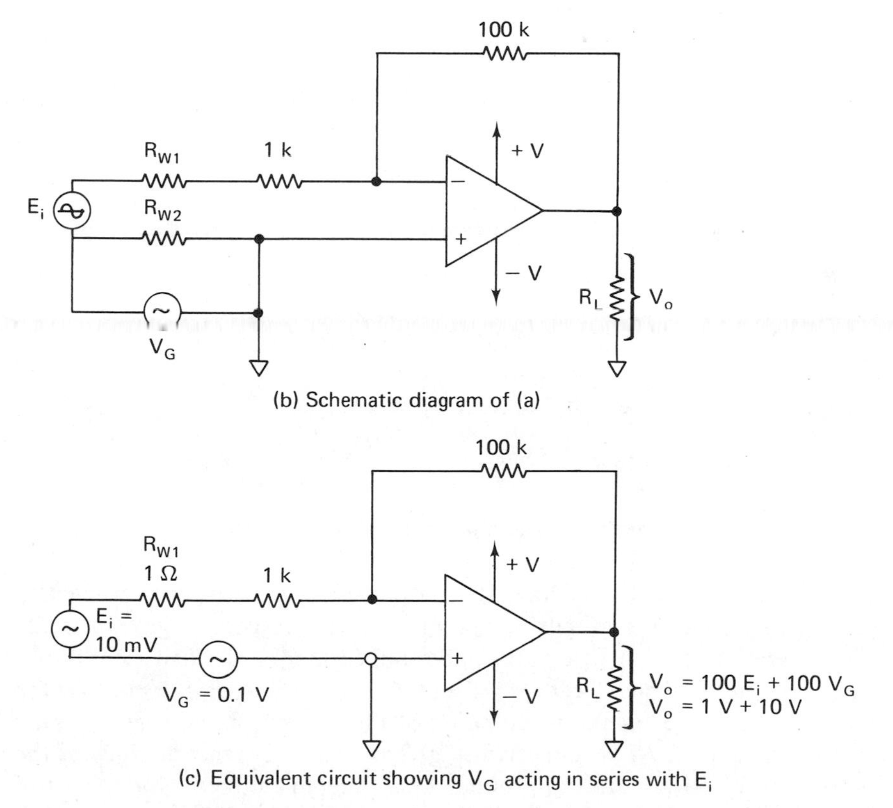

(b) Schematic diagram of (a)

(c) Equivalent circuit showing V_G acting in series with E_i

Figure 8-3 Measuring remote signals incorrectly.

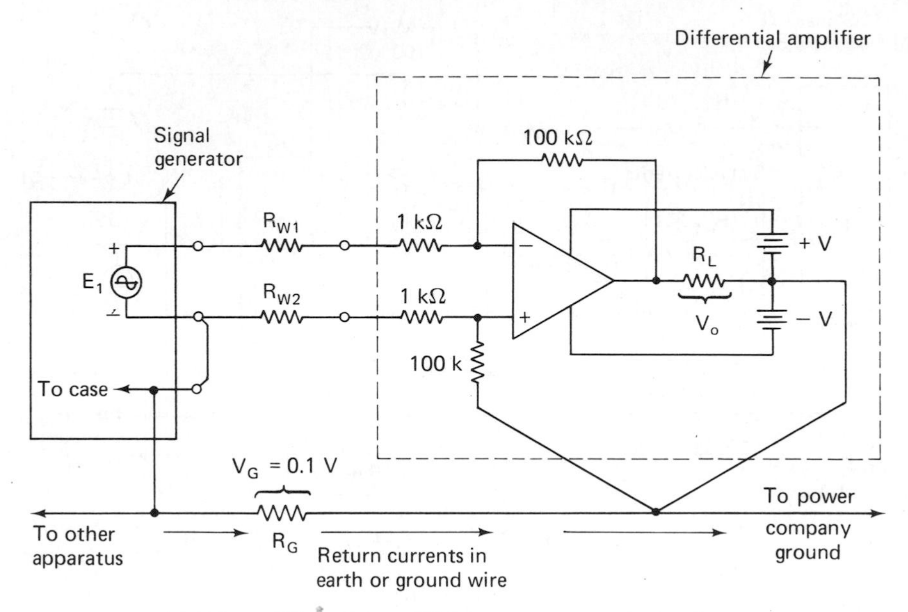

(a) Using a differential amplifier to measure remote signals

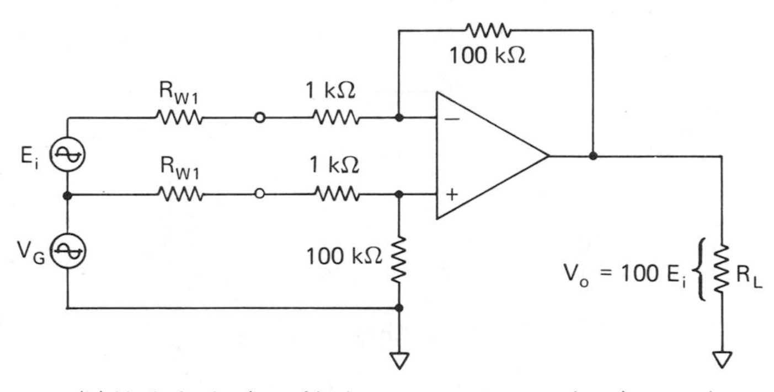

(b) Undesired voltage V_G is now a common-mode voltage and is not amplified by the differential amplifier

Figure 8-4 Measuring remote signal correctly.

voltage followers in Fig. 8-5(a). The output of op amp A_1 with respect to ground is E_1, and the output of op amp A_2 with respect to ground is E_2. The differential output voltage V_o is developed across the load resistor R_L. V_o equals the difference between E_1 and E_2 ($V_o = E_1 - E_2$). Note that the output of the basic differential amplifier of Fig. 8-1 is a single-ended output; that is, one side of R_L is connected to ground, and V_o is measured from the output pin of the op amp to ground. The buffered differential amplifier of Fig. 8-5(a) is a differential output; that is, neither side of R_L is connected to

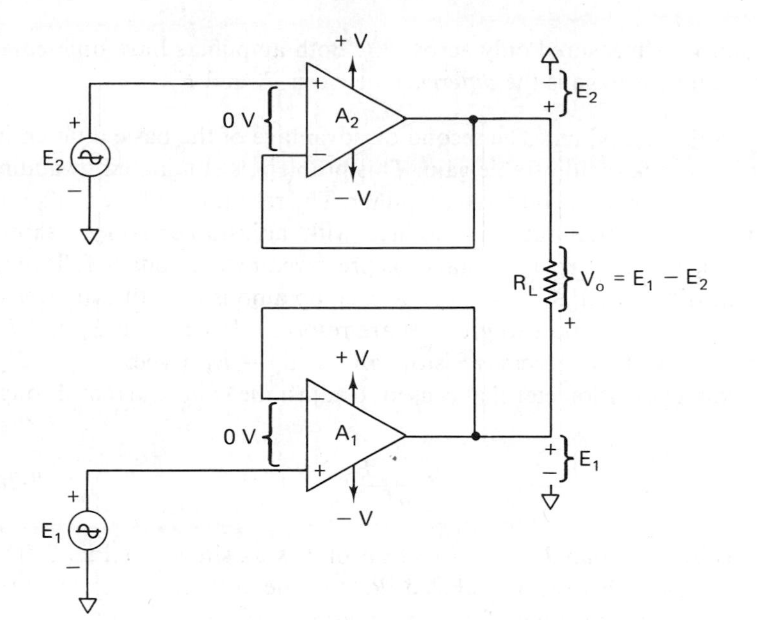

(a) Buffered differential-input to differential-output amplifier

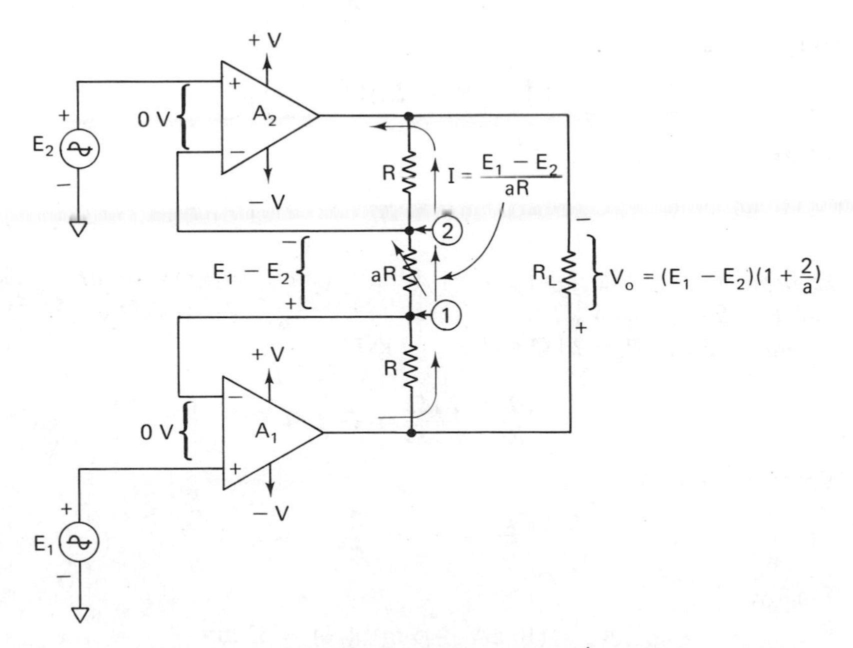

(b) Buffered differential-input to differential-output amplifier with adjustable gain

Figure 8-5 Improving the basic differential amplifier.

ground, and V_o is measured only across R_L. Both amplifiers have differential inputs; the input voltage is the *difference* between E_1 and E_2.

8-3.2 Adjustable Gain. The second disadvantage of the basic differential amplifier is the lack of adjustable gain. This problem is eliminated by adding three more resistors to the buffered amplifier. The resulting buffered differential-input to differential-output amplifier with adjustable gain is shown in Fig. 8-5(b). The high input resistance is preserved by the voltage followers.

Since the differential input voltage of each op amp is 0 V, the voltages at points 1 and 2 (with respect to ground) are respectively equal to E_1 and E_2. Therefore, the voltage across resistor aR is $E_1 - E_2$ (where $E_1 > E_2$). Resistor aR is a potentiometer that is used to adjust the gain. Current through aR is

$$I = \frac{E_1 - E_2}{aR} \tag{8-2a}$$

When E_1 is greater than E_2, the direction of I is as shown in Fig. 8-5(b). I flows through both resistors labeled R, and the voltage across all three resistors establishes the value of V_o. In equation form,

$$V_o = I(aR + 2R) = \frac{E_1 - E_2}{R}(aR + 2R)$$

Simplifying,

$$V_0 = (E_1 - E_2)\left(1 + \frac{2}{a}\right) \tag{8-2b}$$

where

$$a = \frac{aR}{R}$$

Example 8-2: In Fig. 8-5(b), $E_1 = 10$ mV and $E_2 = 5$ mV. If $aR = 2$ kΩ and $R = 9$ kΩ, find V_o.

Solution: Since $aR = 2$ kΩ and $R = 9$ kΩ,

$$\frac{aR}{R} = \frac{2\text{ k}\Omega}{9\text{ k}\Omega} = \frac{2}{9} = a$$

From Eq. (8-2b),

$$1 + \frac{2}{a} = 1 + \frac{2}{\frac{2}{9}} = 10$$

Finally,

$$V_o = (10\text{ mV} - 5\text{ mV})(10) = 50\text{ mV}$$

Conclusion: To change the amplifier gain, only a single resistor aR now has to be adjusted. However, the buffered differential amplifier has one disadvantage: It can only drive floating loads. *Floating loads* are loads that have neither terminal connected to ground. To drive grounded loads, a circuit must be added that converts a differential input voltage to a single-ended output voltage. Such a circuit is the basic differential amplifier. The resulting circuit configuration, to be studied in Section 8-4, is called an *instrumentation amplifier*.

8-4 Instrumentation Amplifier

8-4.1 Circuit Operation. The instrumentation amplifier is one of the most useful, precise, and versatile amplifiers available today. It is made from 3 op amps and 7 resistors as shown in Fig. 8-6. To simplify circuit analysis, note that the instrumentation amplifier is actually made by connecting a buffered amplifier (Fig. 8-5(b)) to a basic differential amplifier (Fig. 8-1). Op amp A_3 and its four equal resistors R form a differential amplifier with a gain of 1. Only the A_3 resistors have to be matched. The primed resistor R' can be made variable to balance out any common-mode voltage, as shown in Fig. 8-2. Only one resistor, aR, is used to set the gain according to

$$\frac{V_o}{E_1 - E_2} = 1 + \frac{2}{a} \tag{8-3}$$

where $a = aR/R$.

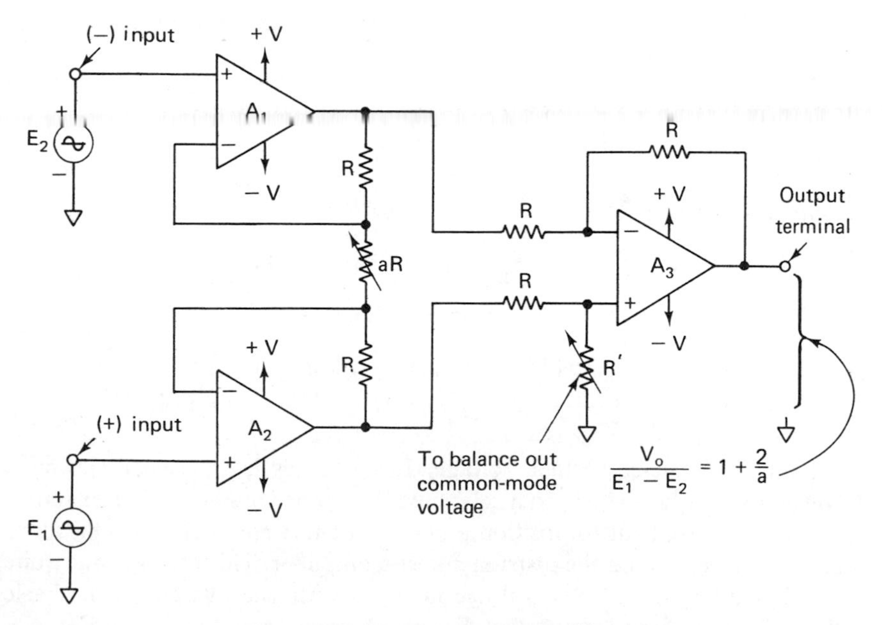

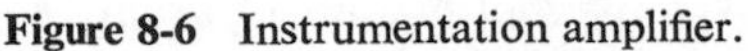

Figure 8-6 Instrumentation amplifier.

E_1 is applied to the (+) input and E_2 to the (−) input. V_o is proportional to the difference between input voltages. Characteristics of the instrumentation amplifier are summarized as follows:

1. The voltage gain, from differential input ($E_1 - E_2$) to single-ended output, is set by *one* resistor.
2. The input resistance of both inputs is very high and does not change as the gain is varied.
3. V_o does *not* depend on the voltage common to both E_1 and E_2 (common-mode voltage), only on their difference.

Example 8-3: In Fig. 8-6, $R = 25\text{ k}\Omega$ and $aR = 50\ \Omega$. Calculate the voltage gain.

Solution: From Eq. (8-3), $\dfrac{aR}{R} = \dfrac{50}{25000} = \dfrac{1}{500} = a$

$$\frac{V_o}{E_1 - E_2} = 1 + \frac{2}{a} = 1 + \frac{2}{\frac{1}{500}} = 1 + (2 \times 500) = 1001$$

Example 8-4: If R is removed in Fig. 8-6 so that $aR = \infty$, what is the voltage gain?

Solution: $a = \infty$, so

$$\frac{V_o}{E_1 - E_2} = 1 + \frac{2}{\infty} = 1$$

Example 8-5: In Fig. 8-6, the following voltages are applied to the inputs. Each voltage polarity is given with respect to ground. Assuming the gain of 1001 from Example 8-3, find V_o for (a) $E_1 = 5.001$ V and $E_2 = 5.002$ V; (b) $E_1 = 5.001$ V and $E_2 = 5.000$ V; and (c) $E_1 = -1.001$ V, $E_2 = -1.002$ V.

Solution: (a)

$$\begin{aligned} V_o &= 1001(E_1 - E_2) = 1001(5.001 - 5.002)\text{ V} \\ &= 1001(-0.001)\text{ V} = -1.001\text{ V} \end{aligned}$$

(b) $V_o = 1001(5.001 - 5.000)\text{ V} = 1001(0.001)\text{ V} = 1.001\text{ V}$
(c) $V_o = 1001[-1.001 - (-1.002)]\text{ V} = 1001(0.001)\text{ V} = 1.001\text{ V}$

8-4.2 Referencing Output Voltage. In some applications, it is desirable to offset the output voltage to a reference level other than 0 V. For example, it would be convenient to position a pen on a chart recorder or oscilloscope trace from a control on the instrumentation amplifier. This can be done quite easily by adding a reference voltage in series with one resistor of the basic differential amplifier. Assume that E_1 and E_2 are set equal to 0 V in Fig. 8-6.

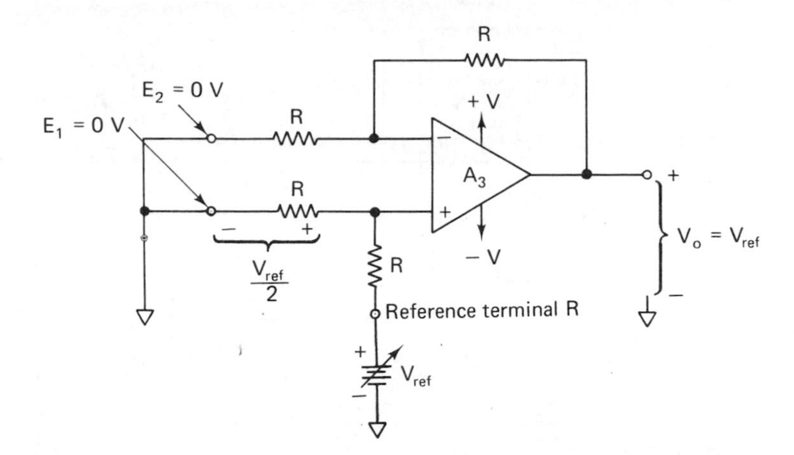

Figure 8-7 Referencing output voltage.

The outputs of A_1 and A_2 will equal 0 V. Thus we can show the inputs of the differential amplifier as 0 V in Fig. 8-7.

A reference voltage V_{ref} is inserted in series with reference terminal R. V_{ref} is divided by 2 and applied to the A_3 op amp's (+) input. Then the noninverting amplifier gives a gain of 2 so that V_o equals V_{ref}. Now V_o can be set to any desired reference value by adjusting V_{ref}. In practice V_{ref} is the output of a voltage follower circuit.

8-5 Sensing and Measuring with the Instrumentation Amplifier

8-5.1 Sense Terminal The versatility and performance of the instrumentation amplifier can be improved by breaking the negative feedback loop around op amp A_3 and bringing out three terminals. As shown in Fig. 8-8, these terminals are *output* terminal O, *sense* terminal S, and *reference* terminal R. If long wires or a current-boost transistor are required between the instrumentation amplifier and load, there will be voltage drops across the connecting wires. To eliminate these voltage drops, the sense terminal and reference terminal are wired directly to the load. Now, wire resistance is added equally to resistors in series with the sense and reference terminals to minimize any mismatch. Even more importantly, by sensing voltage at the load terminals and *not* at the amplifier's output terminal, feedback acts to hold load voltage constant. If only the basic differential amplifier is used, the output voltage is found from Eq. (8-1) with $m = 1$. If the instrumentation amplifier is used, then the output voltage is determined from Eq. (8-3).

This technique is also called *remote voltage sensing*; that is, you sense and control the voltage at a remote load and not at the amplifier's output terminals.

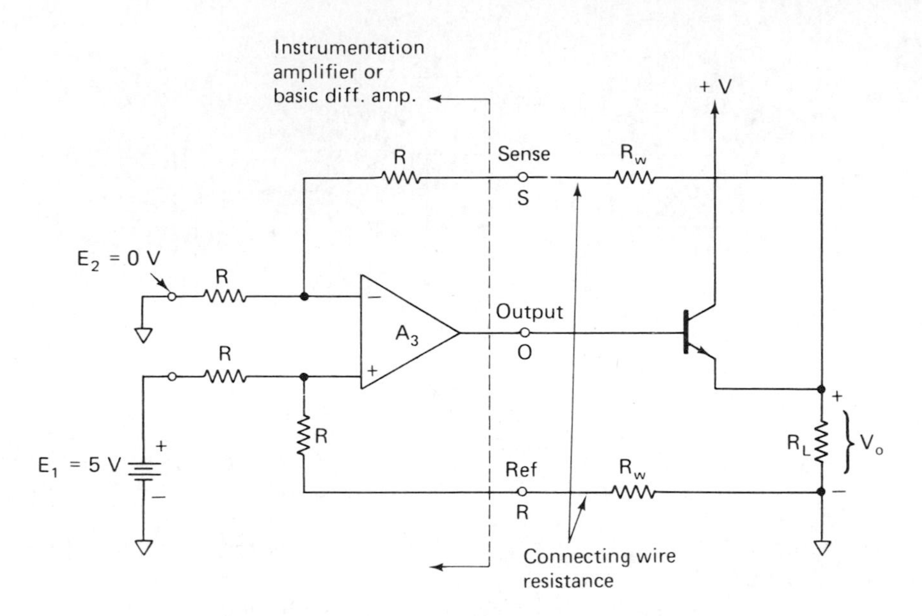

Figure 8-8 Extending the sense and reference terminals to the load makes V_o depend on the amplifier gain and the input voltages, not on the wire resistance.

8-5.2 Current and Differential Voltage Measurements. The schematic drawing of an instrumentation amplifier is shown in Fig. 8-9(a). S is the sense terminal and R is the reference terminal. It is easy to measure the voltage across R_1 with an instrumentation amplifier. Connect the (+) and (−) inputs across R_1 and read V_o with a vacuum tube voltmeter (VTVM). Then calculate the voltage across R_1 or $E_1 - E_2$ from Eq. (8-3):

$$E_1 - E_2 = \frac{V_o}{1 + \frac{2}{a}} \tag{8-3}$$

Figure 8-9(a) can be used to determine the current in a circuit. Let R_1 be placed in the circuit and its value be small enough not to disturb circuit operation but large enough to allow current sensing. That is, if we know the value of R_1 and can measure $E_1 - E_2$ as noted above, then the current I can be found from

$$I = \frac{E_1 - E_2}{R_1} \tag{8-4}$$

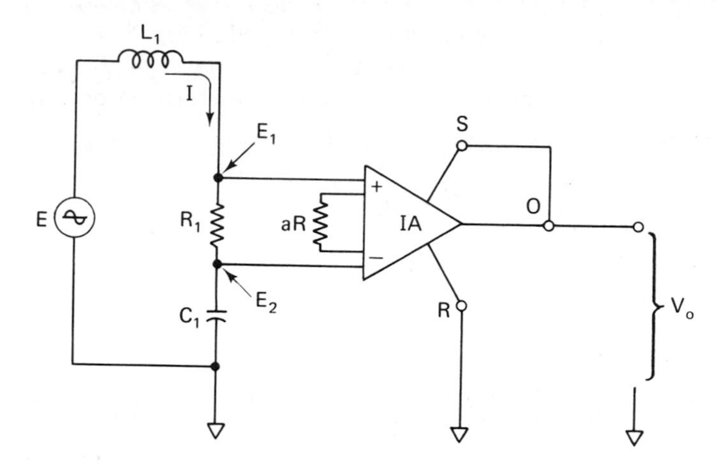

(a) Current measurement with an instrumentation amplifier

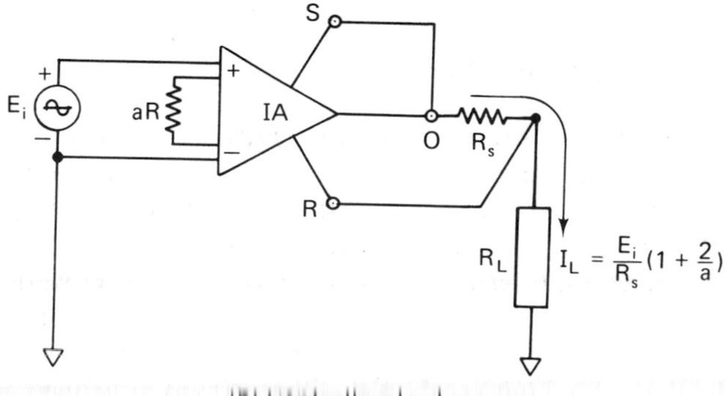

(b) Controlling load current

Figure 8-9 Current measurement and current control using the instrumentation amplifier.

Example 8-6: If $a = \frac{1}{2}$ and $V_o = 10$ V in Fig. 8-9(a), find the differential voltage across resistor R_1.

Solution: By Eq. (8-3),

$$E_1 - E_2 = \frac{V_o}{1 + \dfrac{2}{a}} = \frac{10\text{ V}}{5} = 2\text{ V}$$

Example 8-7: If $R_1 = 1$ kΩ in Example 8-6, find I.

Solution:

$$I = \frac{V_o}{R_1\left(1 + \dfrac{2}{a}\right)} = \frac{10\text{ V}}{1\text{ k}\Omega\,(5)} = \frac{2\text{ V}}{1\text{ k}\Omega} = 2\text{ mA}$$

A current sensing resistor R_S is also employed in Fig. 8-9(b) to sense the load current. The sense and reference terminals draw negligible current with respect to the load current, I_L. Load current flows through both R_S and R_L and is given by

$$I_L = \frac{E_i}{R_S}\left(1 + \frac{2}{a}\right) \qquad (8\text{-}5)$$

Example 8-8: In Fig. 8-9(b), if $E_i = 0.1$ V, $a = 1/2$, and $R_S = 100\ \Omega$, calculate the load current I_L.
Solution: From Eq. (8-5),

$$I_L = \frac{0.1\text{ V}}{100\ \Omega}\left(1 + \frac{2}{\frac{1}{2}}\right) = \frac{(0.1\text{ V})}{100\ \Omega}(5) = 5\text{ mA}$$

E_i controls the load current, and we have a voltage-controlled, *constant current source.*

8-6 Basic Bridge Amplifier

8-6.1 Introduction. An op amp, three equal resistors, and a transducer form the basic bridge amplifier in Fig. 8-10(a). The transducer in this case is any device that converts an environmental change to a resistance change. For example, a *thermistor* is a transducer whose resistance increases as its temperature decreases. A *photoconductive cell* is a transducer whose resistance decreases as light intensity increases. For circuit analysis, the transducer is represented by a resistor R plus a *change* in resistance ΔR. R is the resistance value at the desired reference, and ΔR is the amount of change in R.

For example, a thermistor could have a resistance of $R = 5\text{ k}\Omega$ at a reference of 100°F. Then if a temperature change to 50°F resulted in a net resistance of 7500 Ω, the *change* in resistance ΔR would be $(7500 - 5000) = 2500\ \Omega$. The transducer then would be modeled as $R = 5\text{ k}\Omega$ at 100°F reference and $\Delta R = 2.5\text{ k}\Omega$ for a temperature decrease of 50°F.

To operate the bridge, we need a stable bridge voltage E, which may be either ac or dc. E should have an internal resistance that is small with respect to R. The simplest way to generate E is to use a voltage divider across the stable supply voltages as shown in Fig. 8-10(c). Then connect a simple voltage follower to the divider. For the resistor values shown, E can be adjusted between +5 V and −5V.

8-6.2 Operation. In Fig. 8-10(a), we assumed that the three matched resistors are equal to the transducer's resistance *at the reference condition.*

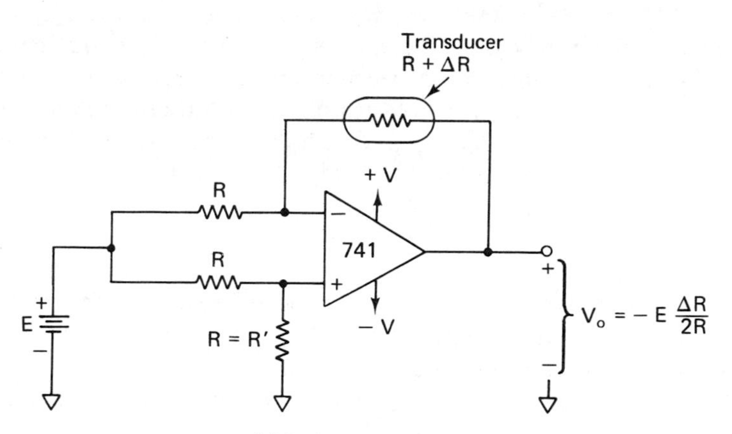

(a) Basic bridge amplifier

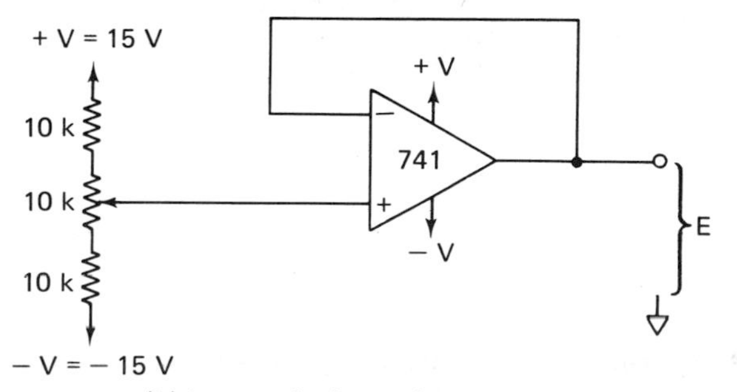

(b) Inexpensive low-resistance voltage source for E

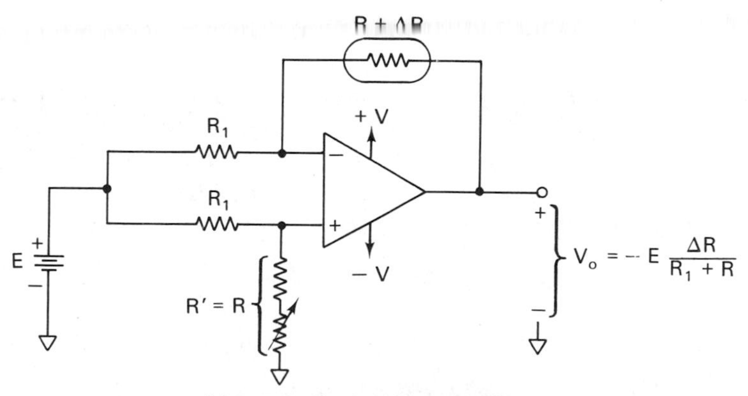

(c) Practical basic bridge amplifier

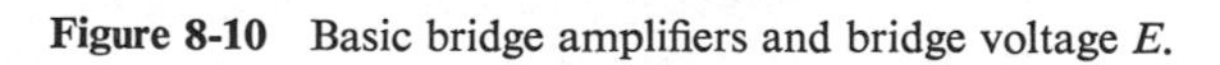

Figure 8-10 Basic bridge amplifiers and bridge voltage *E*.

Unfortunately, this rarely if ever occurs. A more realistic bridge circuit is shown in Fig. 8-10(c). Only two matched resistors, R_1 are required. The primed resistor R' is made up of a series fixed resistor and a variable resistor for the same reasons discussed in Fig. 8-2. The technique of adjusting R' is similar to Fig. 8-2 but must be revised for the bridge circuit as follows.

Zeroing procedure

1. Place the circuit of Fig. 8-10(c) in the reference environment. For example, set it at a temperature of 100°F so that $R \cong 5000\ \Omega$. R is the thermistor discussed in Section 8-6.1.
2. Adjust R' so that $V_o = 0$ V. R' is now exactly equal to the reference value R.
3. E should be set at the largest value allowable for the application. Typically, $E = 5$ to 15 V. The bridge is now calibrated, and the output voltage V_o will be proportional only to the change in transducer resistance ΔR. V_o may be calculated from

$$V_o = -E\frac{\Delta R}{R_1 + R} \tag{8-6}$$

The minus sign means that V_o decreases (goes negative) as ΔR increases.

8-6.3 An IC Thermometer. The circuit of Fig. 8-10(c) can be employed as a thermometer. Assuming that the thermistor is a temperature sensor and the circuit is zeroed, a calibrated meter can easily be used to measure temperature, as shown in the next example.

Example 8-9: In the circuit of Fig. 8-10(c), $R_1 = 10\ \text{k}\Omega$, $R = 5\ \text{k}\Omega$ at 100°F, and $E = -15$ V. Find V_o for the following temperatures: (a) $R = 5\ \text{k}\Omega$ at 100°F; (b) $R = 7500\ \Omega$ at 50°F, $\Delta R = +2500\ \Omega$; (c) $R = 10{,}000\ \Omega$ at 0°F, $\Delta R = +5000\ \Omega$; (d) $R = 4000\ \Omega$ at 110°F, $\Delta R = -1000\ \Omega$.
Solution: From Eq. (8-6),

$$V_o = -(-15\text{ V})\frac{\Delta R}{(10\ \text{k}\Omega + 5\ \text{k}\Omega)}$$

or

$$V_o = \frac{15\text{ V}}{15\ \text{k}\Omega}\Delta R = \frac{1\text{ V}}{1\ \text{k}\Omega} \times \Delta R$$

We tabulate the data.

Problem	ΔR	V_o	*Temperature*
(a)	0	0 V	100°F
(b)	2.5 kΩ	2.5 V	50°F
(c)	5 kΩ	5.0 V	0°F
(d)	−1.0 kΩ	−1.0 V	110°F

Example 8-10: From the results of Example 8-9, show how a −5-V to +5-V voltmeter could be calibrated as a thermometer in Fahrenheit degrees.
Solution: The solution is shown in Fig. 8-11. This voltmeter would be connected to read V_o in Fig. 8-10(c).

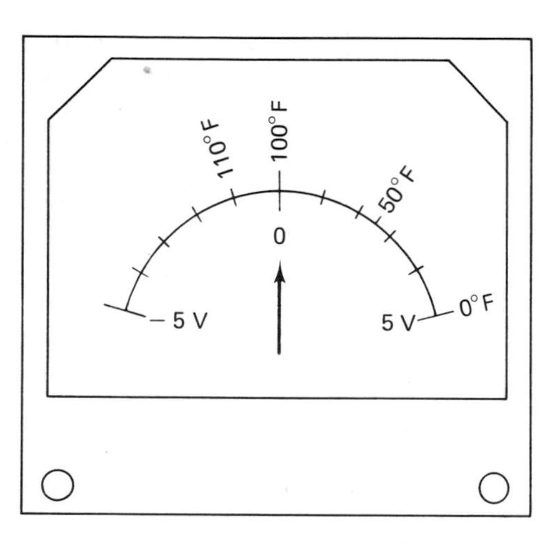

Figure 8-11 Solution to Example 8-10.

8-7 Adding Versatility to the Bridge Amplifier

8-7.1 Grounded Transducers. In some applications, it may be necessary to have the transducer connected to ground. The standard technique is shown in Fig. 8-12(a). Note that V_o will have the same polarity as E for increases in transducer resistance. The resistor R' is made adjustable and set equal to R of the transducer in accordance with the zeroing procedure stated in Section 8-6.2.

8-7.2 High-Current Transducers. If the current required by the transducer is higher than the current capability of the op amp (5 mA), use the circuit of Fig. 8-12(b). Transducer current is furnished from E. Resistors mR are large enough to hold their currents to about 1 mA, typically 10 kΩ. Transducer current and output voltage may be found from the equations

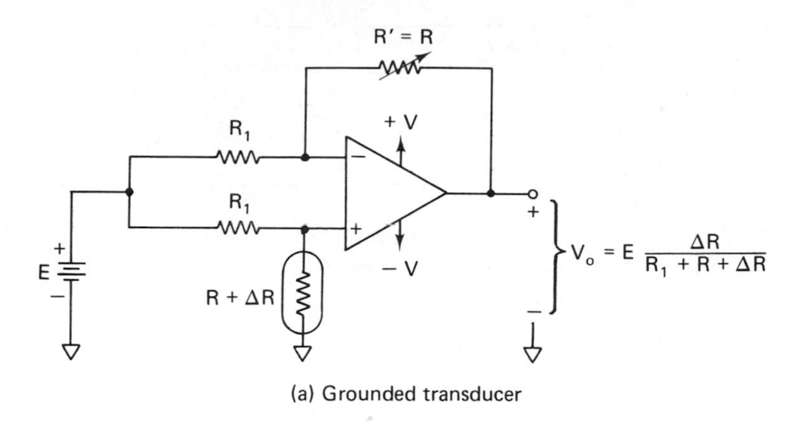

(a) Grounded transducer

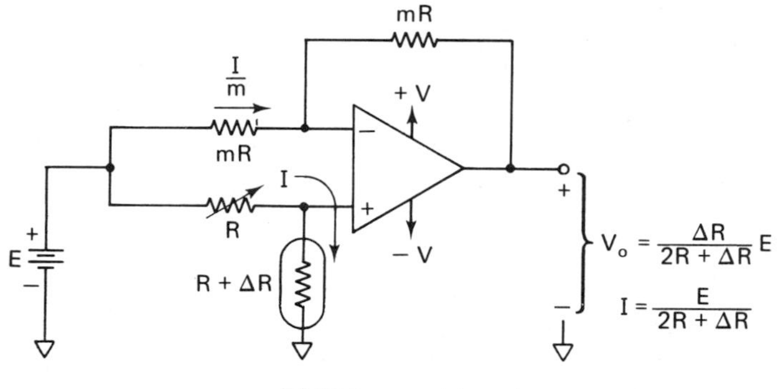

(b) High-current transducer

Figure 8-12 Other ways of using the basic bridge amplifier.

in Fig. 8-12(b). If the transducer current is very small (high-resistance transducers), this same circuit can be used except that the *mR* resistors will be smaller than *R* to hold output current of the op amp at about 1 mA.

8-8 Measuring Small Resistance Changes

Normally, it is difficult to measure resistance *changes* of 1% or less. However, the task is simplified by using a resistance bridge network and an instrumentation amplifier, as in Fig. 8-13. Resistors R_1 are equal wire-wound or metal-film resistors matched to within 1%. The test resistor whose resistance *change* is to be measured (for example under stress, vibration, or shock) is shown as $R + \Delta R$. A 10-turn potentiometer R is adjusted to exactly equal R of the test resistor. This is done by connecting the instrumentation amplifier and adjusting R until $V_o = 0$ V. This sets the reference value of the resistor R

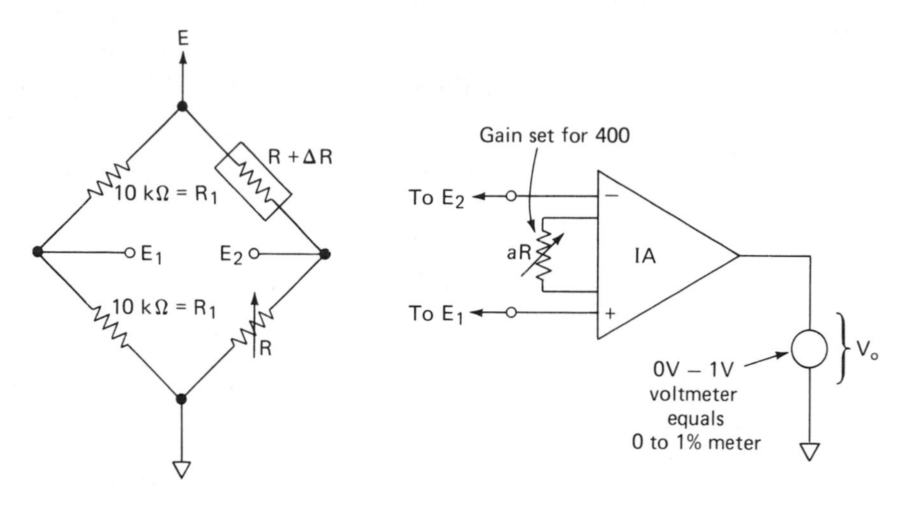

Figure 8-13 Circuit to measure small resistance changes.

and is called *zeroing*. The relationship between V_o and *percentage* of resistance change is developed as follows.

Since the R_1 resistors are matched in Fig. 8-13, $E_1 = E/2$. To find E_2, we note that E divides between R and $R + \Delta R$ so

$$E_2 = \frac{R}{2R + \Delta R}E$$

To find the difference between E_1 and E_2, we subtract to get

$$E_1 - E_2 = \frac{E}{2}\left(\frac{\Delta R}{2R + \Delta R}\right)$$

In this application, the small change ΔR is negligible with respect to $2R$ and can be dropped from the denominator. Thus

$$E_1 - E_2 = \frac{E}{4} \times \frac{\Delta R}{R}$$

The gain of the differential amplifier $V_o/(E_1 - E_2)$ can be set (as in Section 8-4) at 400, or

$$V_o = 400(E_1 - E_2) = 400 \times \frac{E}{4} \times \frac{\Delta R}{R}$$

V_o can now be expressed simply as

$$V_o = 100E \times \frac{\Delta R}{R} \tag{8-7a}$$

Then for $E = 1$ V,

$$V_o = \text{(percentage of change in resistance)} = 100\frac{\Delta R}{R} \quad \text{(8-7b)}$$

Example 8-11: If $R = 10\text{ k}\Omega$, $\Delta R = 100\ \Omega$, and $E_i = 1$ V in Example 8-10, find V_o from Eq. (8-7a).
Solution:

$$\frac{\Delta R}{R} = \frac{100}{10{,}000} = \frac{1}{100} = 1\%\text{ change}$$

By Eq. (8-7a),

$$V_o = 100 \times 1 \times \frac{1}{100} = 1\text{ V}$$

Note from Eq. (8-7b) that $V_o = 1$ V, or 1 V per 1% change in resistance.

8-9 Measuring Force, Pressure, or Acceleration

When force, pressure, or acceleration is applied to an object, it experiences *stress*. The amount of stress is converted into a proportional voltage by a *strain gage*. The strain gage is simply a small-diameter wire, foil or semiconductor material with high resistance. The resistance wire is mounted on a thin plastic carrier sheet for storage and handling.

To measure stress applied to an object with a strain gage, fasten the gage securely to the object with a thin layer of epoxy. The gage will have a typical unstressed resistance value R. When compressed or stretched (force, pressure, or acceleration), resistance respectively decreases or increases. The change in resistance ΔR will be proportional to the amount of stress. ΔR, or the ratio of $\Delta R/R$, can be measured by the circuit of Fig. 8-13 and calculated by measurements of V_o and E from Eq. (8-7a). From manufacturer's data or on-site calibration, ΔR can be related to force, pressure, or acceleration.

Problems

8-1 In Fig. 8-1, $m = 20$, $E_1 = 0.2$ V, and $E_2 = -0.25$ V. Determine V_o.

8-2 If V_o in Fig. 8-1 is 10 V and $E_1 = 7.5$ V and $E_2 = 7.4$ V, calculate m.

8-3 When is the input voltage to a differential amplifier a common-mode voltage?

8-4 If E_{CM} in Fig. 8-2 is 5 V, what is V_o?

8-5 In Fig. 8-4(a), $V_G = 0.15$ V and $E_i = 0.1$ V. What is the output voltage due to (a) V_G and (b) E_i?

8-6 What is the output voltage V_o in Fig. 8-5(a) if $E_1 = 6$ V and $E_2 = 4$ V?

8-7 For the circuit of Fig. 8-5(b), $R = 10$ kΩ and $aR = 2$ kΩ. Determine (a) a, (b) V_o, and (c) I. $E_1 = 3$ V and $E_2 = 2$ V.

8-8 Output voltage V_o in Fig. 8-6 equals 8 V if $E_1 = 1.5$ V and $E_2 = 1$ V. Calculate a.

8-9 (a) If the overall gain $V_o/(E_1 - E_2)$ of Fig. 8-6 is 21, calculate a. (b) If $V_o = 3$ V, determine $(E_1 - E_2)$.

8-10 In Fig. 8-9(b), if $E_i = 0.2$ V, $a = 0.1$, and $R_S = 200$ Ω, calculate the load current I_L.

8-11 Using the values of Problem 8-10, what is the reference voltage (voltage from the R terminal to ground)? $R_L = 10$ kΩ.

8-12 The thermistor in Fig. 8-10(a) has a resistance of 10 kΩ at a reference temperature of 75°. If the temperature changes to 50°, the resistance increases to 15 kΩ. Determine (a) ΔR and (b) V_o. $E = 4$ V.

8-13 In the Fig. 8-10(c), $R_1 = 20$ kΩ, $R = 10$ kΩ at 100°F, and $E = 10$ V; find V_o when $R = 7.5$ kΩ at 125°F.

8-14 If the thermistor of Problem 8-13 is used in the circuit of Fig. 8-12(a), determine V_o.

8-15 In Fig. 8-12(b), $E = 6$ V, $R = 10$ kΩ, and $\Delta R = 5$ kΩ. Find (a) V_o, (b) m such that the current into the output terminal of the op amp is 0.02 mA, and (c) I.

9

Bias, Offsets, and Drift

9-0 Introduction

The op amp is widely used in amplifier circuits to amplify dc or ac signals or combinations of them. In dc amplifier applications, certain electrical characteristics of the op amp can cause large errors in the output voltage. The ideal output voltage should be equal to the product of the dc input signal and the amplifier's closed-loop voltage gain. However, the output voltage may have an added error component. This error is due to differences between an ideal op amp and a real op amp. If the ideal value of output voltage is large with respect to the error component, then we can ignore the op amp characteristic that causes it. But if the error component is comparable to or even larger than the ideal value, we must try to minimize the error. Op amp characteristics that add error components to the dc output voltage are

1. *input bias currents*
2. *input offset current*
3. *input offset voltage*
4. *drift*

When the op amp is used in an ac amplifier, coupling capacitors eliminate dc output-voltage error. Therefore, characteristics (1) to (4) above are usually unimportant in ac applications. However, there are new problems for ac amplifiers. They are

5. *frequency response*
6. *slew rate*

Frequency response refers to how voltage gain varies as frequency changes. The most convenient way to display this data is a plot of voltage gain versus frequency. Op amp manufactures give such a plot for open-loop gain versus frequency. A glance at the plot quickly shows how much gain is obtainable at a particular frequency.

If the op amp has sufficient gain at a particular frequency, there is still a possibility of an error being introduced in V_o. This is because there is a fundamental limit imposed by the op amp (and certain circuit capacitors) on how fast the output voltage can change. If the input signal tells the op amp output to change faster than it can, distortion is introduced in the output voltage. The op amp characteristic responsible for this type of error is its capacitance. This type of error is called *slew-rate limiting*.

Op amp characteristics and the circuit applications that each type of error *may* affect are summarized in Fig. 9-1. The first four characteristics can limit dc performance; the last two can limit ac performance.

Op-amp characteristic that may affect performance	Op-amp application			
	DC amplifier		AC amplifier	
	Small output	Large output	Small output	Large output
1. Input bias current	yes	maybe	no	no
2. Offset current	yes	maybe	no	no
3. Input offset voltage	yes	maybe	no	no
4. Drift	yes	no	no	no
5. Frequency response	no	no	yes	yes
6. Slew rate	no	yes	no	yes

Figure 9-1 Listing of op amp applications and characteristics that affect operation.

Op amp characteristics that cause errors primarily in dc performance will be studied in this chapter. Those that cause errors in ac performance will be studied in Chapter 10. We begin with input bias currents and ways in which they cause errors in the dc output voltage of an op amp circuit.

9-1 Input Bias Currents

Transistors within the op amp must be *biased* correctly before any signal voltage is applied. Biasing correctly means that the transistor has the right value of base and collector current as well as collector-to-emitter voltage. Until now, we have considered that the input terminals of the op amp conduct no current. This is the ideal condition. Practically, however, the input terminals do conduct a small value of dc current to bias the op amps' transistors. (See Appendices 1 and 2.) A simplified diagram of the op amp is shown in Fig. 9-2(a). To discuss the effect of input bias currents, it is conven-

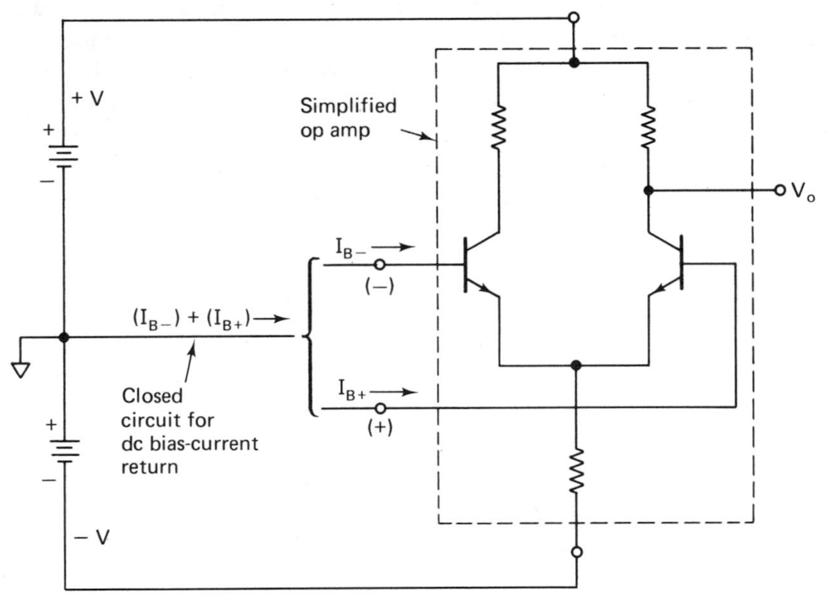

(a) Simplified op amp input circuit

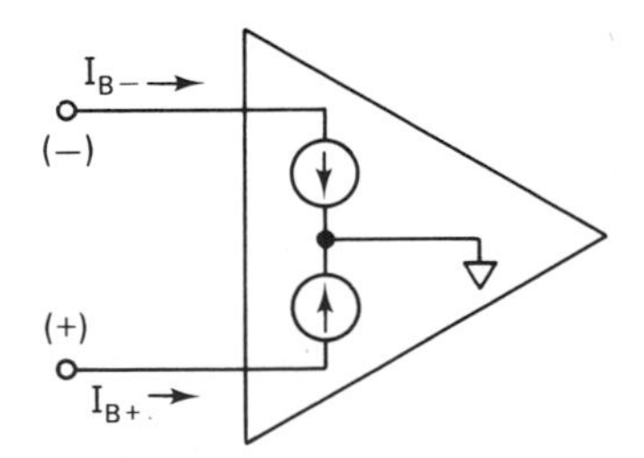

(b) Model for bias currents

Figure 9-2 Origin and model of dc input bias currents.

ient to model them as current sources in series with each input terminal, as shown in Fig. 9-2(b).

The (—) input's bias current, I_{B-}, will usually not be equal to the (+) input's bias current, I_{B+}. Manufacturers specify an *average* input bias current I_B, which is found by adding the *magnitudes* of I_{B+} and I_{B-} and dividing this sum by 2. In equation form,

$$I_B = \frac{|I_{B+}| + |I_{B-}|}{2} \tag{9-1}$$

where $|I_{B+}|$ is the magnitude of I_{B+} and $|I_{B-}|$ is the magnitude of I_{B-}. The range of I_B is from 1 μA or more for general-purpose op amps to 1 pA or less for op amps that have field-effect transistors at the input.

9-2 Input Offset Current

The difference in magnitudes between I_{B+} and I_{B-} is called the *input offset current*, I_{os}

$$I_{os} = |I_{B+}| - |I_{B-}| \tag{9-2}$$

Manufacturers specify I_{os} for a circuit condition where the output is at 0 V and the temperature is 25°C. The typical I_{os} is less than 25% of I_B, the average input bias current. (see Appendices 1 and 2.)

Example 9-1: If $I_{B+} = 0.4\ \mu$A and $I_{B-} = 0.3\ \mu$A, find (a) the average bias current I_B and (b) the offset current. I_{os}.
Solution: (a) By Eq. (9-1),

$$I_B = \frac{(0.4 + 0.3)\ \mu\text{A}}{2} = 0.35\ \mu\text{A}$$

(b) By Eq. (9-2),

$$I_{os} = (0.4 - 0.3)\ \mu\text{A} = 0.1\ \mu\text{A}$$

9-3 Effect of Bias Currents on Output Voltage

9-3.1 Simplification. In this section it is assumed that bias currents are the only op amp characteristic that will cause an undesired component in the output voltage. The effects of other op amp characteristics on V_o will be dealt with individually.

9-3.2 Effect of (—) Input Bias Current. Output voltage should ideally equal 0 V in each circuit of Fig. 9-3, because input voltage E_i is 0 V. The fact that a voltage component will be measured is due strictly to I_{B-}. In Fig.

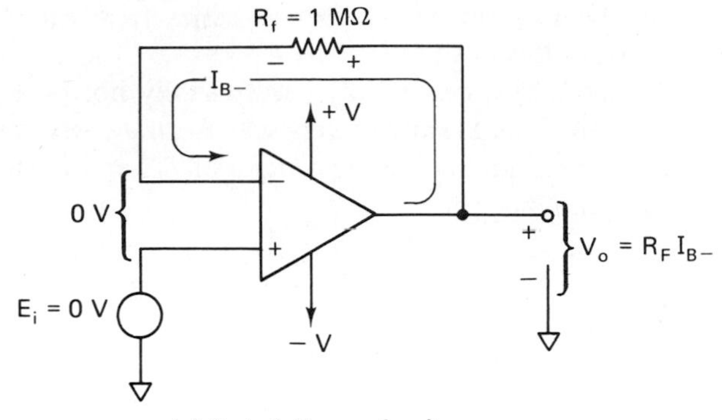

(a) Basic follower circuit

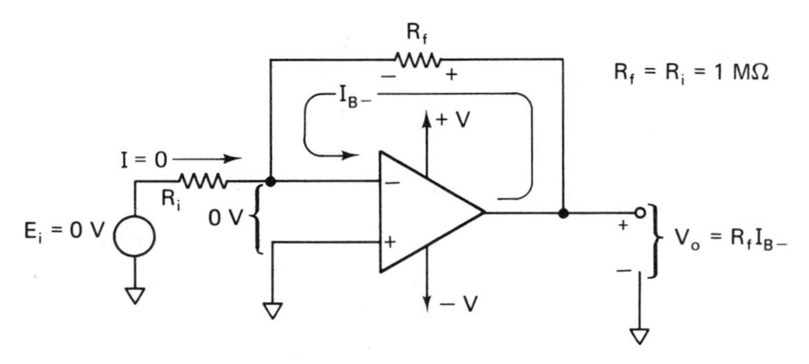

(b) Basic inverting circuit

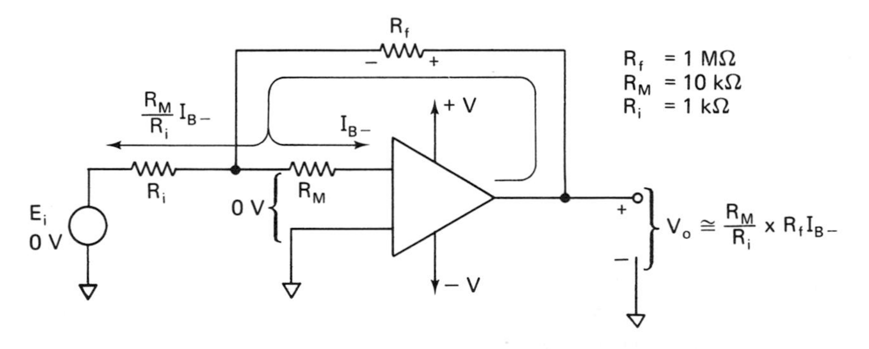

(c) Multiplier resistor R_M increases effect of I_{B-} on V_o

Figure 9-3 Effects of (−) input bias current on output voltage.

9-3(a), the bias current is furnished from the output terminal. Since negative feedback forces the differential input voltage to 0 V, V_o must rise to supply the voltage drop across R_f. Thus, the output voltage error due to I_{B-} is found from $V_o = R_f I_{B-}$. I_{B+} flows through 0 Ω, so it causes no voltage error.

The circuit of Fig. 9-3(b) has the same output-voltage error expression, $V_o = R_f I_{B-}$. No current flows through R_i, because there is 0 V on each side of R_i. Thus all of I_{B-} flows through R_f. (Recall that an ideal amplifier with negative feedback has 0 voltage between the (+) and (−) inputs.)

Example 9-2: In Fig. 9-3(a), $V_o = 0.4$ V. Find I_{B-}.
Solution:

$$I_{B-} = \frac{V_o}{R_f} = \frac{0.4\text{ V}}{1\text{ M}\Omega} = 0.4\ \mu\text{A}$$

Placing a multiplying resistor R_M in series with the (−) input in Fig. 9-3(c) multiplies the effect of I_{B-} on V_o. I_{B-} sets up a voltage drop across R_M that establishes an equal drop across R_i.

Now both the R_i current and I_{B-} must be furnished through R_f. Thus the error in V_o will be much larger. R_M would be undesirable in a normal circuit; however, we want to measure low values of the bias current, and Fig. 9-3(c) is a way of doing it. For the resistor values shown, $V_o \cong 10\ R_f I_{B-}$.

9-3.3 Effect of (+) Input Bias Current. Since $E_i = 0$ V in Fig. 9-4, V_o should ideally equal 0 V. However, the positive input bias current I_{B+} flows through the internal resistance of the signal generator. Internal generator resistance is modeled by resistor R_G in Fig. 9-4. I_{B+} sets up a voltage drop of $R_G I_{B+}$ across R_G and applies it to the (+) input. The differential input voltage is 0 V, so the (−) input is also at $R_G I_{B+}$ in Fig. 9-4. Since there is 0 resistance in the feedback loop, V_o equals $R_G I_{B+}$. (The return path for I_{B+} is through $-V$ supply and back to ground.)

Example 9-3: In Fig. 9-4, $V_o = 0.3$ V. Find I_{B+}.
Solution:

$$I_{B+} = \frac{V_o}{R_G} = \frac{0.3\text{ V}}{1\text{ M}\Omega} = 0.3\ \mu\text{A}$$

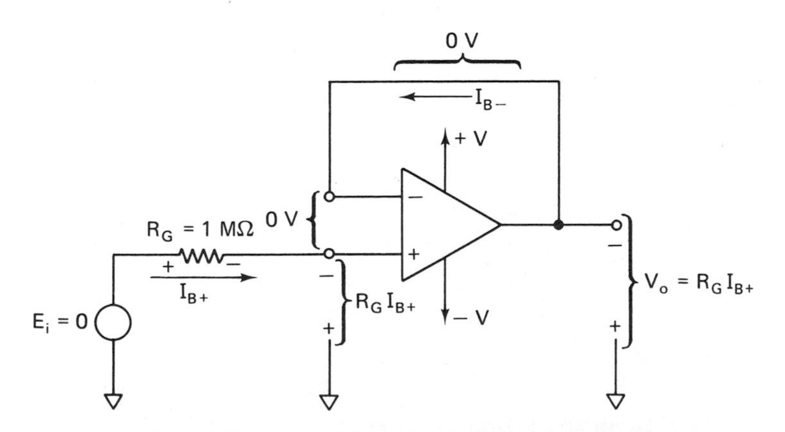

Figure 9-4 Effect of (+) input bias current on output voltage.

9-4 Effect of Offset Current on Output Voltage

9-4.1 Current-Compensating the Voltage Follower. If I_{B+} and I_{B-} were always equal, it would be possible to compensate for their effects on V_o. For example, in the voltage follower of Fig. 9-5(a), I_{B+} flows through the signal generator resistance R_G. If we insert $R_f = R_G$ in the feedback loop, I_{B-} will develop a voltage drop across R_f of $R_f I_{B-}$. If $R_f = R_G$ and $I_{B+} = I_{B-}$, their voltage drops will cancel each other and V_o will equal 0 V when $E_i = 0$ V. Unfortunately, I_{B+} is seldom equal to I_{B-}. V_o will then be equal to R_G times the difference between I_{B+} and $I_{B-}(I_{B+} - I_{B-} = I_{os})$. Therefore, by making $R_f = R_G$, we have reduced the error in V_o from $R_G I_{B+}$ in Fig. 9-4 to

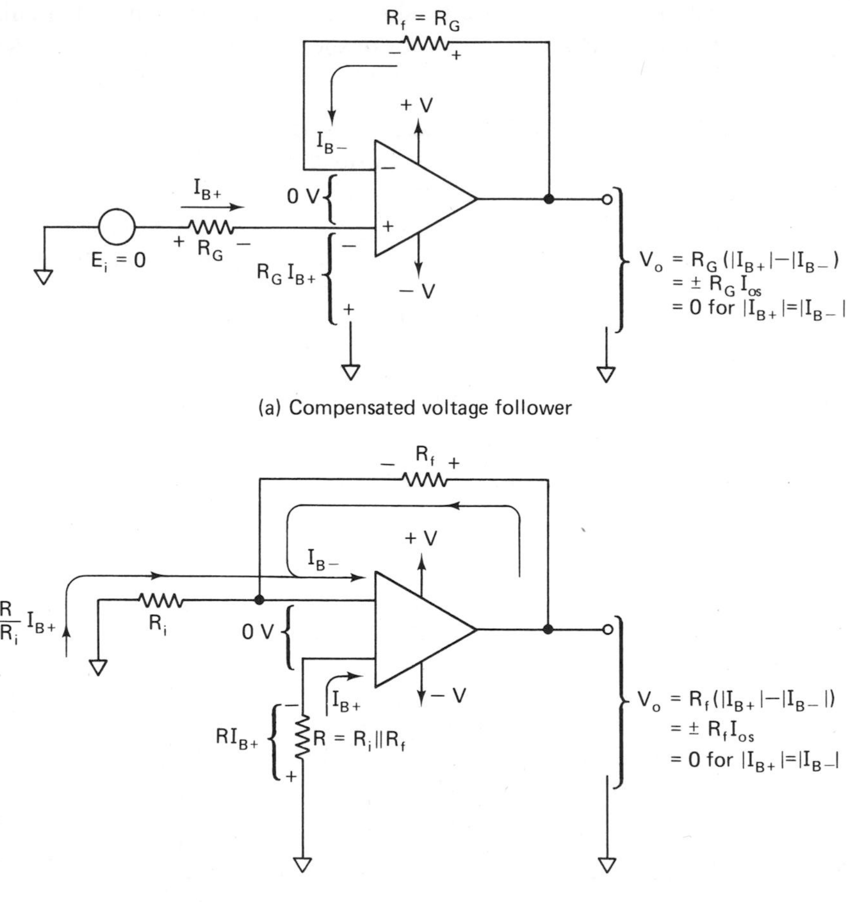

Figure 9-5 Balancing out effects of bias current in V_o.

$R_G I_{os}$ in Fig. 9-5(a). Recall that I_{os} is typically 25% of I_B. If the value of I_{os} is too large, then an op amp with a smaller value of I_{os} is needed.

9-4.2 Current-Compensating Other Amplifiers. To minimize errors in V_o due to bias currents for either inverting or noninverting amplifiers, resistor R as shown in Fig. 9-5(b), must be added to the circuit. With no input signal applied, V_o depends on R_f times I_{os} (where I_{os} is given by Eq. (9-2)). Resistor R is called the *current-compensating resistor* and is equal to the parallel combination of R_i and R_f, or

$$R = R_i \,||\, R_f = \frac{R_i R_f}{R_i + R_f} \tag{9-3}$$

R_i and R should include any signal generator resistance. By insert'ng resistor R, the error voltage in V_o will be reduced more than 25% from $R_f I_{B-}$ in Fig. 9-3(b) to $R_f I_{os}$ in Fig. 9-5(b). In the event that $I_{B-} = I_{B+}$, then $I_{os} = 0$ and $V_o = 0$.

9-4.3 Summary on Bias-Current Compensation. Always add a bias-current compensating resistor R in series with the (+) input terminal (except for FET input op amps). The value of R should equal the parallel combination of all resistance branches connected to the (−) terminal. Any internal resistance in the signal source should also be included in the calculations.

In circuits where more than a single resistor is connected to the (+) input, bias-current compensation is accomplished by observing the following principle. *The dc resistance seen from the* (+) *input to ground should equal the dc resistance seen from the* (−) *input to ground.* In applying this principle, signal sources are replaced by their internal dc resistance and the op amp output terminal is considered to be at ground potential.

Example 9-4: (a) In Fig. 9-5(b), $R_f = 100\text{ k}\Omega$ and $R_i = 10\text{ k}\Omega$. Find R. (b) If $R_f = 100\text{ k}\Omega$ and $R_i = 100\text{ k}\Omega$, find R.

Solution: (a) By Eq. (9-3),

$$R = \frac{(100\text{ k})(10\text{ k})}{100\text{ k} + 10\text{ k}} = 9.1\text{ k}\Omega$$

(b) By Eq. (9-3),

$$R = \frac{(100\text{ k})(100\text{ k})}{100\text{ k} + 100\text{ k}} = 50\text{ k}\Omega$$

9-5 Input Offset Voltage

9-5.1 Definition and Model. In Fig. 9-6(a), the output voltage V_o should equal 0 V. However, there will be a small error-voltage component present in V_o. Its value can range from microvolts to millivolts and is caused by very small but unavoidable unbalances inside the op amp. The easiest way to

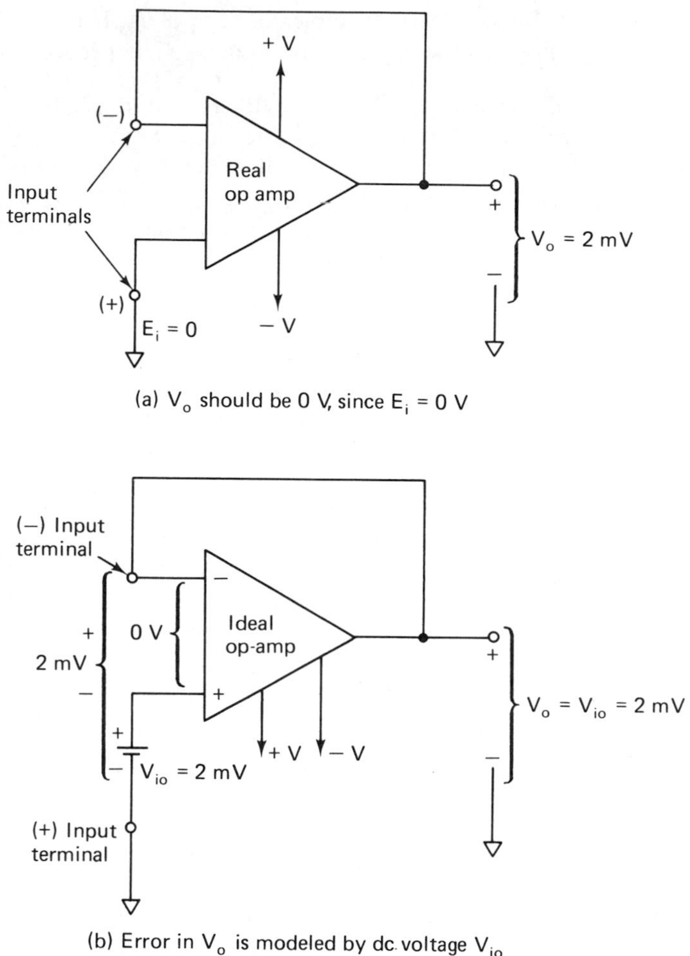

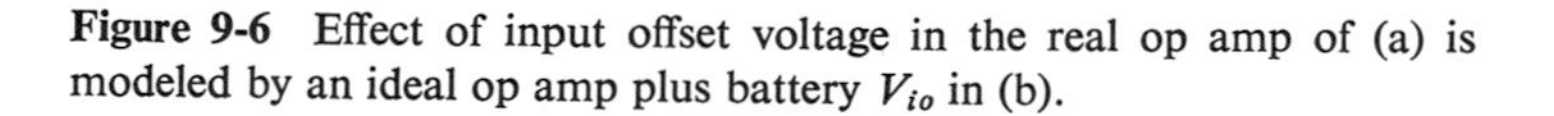

Figure 9-6 Effect of input offset voltage in the real op amp of (a) is modeled by an ideal op amp plus battery V_{io} in (b).

study the *net effect* of all these internal unbalances is to visualize a small dc voltage in *series* with one of the input terminals. This dc voltage is modeled by a battery in Fig. 9-6(b) and is called *input offset voltage*, V_{io}. See Appendices 1 and 2 for typical values. Note that V_{io} is shown in series with the (+) input terminal of the op amp. It makes no difference whether V_{io} is modeled in series with the (−) input or the (+) input. But it is easier to determine the polarity of V_{io} if it is placed in series with the (+) input. For example, if the output terminal is positive (with respect to ground) in Fig. 9-6(b), then V_{io} should be drawn with its (+) battery terminal connected to the ideal op amp's (+) input.

9-5.2 Effect of Input Offset Voltage on Output Voltage. Fig. 9-7(a) shows that V_{io} and the large value of the open-loop gain of the op amp act to drive V_o to negative saturation. Contrast the polarity of V_{io} in Figs. 9-6(b) and 9-7(a). If you buy several op amps and plug them into the test circuit of Fig. 9-7(a), some will drive V_o to $+V_{sat}$ and the remainder will drive V_o to $-V_{sat}$. Therefore the magnitude and polarity of V_{io} varies from op amp to op amp. We also conclude that this is no way to measure V_{io}. To learn how V_{io} affects amplifiers with negative feedback, we study how to measure V_{io}.

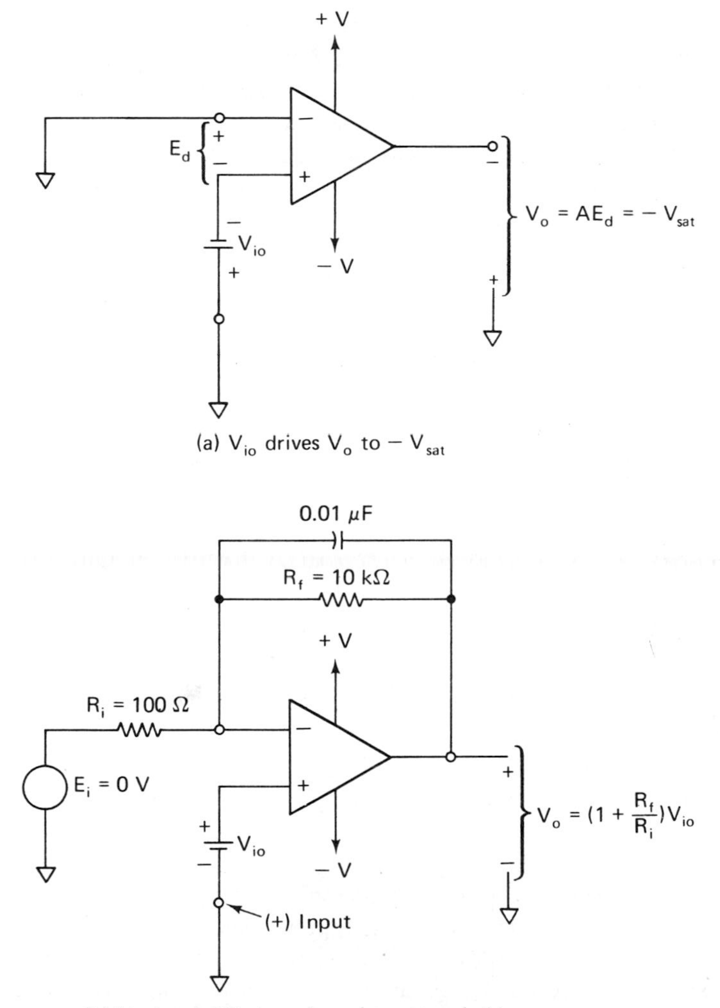

Figure 9-7 V_o should be 0 V in (a) and (b) but contains a dc error voltage due to V_{io}. (Error component due to bias current is neglected.)

9-5.3 Measurement of Input Offset Voltage. For simplicity, the effects of bias currents are neglected in the following discussion. Figure 9-7(b) shows how to measure V_{io}. It also shows how to predict the magnitude of error that V_{io} will cause in the output voltage. Since $E_i = 0$ V, V_o should equal 0 V. But V_{io} acts exactly as would a signal in series with the noninverting input. Therefore, V_{io} is amplified exactly as any signal applied to the (+) input of a noninverting amplifier (see Section 3-6). The error in V_o due to V_{io} is given by

$$\text{Error voltage due to } V_{io} = V_{io}\left(1 + \frac{R_f}{R_i}\right) \tag{9-4}$$

The output error voltage in Fig. 9-7(b) is given by Eq. (9-4) whether the circuit is used as an inverting or as a noninverting amplifier. That is, E_i could be inserted in series with R_i (inverting amplifier) for a gain of $-(R_f/R_i)$ or in series with the (+) input (noninverting amplifier) for a gain of $1 + (R_f/R_i)$. A bias-current compensating resistor (a resistor in series with the (+) input) has no effect on this type of error.

Conclusion: To measure V_{io}, set up the circuit of Fig. 9-7(b). The capacitor is installed across R_f to minimize noise in V_o. Measure V_o, R_f and R_i. Calculate V_{io} from

$$V_{io} = \frac{V_o}{1 + \dfrac{R_f}{R_i}} \tag{9-5}$$

Note that R_f is made small to minimize the effect of input bias current.

Example 9-5: V_{io} is specified to be 1 mV for a 741-type op amp. Predict the value of V_o that would be measured in Fig. 9-7(b).
Solution: From Eq. (9-5),

$$V_o = \left(1 + \frac{10{,}000}{100}\right)(1 \text{ mV}) = 101 \text{ mV}$$

9-6 Input Offset Voltage for the Adder Circuit

9-6.1 Comparison of Signal Gain and Offset Voltage Gain. In both inverting and noninverting amplifier applications, the input offset voltage V_{io} is multiplied by $(1 + R_f/R_i)$. The input signal in either circuit is also multiplied by approximately the same value (magnitude only). R_f/R_i is the gain for the inverter and $(1 + R_f/R_i)$ for the noninverter. In the inverting adder circuit of Fig. 9-8(a) (neglecting bias currents), V_{io} is multiplied by a larger number than the signal at each input.

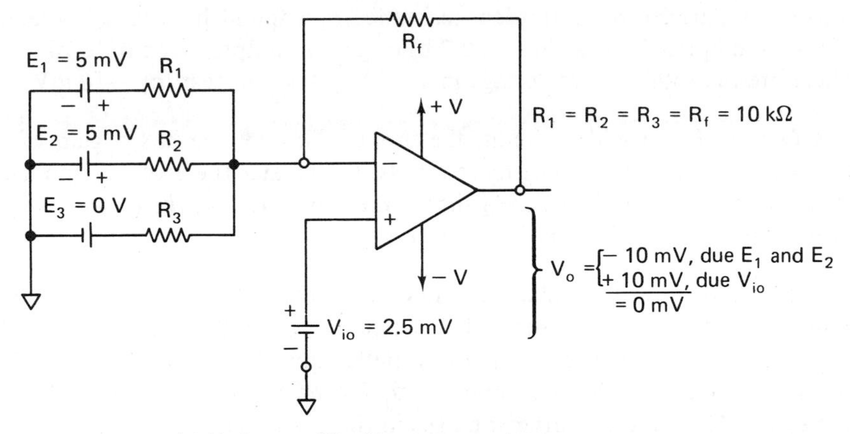

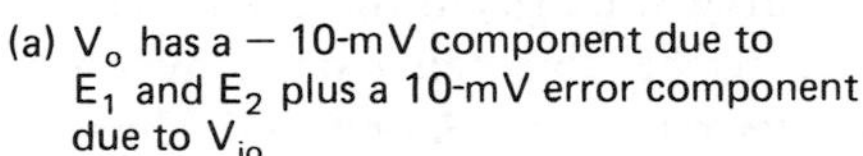
(a) V_o has a − 10-mV component due to E_1 and E_2 plus a 10-mV error component due to V_{io}

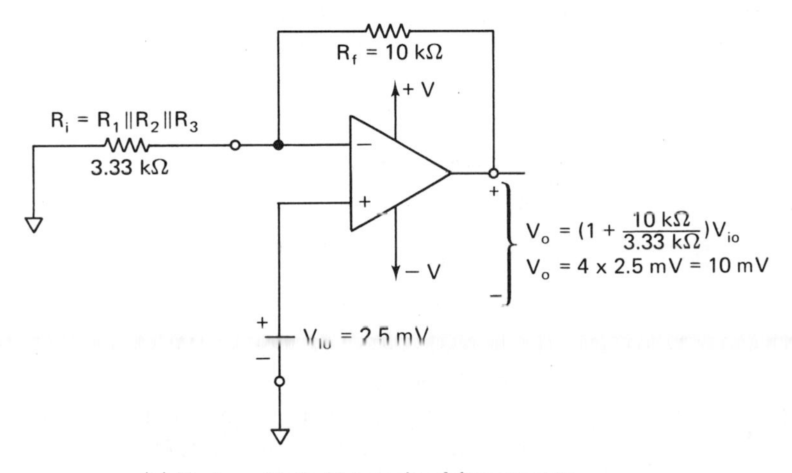

(b) V_{io} is multiplied by a gain of 4 to generate a 10-mV error component in V_o

Figure 9-8 Each input voltage of the inverting adder in (a) is multiplied by a gain of −1. V_{io} is multiplied by a gain of +4.

For example, in Fig. 9-8(a) signals E_1 and E_2 are each larger than V_{io} but E_1 is multiplied by $-R_f/R_1 = -1$ and develops a component of −5 mV in V_o. E_2 is likewise multiplied by −1 and adds a −5-mV component to V_o. Thus the correct value of V_o should be −10 mV. Since E_3 is 0 its contribution to V_o is 0. See Section 3-2.

If we temporarily let E_1 and $E_2 = 0$ V in Fig. 9-8(a), then the (−) input sees three equal resistors forming parallel paths to ground. The single equivalent series resistance, R_i, is shown in Fig. 9-8(b). For three equal 10-kΩ resistors in parallel, the equivalent resistance R_i is found by 10 kΩ/3 = 3.33 kΩ. V_{io} is amplified just as in Fig. 9-7(b) to give an output error of +10 mV. Therefore the total output voltage in Fig. 9-8(a) is 0 instead of −10 mV.

Conclusions: In an adder circuit, the input offset voltage has a gain of *1 plus the number of inputs*. The more inputs, the greater the error component in the output voltage. Since the gain for the inputs is −1, *the offset voltage gain always exceeds the signal voltage gain.*

9-6.2 How Not to Eliminate the Effects of Offset Voltage. One might be tempted to add an input to the adder such as E_3 in Fig. 9-8(a) to balance out the effect of V_{io}. For example, if E_3 is made equal to 10 mV, then E_3, R_3, and R_f will add a −10-mV component to V_o and balance out the +10 mV due to V_{io}. There are two disadvantages to this approach. First, such a small value of E_3 would have to be obtained from a resistor-divider network between the power supply terminals of $+V$ and $-V$. The second disadvantage is that any resistance added between the (−) input and ground raises the *noise gain*. This situation will be treated in Sections 10-5.3 and 10-5.4.

Section 9-7 shows how to minimize the output voltage errors caused by both bias currents and input offset voltage.

9-7 Nulling Out Effects of Bias Currents and Offset Voltage

9-7.1 Design or Analysis Sequence. To minimize dc error voltages in the output voltage, follow this sequence:

1. Select a bias-current compensating resistor in accordance with the principles set forth in Section 9-4.3.
2. Get a circuit for minimizing effects of the input offset voltage from the manufacturer's data sheet. This principle is treated in more detail in Section 9-7.2 and in Appendices 1 and 2.
3. Go through the output-voltage nulling procedure given in Section 9-7.3.

9-7.2 Null Circuits for Offset Voltage. It is possible to imagine a fairly complex resistor-divider network that would inject a small variable voltage into the (+) or (−) input terminal. This would compensate for the effects of both input offset voltage and offset current. However, the extra components are more costly and bulky than necessary. It is far better to go to the op amp manufacturer for guidance. The data sheet for your op amp will have a *voltage offset null circuit* recommended by the manufacturer. Experts have designed the null circuit to minimize offset errors at the lowest cost to the user. See Appendices 1 and 2.

Some typical output-voltage null circuits are shown in Fig. 9-9. In Fig. 9-9(a), one variable resistor is connected between the $+V$ supply and a *trim* terminal. For an expensive op amp, the manufacturer may furnish a metal film resistor selected especially for that op amp. In Fig. 9-9(b), a 10-kΩ pot is connected between terminals called *offset null*. More complicated null circuits are shown in Figs. 9-9(c) and (d). Note that only the offset-voltage compensating resistors are shown by the manufacturer. They assume that a current-compensating resistor will be installed in series with the (+) input.

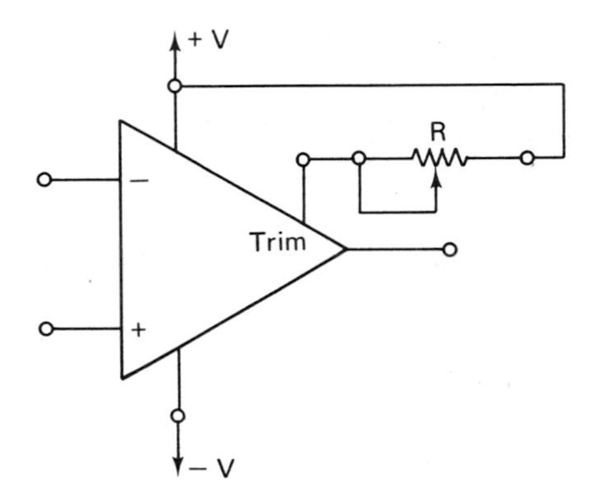

(a) Trim resistor 0 to 50 kΩ or 25 kΩ fixed for discrete op amp

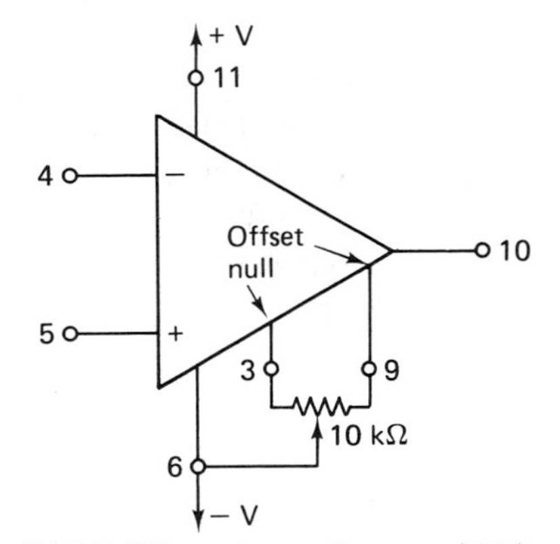

(b) 741 Offset voltage adjustment (DIP)

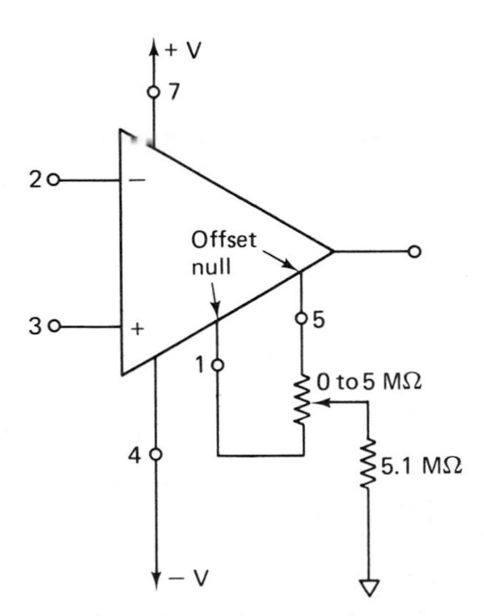

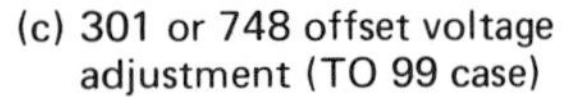

(c) 301 or 748 offset voltage adjustment (TO 99 case)

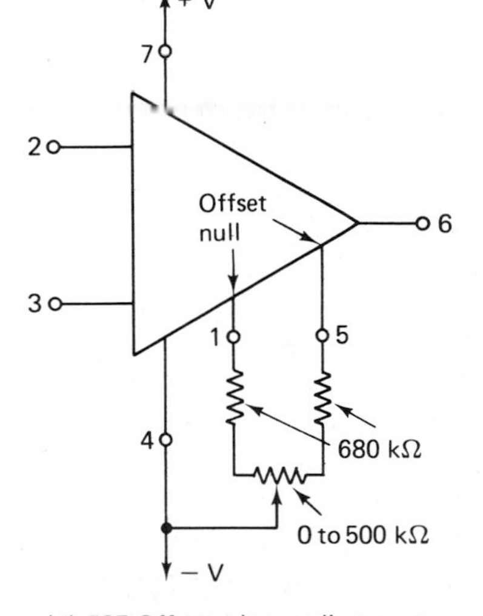

(d) 537 Offset voltage adjustment

Figure 9-9 Typical circuits to minimize error in output voltage due to input offset voltage. (See also Appendices 1 and 2.)

9-7.3 Nulling Procedure for Output Voltage.

1. Build the circuit. Include (a) the current-compensating resistor (see Section 9-4.3) and (b) the voltage offset null circuit (see Section 9-7.2).
2. Reduce all generator signals to 0. If their output cannot be set to 0, replace them with resistors equal to their internal resistance. This step is unnecessary if their internal resistance is negligible with respect to (more than about 1% of) any series resistor R_i connected to the generator.
3. Connect the load to the output terminal.
4. Turn on the power and wait a few minutes for things to settle down.
5. Connect a dc voltmeter or a CRO (dc coupled) across the load to measure V_o. (The voltage sensitivity should be capable of reading down to a few millivolts).
6. Vary the offset voltage adjustment resistor until V_o reads 0 V. Note that output voltage errors due to both input offset voltage and input offset current are now minimized.
7. Install the signal sources and do *not* touch the offset-voltage adjustment resistor again.

9-8 Drift

It has been shown in this chapter that dc error components in V_o can be minimized by installing a current-compensating resistor in series with the (+) input and by trimming the offset-voltage adjustment resistor. It must also be emphasized that the zeroing procedure holds only at one temperature and at one time.

The offset current and offset voltage change with time because of aging of components. The offsets will also be changed by temperature changes in the op amp. In addition, if the supply voltage changes, bias currents, and consequently the offset current, change. By use of a well-regulated power supply, the output changes that depend on supply voltage can be eliminated. However the offset changes with temperature can only be minimized by (1) holding the temperature surrounding the circuit constant or (2) selecting op amps with offset current and offset voltage ratings that change very little with temperature changes.

The changes in offset current and offset voltage due to temperature are described by the term *drift*. Drift is specified for offset current in nA/°C (nanoamperes per degree Centigrade). For offset voltage, drift is specified in μV/°C (microvolts per degree Centigrade). Drift rates may differ at different temperatures and may even reverse; that is, at low temperatures V_{io} may drift by +20 μV/°C (increase), and at high temperatures V_{io} may change by −10 μV/°C (decrease). For this reason, manufacturers may specify either an average or maximum drift between two temperature limits. Even better is to have a plot of drift vs temperature. An example is shown to calculate the effects of drift.

Example 9-6: A 301 op amp in the circuit of Fig. 9-10 has the following drift specifications. As temperature changes from 25°C to 75°C, I_{os} changes by a *maximum* of 0.3 nA/°C and V_{io} changes by a *maximum* of 30 μV/°C. Assume that V_o has been zeroed at 25°C; then the surrounding temperature is raised to 75°C. Find the maximum error in output voltage due to drift in (a) V_{io} and (b) I_{os}.

Solution: (a) V_{io} will change by

$$\pm\frac{30\ \mu\text{V}}{°\text{C}} \times (75 - 25)°\text{C} = \pm 1.5\ \text{mV}$$

From Fig. 9-8(b), the change in V_o due to the change in V_{io} is

$$1.5\ \text{mV}\left(1 + \frac{R_f}{R_i}\right) = 1.5\ \text{mV}(101) \cong \pm 150\ \text{mV}$$

(b) I_{os} will change by

$$\pm\frac{0.3\ \text{nA}}{°\text{C}} \times 50°\text{C} = \pm 15\ \text{nA}$$

From Fig. 9-4(b), the change in V_o due to the change in I_{os} is ± 15 nA $\times R_f$ $= \pm 15$ nA(1 MΩ) $= \pm 15$ mV.

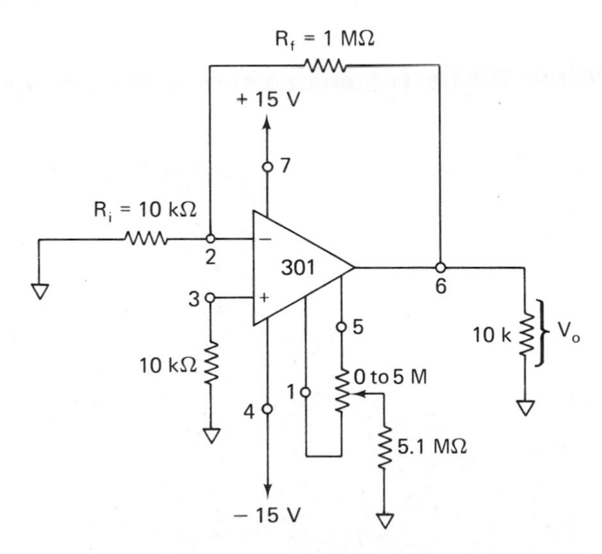

Figure 9-10 Circuit for Example 9-6.

The changes in V_o due to both V_{io} and I_{os} can either add or subtract from one another. Therefore, the worst possible change in V_o is either +165 mV or −165 mV, from the 0 value at 25°C.

Problems

9-1 Which op amp characteristics normally have the most effect on (a) dc amplifier performance (b) ac amplifier performance?

9-2 If $I_{B+} = 0.2\ \mu A$ and $I_{B-} = 0.1\ \mu A$, find (a) the average bias current I_B and (b) the offset current I_{os}.

9-3 In Example 9-2, $V_o = 0.2$ V. Find I_{B-}.

9-4 In Example 9-3, $V_o = 0.2$ V. Find I_{B+}.

9-5 I_{B-} is 0.2 μA in Fig. 9-3(c). Find V_o.

9-6 In Fig. 9-5(a), $R_f = R_G = 100\ k\Omega$. $I_{B+} = 0.3\ \mu A$ and $I_{B-} = 0.2\ \mu A$. Find V_o.

9-7 In Fig. 9-5(b), $R_i = 25\ k\Omega$ and $R = 12.5\ k\Omega$. If $I_{os} = 0.1\ \mu A$, find V_o.

9-8 In Fig. 9-5(b), $R_i = R_f = 25\ k\Omega$ and $R = 12.5\ k\Omega$. If $I_{os} = 0.1\ \mu A$, find V_o.

9-9 $V_o = 200$ mV in Fig. 9-7(b). Find V_{io}.

9-10 Resistors R_1, R_2, R_3, and R_f all equal 20 kΩ in Fig. 9-8(a). $E_1 = E_2 = E_3 = V_{io} = 2$ mV. Find (a) the actual value of V_o and (b) V_o assuming that $V_{io} = 0$.

9-11 What value of current-compensating resistor should be added in Problem 9-10?

9-12 What is the general procedure to null the output voltage of an op amp to 0 V?

9-13 In Fig. 9-10, V_{io} changes by ±1 mV when the temperature changes by 50°C. What is the *change* in V_o due to the *change* in V_{io}?

9-14 I_{os} changes by ±20 nA in Fig. 9-10 for a temperature change of 50°C. What is the resulting change in V_o?

10

Band Width, Slew Rate, Noise, and Frequency Compensation

10-0 Introduction

When the op amp is used in a circuit that amplifies only ac signals, we must consider whether ac output voltages will be small signals (below about 1 V peak) or large signals (above 1 V peak). If only small ac output signals are present, the important op amp characteristics that limit performance are *noise* and *frequency response*. If large ac output signals are expected, then an op amp characteristic called *slew-rate limiting* determines whether distortion will be introduced by the op amp.

Bias currents and offset voltages affect dc performance and usually do not have to be considered with respect to ac performance. This is true because a coupling capacitor is usually in the circuit to pass ac signals and block dc currents and voltages. We begin with an introduction to the frequency response of an op amp.

10-1 Frequency Response of the Op Amp

10-1.1 Internal Frequency Compensation. Many types of general-purpose op amps and specialized op amps are *internally compensated*; that is, the manufacturer has installed within such op amps a small capacitor, usually 30 pF. This *internal frequency compensation capacitor* prevents the op amp from oscillating at high frequencies. Oscillations are prevented by decreasing the op amp's gain as frequency increases. Otherwise, there would be sufficient gain and phase shift at some high frequency where enough output signal could be fed back to the input and cause oscillations. (See Appendix 1.)

From basic circuit theory it is known that the reactance of a capacitor goes down as frequency goes up: $X_C = 1/(2\pi fC)$. For example, if the frequency is increased by 10, the capacitor reactance decreases by 10. Thus, it is no accident that the voltage gain of an op amp goes down by 10 as the frequency of the input signal is increased by 10. A change in frequency of 10 is called a *decade*. Manufacturers show how the open-loop gain of the op amp is related to the frequency of the differential input signal by a curve called *Open-Loop Voltage Gain vs Frequency*. The curve may also be called *Small-Signal Response*.

10-1.2 Frequency-Response Curve. A typical curve is shown in Fig. 10-1 for internally compensated op amps such as the 741 and 747. At low frequencies (below 0.1 Hz), the open-loop voltage gain is very high. A typical value is 200,000 (106 db), and it is this value that is specified on data sheets

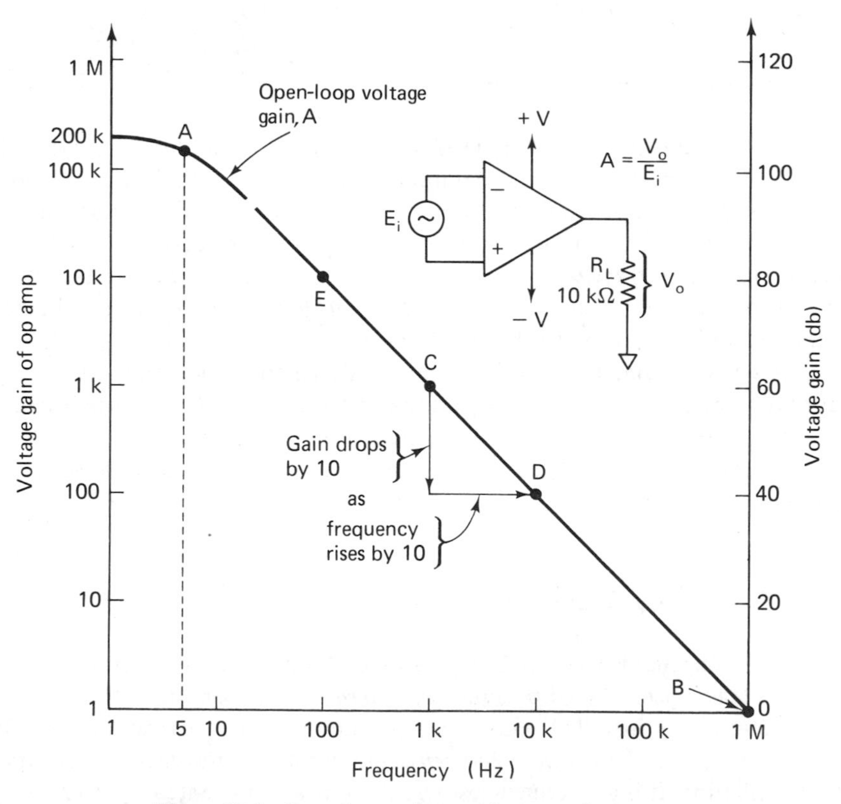

Figure 10-1 Open-loop voltage gain versus frequency.

where a curve is not given. See also Appendix 1, p. 272 and Appendix 2, p. 279.

Point A in Fig. 10-1 locates the *break frequency* where the voltage gain is 0.707 times its value at very low frequencies. Therefore, the voltage gain at point A (where the frequency of E_i is 5 Hz) is about 140,000, or $0.707 \times 200{,}000$.

Points C and D show how gain drops by a factor of 10 as frequency rises by a factor of 10. Changing frequency or gain by a factor of 10 is expressed more efficiently by the term *per decade* ("decade" signifies ten). The right-hand vertical axis of Fig. 10-1 is a plot of voltage gain in decibels (db). The voltage gain decreases by 20 db for an increase in frequency of one decade. This explains why the frequency-response curve from A to B is described as *rolling off at* 20 *db per decade*. An alternative description is 6 *db per octave rolloff* ("octave" signifies a frequency change of 2). Therefore, each time the frequency doubles the voltage gain decreases by 6 db.

10-1.3 Unity-Gain Band Width. Point B in Fig. 10-1 defines the *small-signal unity-gain band width* of the op amp. It is located at that frequency where the open-loop voltage gain is unity, or 1.

Some data sheets do not give a specification called unity-gain band width or a curve like Fig. 10-1. Instead, they give a specification called *Transient Response Rise Time* (*Unity Gain*). For a 741 op amp it is typically 0.25 μs and 0.8 μs at maximum. The band width B is calculated from the rise time specification by

$$B = \frac{0.35}{\text{rise time}} \tag{10-1}$$

where B is in Hz and rise time is in seconds. Rise time is defined in Section 10-1.4. (See Appendix 1, p. 269.)

Example 10-1: A 741 op amp has a rise time of 0.35 μs. Find the small-signal or unity-gain band width.
Solution: From Eq. (10-1),

$$B = \frac{0.35}{0.35\ \mu\text{s}} = 1\ \text{MHz}$$

Example 10-2: What is the open-loop voltage gain for the op amp of Example 10-1 at 1 MHz?
Solution: From the definition of B, the voltage gain is 1.

Example 10-3: What is the open-loop voltage gain at 100 kHz for the op amp in Examples 10-1 and 10-2?

Solution: By inspection of Fig. 10-1, if the frequency goes down by 10, the gain goes up by 10. Therefore, since the frequency goes down a decade (from 1 MHz to 100 kHz), the gain must go up by a decade from 1 at 1 MHz to 10 at 100 kHz.

Example 10-3 leads to the conclusion that if you divide the frequency of the signal, f, into the unity-gain band width, B, the result is the op amp's gain at the signal frequency. Expressed mathematically.

$$\text{open-loop gain at } f = \frac{\text{band width}}{\text{input signal frequency}} \tag{10-2}$$

Example 10-4: What is the open-loop gain of an op amp that has a unity-gain band width of 1.5 MHz for a signal of 1 kHz?
Solution: From Eq. (11-2), the open-loop gain at 1 kHz is

$$\frac{1.5 \text{ MHz}}{1 \text{ kHz}} = 1500$$

The data shown in Fig. 10-1 are useful for learning but will probably not apply to your op amp. For example, while 200,000 is a specified typical open-loop gain, the manufacturer guarantees only a minimum gain of 20,000 for inexpensive op amps. Still, 20,000 may be enough to do the job. Section 10-2 deals with this question.

10-1.4 Rise Time. Assume that the input voltage E_i of a unity-gain amplifier is changed very rapidly by a square wave or pulse signal. Ideally, E_i should be changed from 0 V to + 20 mV in 0 time; practically, a few nanoseconds are required to make this change. (See Appendix 1, pp. 269 and 273.) At unity gain, the output should change from 0 to +20 mV in the same few nanoseconds. However, it takes time for the signal to propagate through all the transistors in the op amp. It also takes time for the output voltage to rise to its final value. *Rise time* is defined as the time required for the output voltage to rise from 10% of its final value to 90% of its final value. From Section 10-1.3, the rise time of a 741 is 0.35 μs. Therefore, it would take 0.35 μs for the output voltage to change from 2 mV to 18 mV.

10-2 Amplifier Frequency Response to Small Signals

10-2.1 Closed-Loop and Open-Loop Gain. Ideally, the closed-loop gain of the amplifier in Fig. 10-2 should (a) be determined only by resistors R_f and R_i, (b) be the same for all frequencies, and (c) not depend on the op amp. But as shown in Section 10-1, the op amp's open-loop gain does depend

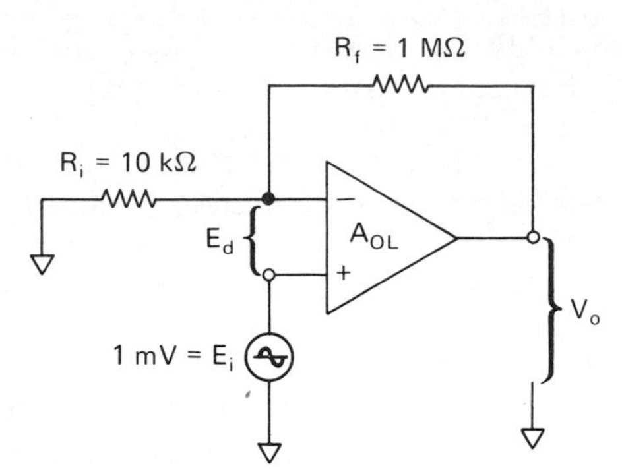

Figure 10-2 Circuit for Example 10-5.

on frequency. Therefore, any amplifier circuit that contains an op amp will have a *closed-loop* or amplifier gain that depends in some frequency range on the op amp's *open-loop gain*. Therefore, one must understand about each gain term. The gain of the op amp is called the open-loop gain, A_{OL}. The gain of the amplifier (op amp plus feedback resistors) is called the closed-loop gain, A_{CL}, because the feedback resistors close a loop from the op amp output terminal to the (—) input terminal. To summarize: *amplifier or circuit gain is closed-loop gain*, and *op amp gain is open-loop gain.*

10-2.2 Effect of Open-Loop Gain on Closed-Loop Gain. In Fig. 10-2, the open-loop gain of the op amp is defined by $A_{OL} = V_o/E_d$. The closed-loop gain of the amplifier is $A_{CL} = V_o/E_i$. The closed-loop gain for a non-inverting amplifier shoud be set by $1 + R_f/R_i$ and for Fig. 10-2 equals 101. Therefore, when $E_i = 1$ mV, V_o should equal 101 mV. This conclusion is based on the assumption that E_d is approximately 0. But if A_{OL} is *not* very large, E_d will not be 0. If E_d is within 1% or more of E_i, V_o depends not only on $1 + R_f/R_i$ but also on A_{OL}. To illustrate this point, consider the following example.

Example 10-5: Let the open-loop gain A_{OL} of the op amp in Fig. 10-2 be set at 10,000 and 1,000. The resulting closed-loop gain A_{CL} and value of E_i for each open-loop gain value are given as follows:

Open-Loop Gain A_{OL}	*Op-amp Input Voltage* E_d *(mV)*	*Actual Gain* A_{CL}	*Ideal Gain* $1 + \frac{R_f}{R_i}$	*Percentage of Error*	*Fig. 10-1 Point*
10,000	0.0099	99.9	101	1	*E*
1,000	0.091	91.8	101	10	*C*

These results are interpreted as follows. When A_{OL} is ten times greater than $1 + R_f/R_i$, the actual gain A_{CL} will be in error by about 10%. When

A_{OL} is 100 times greater than $1 + R_f/R_i$, the actual gain A_{CL} will be within 1% of what the ideal gain should be, namely $1 + R_f/R_i$.

By comparing these interpretations with Fig. 10-1, we can relate them to frequency. Note that the open-loop gains in Example 10-5 correspond to points *E* and *C*. A simple method to predict the range of frequencies for 1% error or 10% error with either inverting or noninverting amplifiers is given in Section 10-2.3.

10-2.3 Frequency Range for Precise Amplifier Gain, 1% Accuracy. A three-step graphical procedure locates the frequency range where the actual amplifier closed-loop gain is within 1% of the ideal closed-loop gain of R_f/R_i or $1 + R_f/R_i$. The procedure is as follows.

1. Locate the unity-gain band width point *B* on the manufacturer's open-loop gain versus frequency curve for your op amp. (See Fig. 10-3 or Appendices 1 and 2.) Multiply band width *B* by 1%, or $\frac{1}{100}$, to obtain frequency $f_1 = B/100$. Locate on the horizontal axis f_1 which is point (1) in Fig. 10-3.
2. Sketch a line from point (1) parallel to the open-loop gain curve. This is called *maximum frequency for* 1% *gain error*. Note that every point on this line is 2 decades below (or $\frac{1}{100}$ of) the gain of a corresponding frequency on the open-loop curve.
3. Locate the ideal closed-loop gain for the amplifier on the vertical gain axis (R_f/R_i for inverting amplifiers or $1 + R_f/R_i$ for noninverting amplifiers). Draw a horizontal line of ideal closed-loop gain to the curve of maximum frequency for 1% gain error, which is point (3) in Fig. 10-3. The maximum frequency for 1% accuracy is directly below point (3) on the frequency scale.

Example 10-6: The amplifier of Fig. 10-2 has a closed-loop gain of $1 + R_f/R_i = 1 + 1\ \text{M}\Omega/10\ \text{k}\Omega = 101$. The op amp has an open-loop gain versus frequency curve as shown in Fig. 10-3. In what frequency range will small input signals (E_i) have a gain that is within 1% of 101?
Solution: In Fig. 10-3, draw the curve of maximum frequency for 1% gain error. Draw the dashed line of $A_{CL} = 101$ to locate point (3). Drop a vertical line from point (3) to read 100 Hz. The amplifier will amplify signals by a gain of 101 (within 1%) up to 100 Hz.

The range of frequencies where the amplifier gain is within 1% of the ideal closed-loop gain (set by feedback resistors) is sometimes called the *precision frequency range*. By inspection of Fig. 10-3, the precision frequency range is 1 kHz for an amplifier with a gain of 10 and 10 Hz for an amplifier with a gain of 1,000.

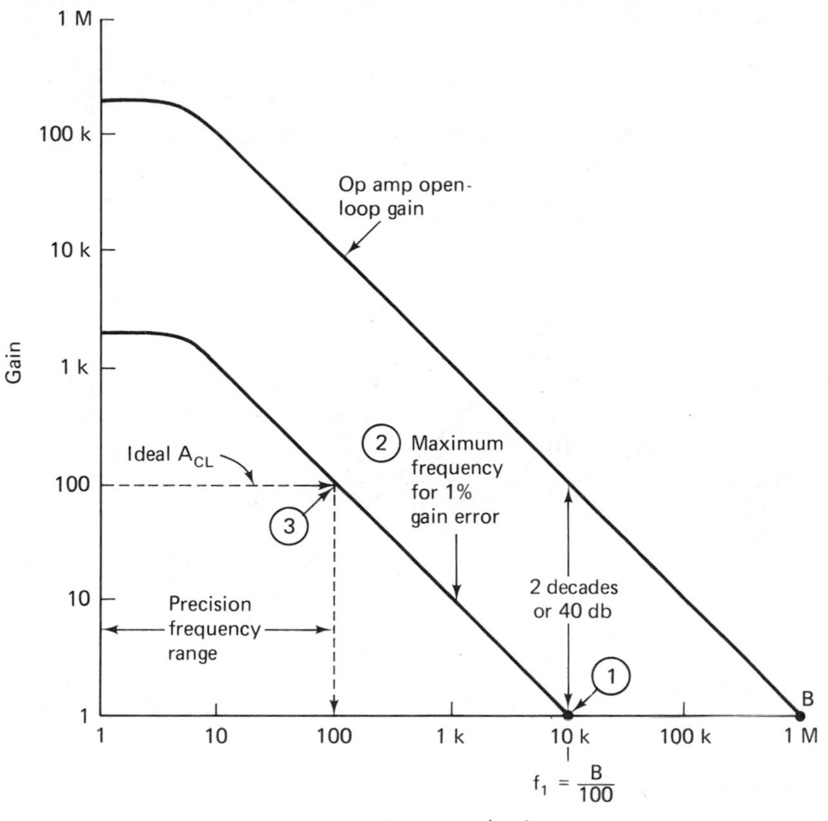

Figure 10-3 Procedure to determine precision frequency range where gain depends on feedbeck resistors to within 1%.

10-2.4 Frequency Range for 10% Gain Accuracy. To find the frequency range where the closed-loop gain A_{CL} is within 10% of the ideal value set by R_f and R_i, the following should be used. (From Example 10-5, we learn that the 10% error occurs when open-loop gain A_{OL} is only ten times ideal A_{CL}. The procedure is similar to that developed in Section 10-2.3.)

1. Evaluate f_{10} from $f_{10} = B/10$. Locate f_{10} as point (1) on the horizontal axis of the open-loop frequency response curve. See Fig. 10-4.
2. Draw the curve of 10% gain error from f_{10} parallel to the open-loop curve as in Fig. 10-4.
3. Draw a horizontal line of ideal closed-loop gain to intersect with the 10% gain error curve, which is point (3) in Fig. 10-4.
4. Read the maximum frequency or frequency range for 10% gain error directly below point (3) on the horizontal axis.

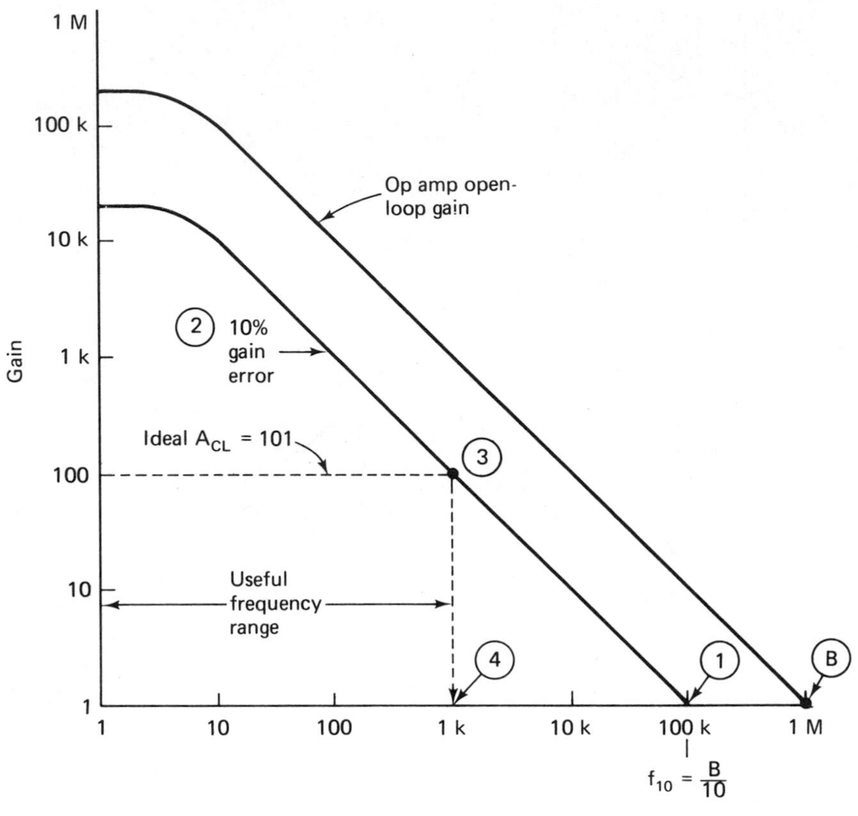

Figure 10-4 Procedure to find *useful* frequency range where amplifier gain depends on feedback resistors to within 10%. Compare with Fig. 10-3.

Example 10-7: What is the frequency range with 10% gain accuracy for the amplifier in Example 10-6?
Solution: Ideal $A_{CL} = 1 + (R_f/R_i) = 101$. Its intersection with 10% gain error is at point (3) in Fig. 10-4. Below point (3), read $f = 1$ kHz. Thus the amplifier of Fig. 10-2 amplifies small signals by a gain of 101 with an error of less than 10% from dc to 1 kHz.

Extending Example 10-7, we see from Fig. 10-4 that an amplifier with a gain of 10 will amplify signals up to 10 kHz with an accuracy of 10%. For an amplifier with unity gain, the 10% frequency range would be 100 kHz.

10-2.5 Small-Signal Band Width. The useful frequency range of any amplifier (closed- or open-loop) is defined by a high-frequency limit f_H and a low-frequency limit f_L. At f_L and f_H, the voltage gain is down to 0.707

times its maximum value in the middle of the useful frequency range. In terms of decibels, the voltage gain is down 3 db at both f_L and f_H. These statements are summarized on the general frequency-response curve in Fig. 10-5 and in Appendices 1 and 2.

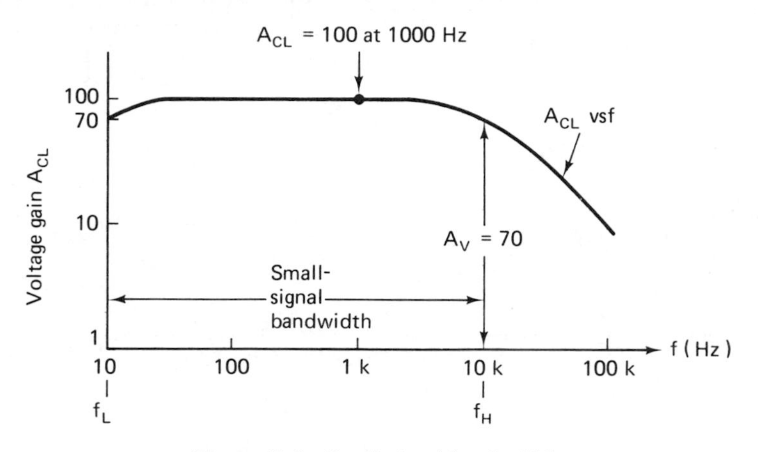

Figure 10-5 Small-signal band width.

Small-signal band width is the difference between f_H and f_L. Often, f_L is very small with respect to f_H or f_L is 0 for a dc amplifier. Therefore, the small-signal band width approximately equals the high-frequency limit f_H. From point *A* of Fig. 10-1, we see that the small-signal band width of an op amp is 5 Hz.

The small-signal band width for closed-loop amplifiers is determined by both the unity-gain band width (Section 10-1.3) *B* and the closed-loop gain A_{CL}. The relationship is simply

$$\text{closed-loop small-signal band width} = \frac{B}{A_{CL}} \tag{10-3}$$

Equation (10-3) can be shown graphically as in Fig. 10-6. Calculate the ideal A_{CL} from $R_f/R_i = 1000$. Extend the horizontal lines of Ideal A_{CL} to intersect the open-loop curve at point f_H. Read the small-signal band width directly below f_H as 1 kHz. Observe that the closed-loop gain at f_H is $0.707 \times$ ideal gain $= 0.707 \times 1000 = 700$. For frequencies above f_H, the gain is determined not by R_f and R_i but by the op amp.

Example 10-8: The unity-gain band width of a 741 is 1 MHz. When it is used in an amplifier designed for a closed-loop gain of 100, find (a) the small-signal band width, (b) A_{CL} at f_H, and (c) the same quantities for an amplifier with a closed-loop gain of 10.

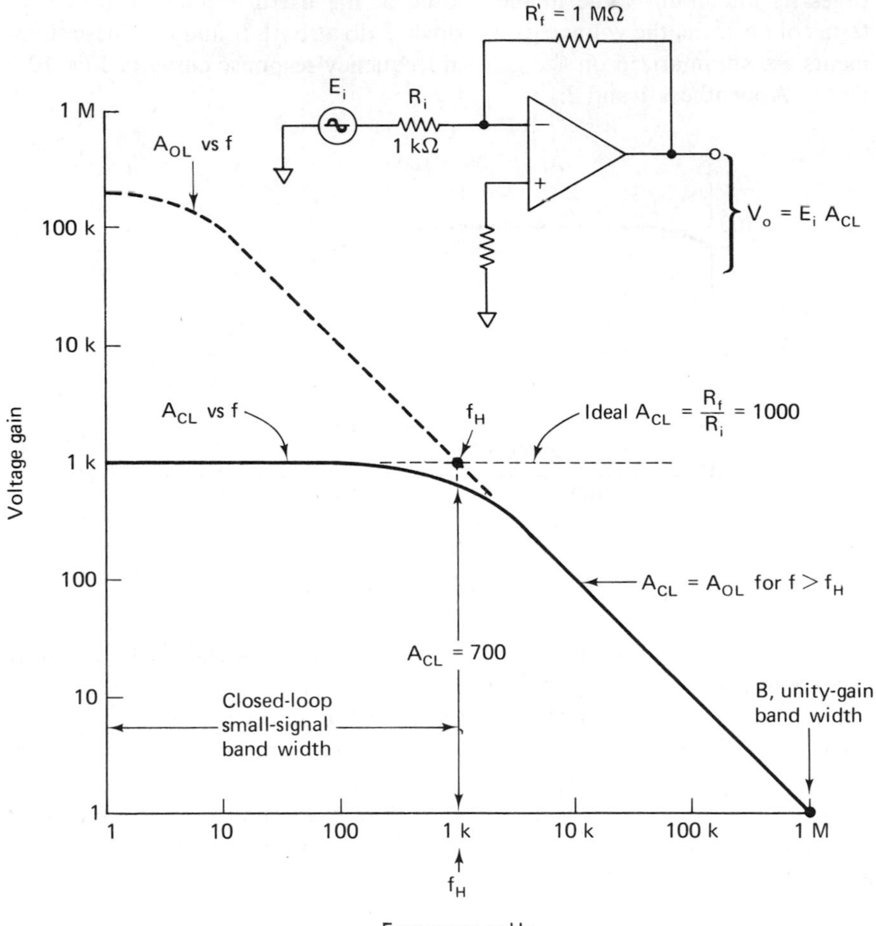

Figure 10-6 Small-signal band width and closed-loop gain.

Solution: (a) By Eq. (10-3),

$$\text{band width} = \frac{10^6\ \text{Hz}}{100} = 10\ \text{kHz}$$

(b) A_{CL} at $f_H = 100 \times 0.707 = 70$
(c) By Eq. (10-3),

$$\text{band width} = \frac{10^6\ \text{Hz}}{10} = 100\ \text{kHz}$$

A_{CL} at $f_H = 10 \times 0.707 \cong 7$

Example 10-8 and Eq. (10-3) show that there is a direct trade-off between small-signal gain and band width. If you increase the gain by 10, the band width decreases by 10. Notice that the product of closed-loop gain and small-signal band width always equals the unity-gain band width B. For this reason, the unity-gain band width is also called *gain-band width product* and is a figure of merit for the op amp.

10-3 Slew Rate and Output Voltage

10-3.1 Definition of Slew Rate. The slew rate of an op amp tells how fast its output voltage can change. For a general-purpose op amp such as the 741, the maximum slew rate is 0.5 V/μs. This means that the output voltage can change a maximum of $\frac{1}{2}$ V in 1 μs. Slew rate depends on many factors: the amplifier gain, compensating capacitors, and even whether the output voltage is going positive or negative. The worst case, or slowest slew rate, occurs at unity gain. Therefore, slew rate is usually specified at unity gain. (See Appendices 1 and 2.)

10-3.2 Cause of Slew-Rate Limiting. Either within or outside the op amp there is at least one capacitor required to prevent oscillation; See Section 10-1.1. Connected to this capacitor is a portion of the op amp's internal circuitry that can furnish a maximum current that is limited by op amp design. The ratio of this maximum current I to the compensating capacitor C is the *slew rate*. For example, a 741 can furnish a maximum of 15 μA to its 30-pF compensating capacitor. (See Appendix 1, p. 269.) Therefore,

$$\text{slew rate} = \frac{\text{output voltage change}}{\text{time}} = \frac{I}{C} = \frac{15\ \mu\text{A}}{30\ \text{pF}} = 0.5\,\frac{\text{V}}{\mu\text{s}} \qquad (10\text{-}4)$$

From Eq. (10-4), a faster slew rate requires the op amp to have either a higher maximum current or a smaller compensating capacitor. For example, the AD518 has a slew rate of 80 V/μs with $I = 400\ \mu$A and $C = 50$ pF.

Example 10-9: An instantaneous input change of 10 V is applied to a unity-gain inverting amplifier. If the op amp is a 741, how long will it take for the output voltage to change by 10 V?
Solution: By Eq. (10-4),

$$\text{slew rate} = \frac{\text{output voltage change}}{\text{time}}$$

$$\frac{0.5\ \text{V}}{\mu\text{s}} = \frac{10\ \text{V}}{\text{time}} \qquad \text{time} = \frac{10\text{V} \times \mu\text{s}}{0.5\ \text{V}} = 20\ \mu\text{s}$$

10-3.3 Slew-Rate Limiting of Sine Waves. In the voltage follower of Fig. 10-7, E_i is a sine wave with peak amplitude E_p. The maximum rate of change of E_i depends on both its frequency f and the peak amplitude as given

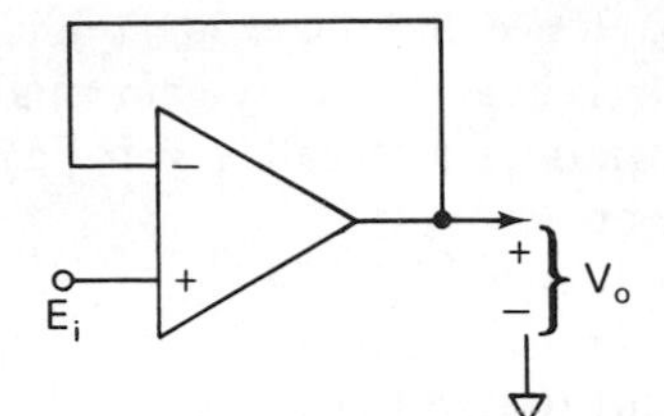

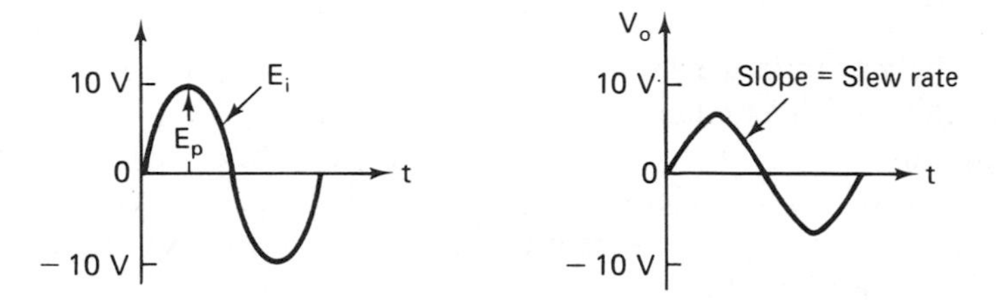

Figure 10-7 An example of slew-rate limiting of output voltage V_o.

by $2\pi f E_p$. If this rate of change is larger than the op amp's slew rate, the output V_o will be distorted. That is, output V_o tries to follow E_i but cannot do so because of slew-rate limiting. The result is distortion, as shown by the triangular shape of V_o in Fig. 10-7. The maximum frequency f_{max} at which we can obtain an undistorted output voltage with a peak value of V_{op} is determined by the slew rate in accordance with

$$f_{max} = \frac{\text{slew rate}}{6.28 \times V_{op}} \tag{10-5}$$

where f_{max} is the maximum frequency in Hz, V_{op} is the maximum undistorted output voltage in volts, and the slew rate is in volts per microsecond.

Example 10-9: The slew rate for a 741 is 0.5 V/μs. At what maximum frequency can you get an undistorted output voltage of (a) 10 V peak (b) 1 V peak?

Solution: (a) From Eq. (10-5),

$$f_{max} = \frac{1}{6.28 \times 10\text{ V}} \times \frac{0.5\text{ V}}{\mu\text{s}} = 8\text{ kHz}$$

(b) From Eq. (10-5)

$$f_{max} = 80\text{ kHz}$$

Example 10-10: The peak output voltage that can be gotten out of a 741 is 13 V with a ±-15 V supply. This output voltage is described as *full power output*. What is the maximum frequency for full power output of a 741? Note this *full power output frequency* specification is often supplied by the manufacturer. See Appendix 1, "output voltage swing as a function of frequency."

Solution: From Eq. (10-5),

$$f_{\text{max full power}} = \frac{1}{6.28 \times 13 \text{ V}} \times \frac{0.5 \text{ V}}{\mu\text{s}} = 6 \text{ kHz}$$

Examples 10-9 and 10-10 show that the slew rate limits the upper frequency of large-amplitude output voltages. As the peak output voltage required from the op amp is reduced, the upper frequency limitation imposed by the slew-rate increases.

Recall that the upper-frequency limitation imposed by small-signal response increases as the closed-loop gain decreases. For each amplifier application, the upper-frequency limit imposed by slew-rate limiting (Section

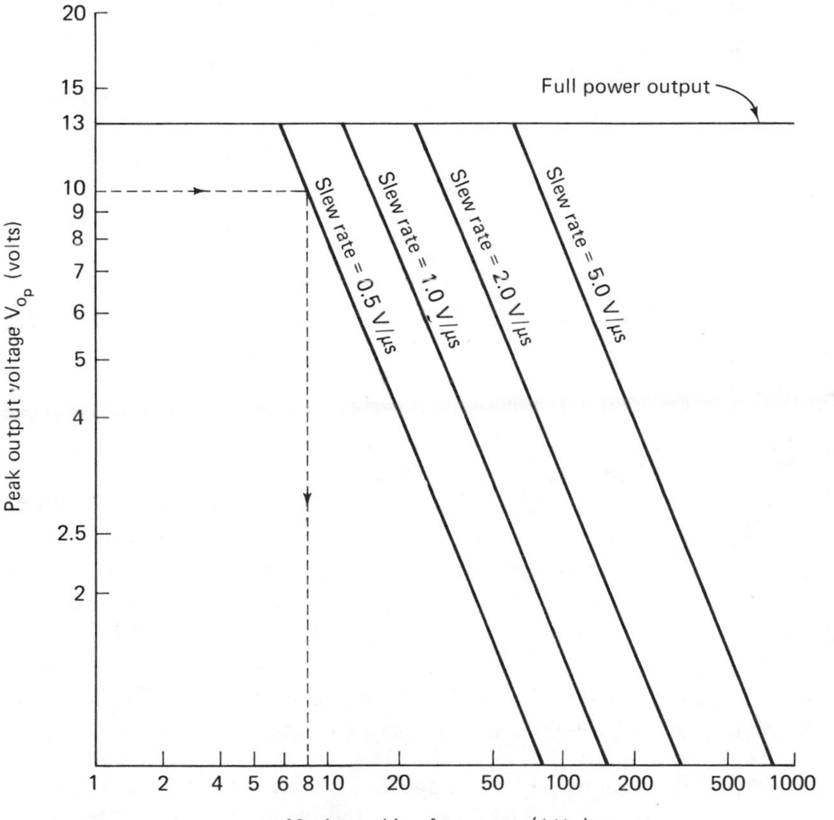

Figure 10-8 Slew rate made easy. Any point on a slew-rate line shows the maximum sinusoidal frequency allowed for the corresponding peak output voltage.

10-3.3) and small-signal band width (Section 10-2.5) must be calculated. The smaller value determines the actual upper-frequency limit. In general, the slew rate is a large-signal frequency limitation and small-signal frequency response is a small-signal frequency limitation.

10-3.4 Slew Rate Made Easy. Figure 10-8 simplifies the problem of finding f_{max} at any peak output voltage for slew rates between 0.5 V/μs and 5 V/μs. For example, to do part (a) of Example 10-9, locate where the horizontal line of $V_{op} = 10$ V intersects the slew-rate line 0.5 V/μs. Below the intersection, read $f_{max} = 8$ kHz.

10-4 Noise in the Output Voltage

10-4.1 Introduction. Undesired electrical signals present in the output voltage are classified as *noise*. Drift (see Chapter 9) and offsets can be considered as very low-frequency noise. If you view the output voltage of an op amp amplifier with a sensitive CRO (1 mV/cm), you will see a random display of noise voltages called *hash*. The frequencies of these noise voltages range from 0.01 Hz to megahertz.

Noise is generated in any material that is above absolute zero (−273°C). Noise is also generated by all electrical devices and their controls. For example, in an automobile, the spark plugs, voltage regulator, fan motor, air conditioner, and generator all generate noise. Even when headlights are switched on (or off), there is a sudden change in current that generates noise. This type of noise is external to the op amp. Effects of external noise can be minimized by proper construction techniques and circuit selection. (See Sections 10-4.3 to 10-4.5).

10-4.2 Noise in Op Amp Circuits. Even if there were no external noise, there would still be noise in the output voltage caused by the op amp. This internal op amp noise is modeled most simply by a noise voltage source E_n. As shown in Fig. 10-9, E_n is placed in series with the (+) input. On data sheets, noise voltage is specified in microvolts (rms) for different values of source resistance over a particular frequency range. For example, the 741 op amp has 2 μV of *total* noise over a frequency of 10 Hz to 10 kHz. This noise voltage is valid for source resistors (R_i) between 100 Ω and 20 kΩ. The noise voltage goes up directly with R_i, once R_i exceeds 20 kΩ. Thus R_i should be kept below 20 kΩ to minimize noise in the output. (See Appendix 1, p.272.)

10-4.3 Noise Gain. Noise voltage is amplified just as offset voltage is. That is, *noise voltage gain* is the same as the gain of an inverting amplifier, namely

$$\text{noise gain} = 1 + \frac{R_f}{R_i}$$

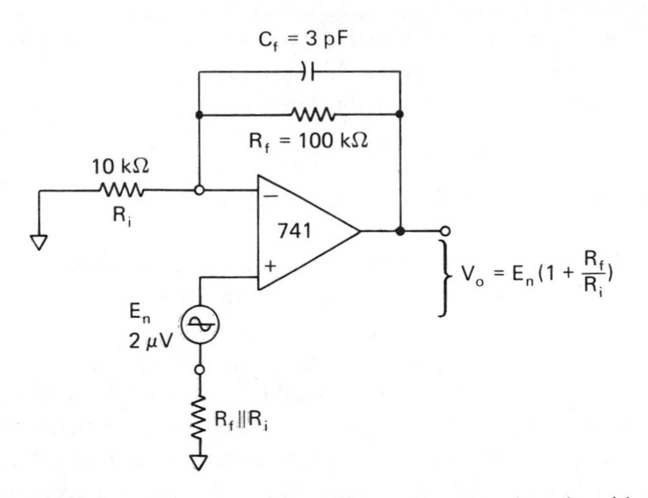

Figure 10-9 Op amp noise is modeled by a noise voltage in series with the (+) input.

What can you do about minimizing output voltage errors due to noise? First, avoid, if possible, large values of R_f. Install a small capacitor (3 pF) across R_f to shunt it at high noise frequencies. Then the higher noise frequencies will not be amplified so much. Next, do not shunt R_i with a capacitor; otherwise, the R_iC combination will have a smaller impedance at higher noise frequencies than R_i alone, and gain will increase with frequency and aggravate the situation. Finally, try to keep R_i at about 10 kΩ.

Noise currents, like bias currents, are also present at each op amp input terminal. If a bias-current compensation resistor is installed, (see Chapter 9), the effect of noise currents on output voltage will be reduced.

As with offset current, the effects of noise currents also depend on the feedback resistor. So if possible, reduce the size of R_f to minimize the effects of noise currents.

10-4.4 Noise in the Inverting Adder. In the inverting adder (see Section 3-2), each signal input voltage has a gain of 1. However, the noise gain will be 1 plus the number of inputs; for example, a four-input adder would have a noise gain of 5. Thus, noise voltage has 5 times as much gain as each input signal. Therefore, low amplitude signals should be preamplified before connecting them to an adder.

10-4.5 Summary. To reduce the effects of op amp noise,

1. *Never* connect a capacitor across the input resistor or from (—) input to ground. There will always be a few pF of stray capacitance from (—) input to ground due to wiring, so

2. *Always* connect a small capacitor (3 pF) across the feedback resistor. This reduces the noise gain at high frequencies.
3. If possible, avoid large resistor values.

10-5 External Frequency Compensation

10-5.1 The Need for External Frequency Compensation. Op amps with internal frequency compensation (see Section 10-2 and Appendix 1) are very stable with respect to signal frequencies. They do not burst into spontaneous oscillation or wait to oscillate occasionally when a signal is applied. However, the tradeoffs for frequency stability are limited small-signal band width, slow slew rate, and reduced-power band width. Internally compensated op amps are useful at audio frequencies but not at higher frequencies.

A 741 that has a 1-MHz gain band width product will give a useful gain of 1000 only up to frequencies of about 1 kHz. To obtain more gain from the op amp at higher frequencies, the internal frequency-compensating capacitor of the op amp must be removed. If this is done, the resulting op amp structure has a higher slew rate and greater power band width. But these improvements would be cancelled out because the op amp probably would oscillate continually. As usual, there is a tradeoff: frequency stability for a larger band width and slew rate.

In order to be able to make these tradeoffs, manufacturers of op amps bring out from 1 to 3 *frequency compensating terminals*. Such terminals allow the user to choose the best allowable combination of stability and band width. This choice is made by connecting external capacitors and resistors to the compensating terminals. Accordingly, this versatile type of op amp is classified as *externally frequency-compensated.* (See Appendix 2.)

10-5.2 Single Capacitor Compensation. The frequency response of the 101 general-purpose op amp can be tailored by connecting a single capacitor, C_1, to pins 1 and 8. As shown in Fig. 10-10 and Appendix 2, by making $C_1 = 3$ pF, the 101 has an open-loop frequency response curve with a small-signal band width of 10 MHz. Increasing C_1 by a factor of 10 (to 30 pF) reduces the small-signal band width by a factor of 10 (to 1 MHz). Therefore, the 101 can be externally compensated to have the same small-signal band width as the 741.

When the 101 is used in an amplifier circuit, the amplifier's useful frequency range now depends on the compensating capacitor. For example, with R_f/R_i set for an amplifier gain of 100, its small-signal band width would be 10 kHz for $C_1 = 30$ pF. By reducing C_1 to 3 pF, the small-signal band width is increased to 100 kHz. The full-power band width is also increased from about 6 kHz to 60 kHz.

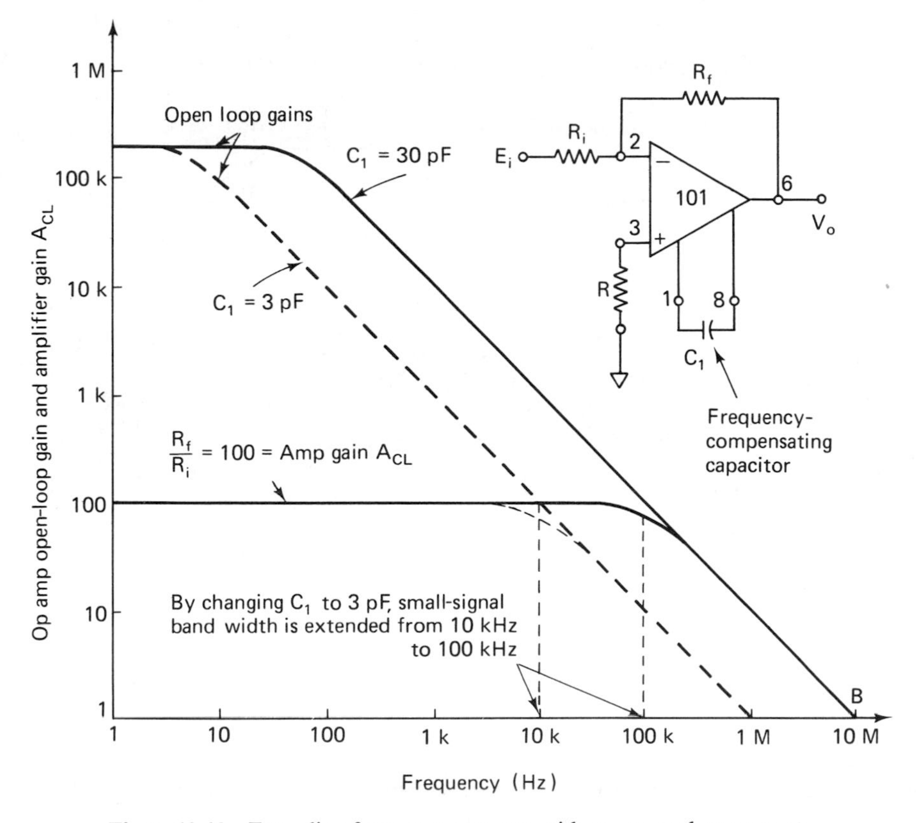

Figure 10-10 Extending frequency response with an external compensating capacitor.

10-5.3 Feed-Forward Frequency Compensation. There are many other types of frequency compensation. Among the more popular are *two-capacitor, or two-pole, compensation* and *feed-forward compensation*. Manufacturer's data sheets give precise instructions on the type best suited for your application.

Feed-forward compensation for the 101 is illustrated in Fig. 10-11. Feed-forward capacitor C is wired from the (—) input to compensating terminal 1. A small capacitor C_f is needed across R_f to insure frequency stability. The slew rate is increased to 10 V/μs and the full-power band width to over 200 kHz. Of course, the added high-frequency gain will also amplify high-frequency noise.

We should conclude that frequency compensation techniques must be applied only to the extent required for the circuit. Do not use any more high-frequency gain than is absolutely necessary; otherwise, there will be a needless amount of high-frequency noise in the output.

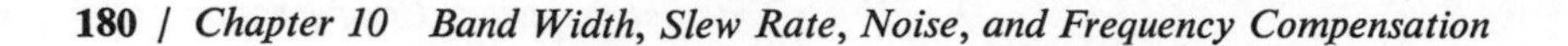

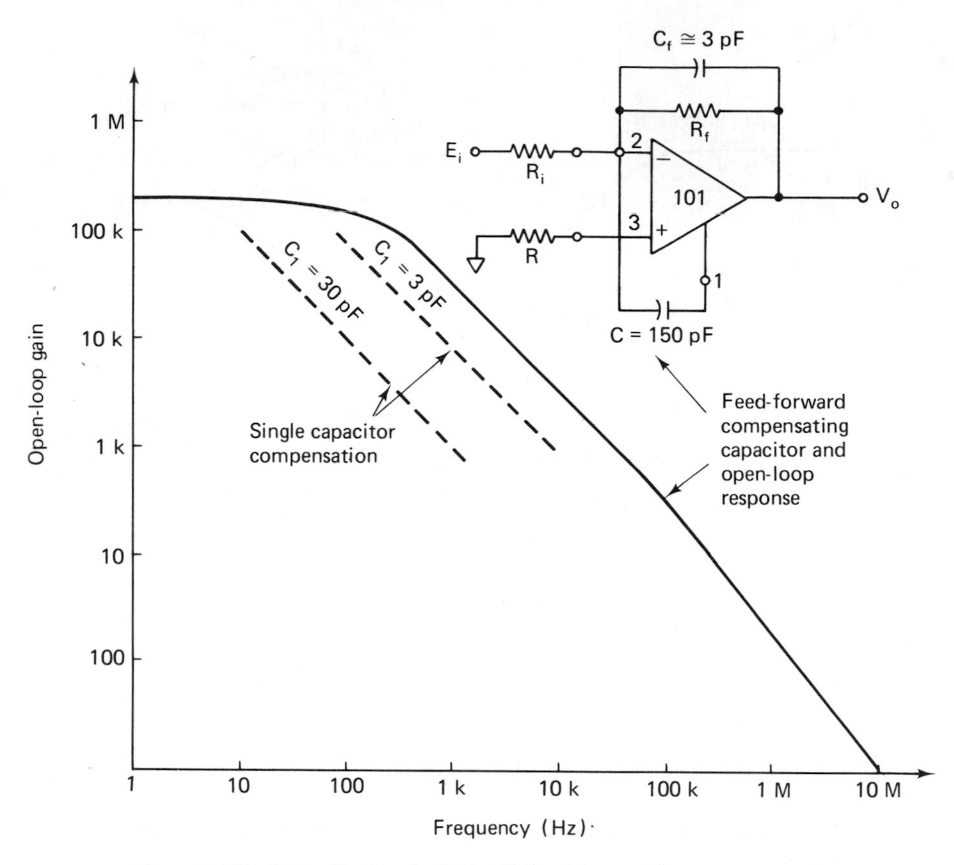

Figure 10-11 Extending band width with feed-forward frequency compensation.

Problems

10-1 What is the typical open-loop gain of an op amp at very low frequencies?

10-2 The dc open-loop gain of an op amp is 100,000. What is the open-loop gain at its break frequency?

10-3 The transient-response rise time at unity gain is given as 0.07 μs. Find the small-signal unity-gain band width.

10-4 What is the open-loop voltage gain for the op amp in Problem 10-3, at 5 MHz?

10-5 An op amp has a unity-gain band width of 2 MHz. What is its open-loop gain at 200 kHz?

10-6 What is the difference between open-loop gain and closed-loop gain?

10-7 In Example 10-6, closed-loop amplifier gain is 50. What is the maximum frequency for 1% gain error?

10-8 In Problem 10-7, what is the maximum frequency for 10% gain accuracy?

10-9 The closed-loop gain of a dc amplifier is 10, and the op amp's unity-gain band width is 10 MHz. Find the small-signal band width of the amplifier.

10-10 How fast can the output of an op amp change by 10 V if its slew rate is 1 V/μs?

10-11 What is the maximum frequency for an output voltage of 10 V peak for the op amp in Problem 10-10?

10-12 Use Fig. 10-8 to estimate the maximum frequency for a 5-V output with a 5-V/μs slew rate.

10-13 What is the noise gain for an inverting amplifier with a gain of $R_f/R_i = 10$?

10-14 What is the noise gain for a six-input inverting adder?

10-15 Does increasing the compensating capacitor size increase or decrease the unity-gain band width?

11

Modulating, Demodulating and Frequency Changing with the Multiplier

11-0 Introduction

Analog multipliers are complex arrangements of op amps and other circuit elements now available in either integrated circuit or functional module form. Multipliers are easy to use; some of their applications are (1) measurement of power, (2) frequency doubling and shifting, (3) detecting phase angle difference between two signals of equal frequency, (4) multiplying two signals, (5) dividing one signal by another, (6) taking the square root of a signal, and (7) squaring a signal. Another use for multipliers is to demonstrate the principles of amplitude modulation and demodulation. The schematic of a typical multiplier is shown in Fig. 11-1(a). There are two input terminals, x and y, which are used for connecting the two voltages to be multiplied. Typical input resistance of each input terminal is 10 kΩ or greater. One output terminal furnishes about the same current as an op amp to a grounded load (5 mA to 10 mA). The output voltage equals the product of the input voltages reduced by a scale factor. The *scale factor* is explained in Section 11-1.

11-1 Multiplying dc Voltages

11-1.1 Multiplier Scale Factor. The schematic of a multiplier shown in Fig. 11-1(a) may have a $\times$ to symbolize multiplication. Another type of schematic shows the inputs and the output voltage equation, as in Fig. 11-1b.

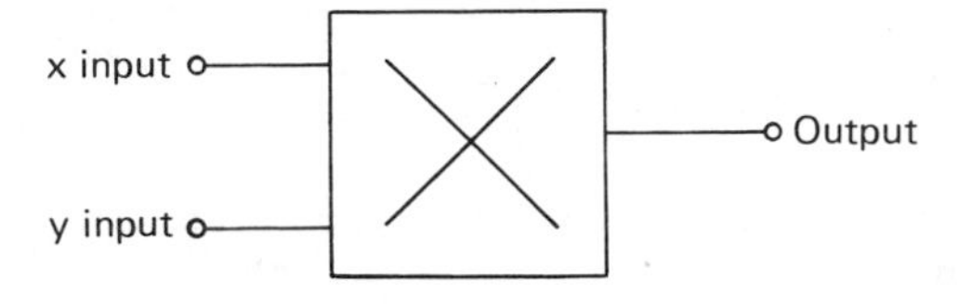

(a) Multiplier schematic

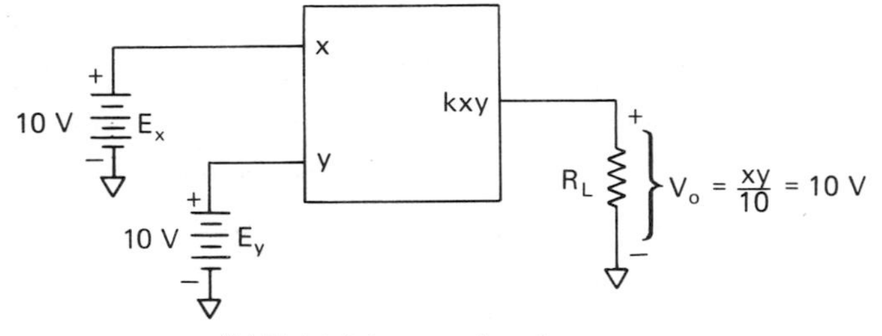

(b) Multiplying two dc voltages

Figure 11-1 Introduction to the multiplier.

In general terms, the output voltage V_o is the product of input voltages x and y and is expressed by

$$V_o = kxy \tag{11-1a}$$

The constant k is called a *scale factor* and is usually equal to $\frac{1}{10}$. This is because multipliers are designed for the same type of power supplies used for op amps, namely ± 15 V. For best results, the voltages applied to either x or y inputs should not exceed $+10$ V or -10 V with respect to ground. This ± 10 V limit also holds for the output, so the scale factor is usually the reciprocal of the voltage limit, or $\frac{1}{10}$. If both input voltages are at their positive limits of $+10$ V, the output will be at its positive limit of 10 V. Thus Eq. (11-1a) is expressed for most multipliers by

$$V_o = \frac{xy}{10} = \frac{E_x E_y}{10} \tag{11-1b}$$

11-1.2 Multiplier Quadrants. Multipliers are classified by quadrants; for example, there are one-quadrant, two-quadrant and four-quadrant multipliers. The classification is explained in two ways in Fig. 11-2. In Fig. 11-2(a), the input voltages can have four possible polarity combinations. If both x and y are positive, operation is in Quadrant 1, since x is the horizontal and y the vertical axis. If x is positive and y is negative, Quadrant 4 operation results, and so forth.

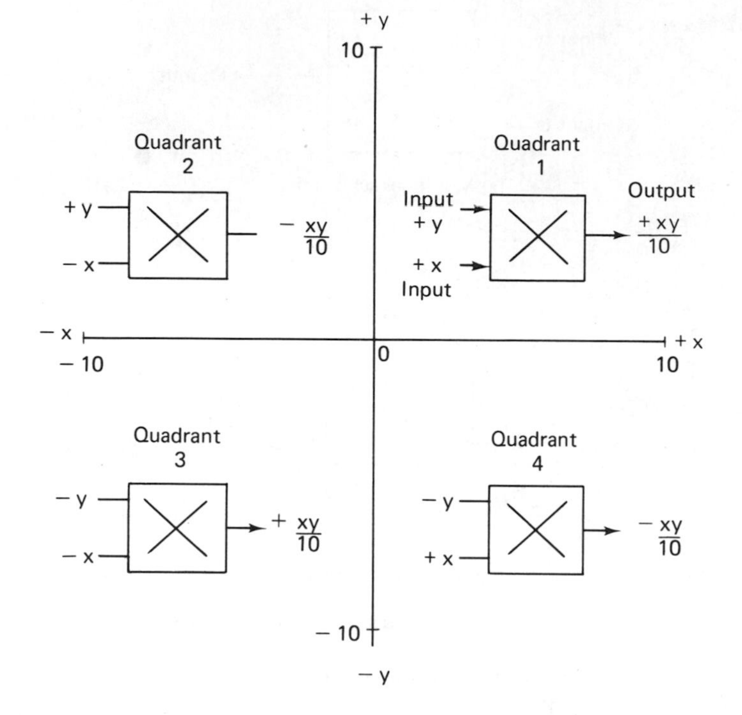

(a) y vs x plot shows location of input operating point in one of four quadrants

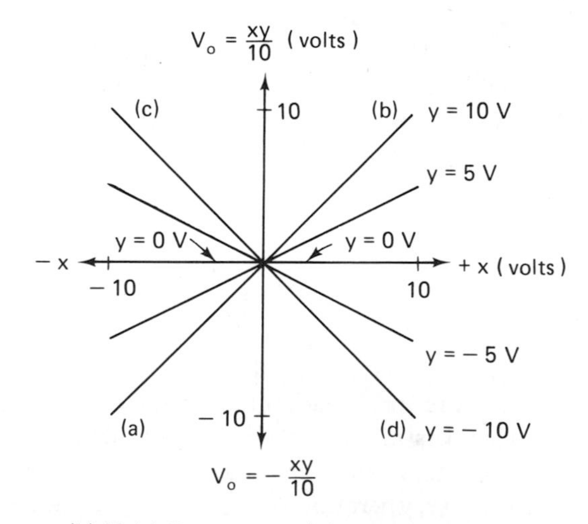

(b) Multiplier output xy/10 versus input x

Figure 11-2 Multiplying two dc voltages, x and y.

Example 11-1: Find V_o for the following combination of inputs. (a) $x = 10$ V, $y = 10$ V; (b) $x = -10$ V, $y = 10$ V; (c) $x = 10$ V, $y = -10$ V, (d) $x = -10$ V, $y = -10$ V.

Solution: From Eq. (11-1b),

(a) $V_o = \dfrac{(10)(10)}{10} = 10\text{ V}$

(b) $V_o = \dfrac{(-10)(10)}{10} = -10\text{ V}$

(c) $V_o = \dfrac{(10)(-10)}{10} = -10\text{ V}$

(d) $V_o = \dfrac{(-10)(-10)}{10} = 10\text{ V}$

In Fig. 11-2(b), V_o is plotted on the vertical axis and x on the horizontal axis. If we apply 10 V to the y input and vary x from -10 V to $+10$ V, we plot the line *ab* labeled $y = 10$ V. If y is changed to -10 V, the line *cd* labeled $y = -10$ V results. These lines can be seen on a cathode-ray oscilloscope (CRO) by connecting V_o of the multiplier to the y input of the CRO and x of the multiplier to the $+x$ input of the CRO.

For accuracy, V_o should be 0 V when either multiplier input is 0 V. If this is not the case, a zero trim adjustment should be made as shown in Section 11-1.3.

11-1.3 Zero Trim Adjustment. To insure that the multiplier output voltage is 0 V when either input is 0 V, a zero trim (output offset voltage) adjustment may be required. This capability is designed into the multiplier, as shown for the Teledyne Philbrick Nexus model 4450 multiplier/divider in Fig. 11-3(a). The user connects a 50 kΩ trimpot as shown, grounds both inputs, and adjusts the pot for $V_o = 0$ V. The adjustment range is typically from -0.5 V to $+0.5$ V. For even more precision, scale factor adjustment should be made as shown in Section 11-1.4.

11-1.4 Scale Factor Trim. Assume that $+10.0$ V is applied to the y input of the multiplier in Fig. 11-3(b). Then if the x input voltage is either $+10.0$ V or -10.0 V, V_o should be either $+10.0$ V or -10.0 V respectively. In reality, V_o may fail to be symmetrical at $+10.0$ V when E_x and E_y are equal to $+10.0$ V, and at -10.0 V when $E_x = -10.0$ V and $E_y = +10.0$ V. Therefore, a scale factor trim adjustment is provided with the multiplier. As shown in Fig. 11-3, the user places a 1-kΩ *scale factor trim resistor* in series with the *trim* input.

The x input voltage is now applied to the trim resistor. As E_x is switched between $+10.0$ V and -10.0 V, the trim resistor is adjusted until $V_o = +10.0$ V when $E_x = +10.0$ V and $V_o = -10.0$ V when $E_x = -10.0$ V.

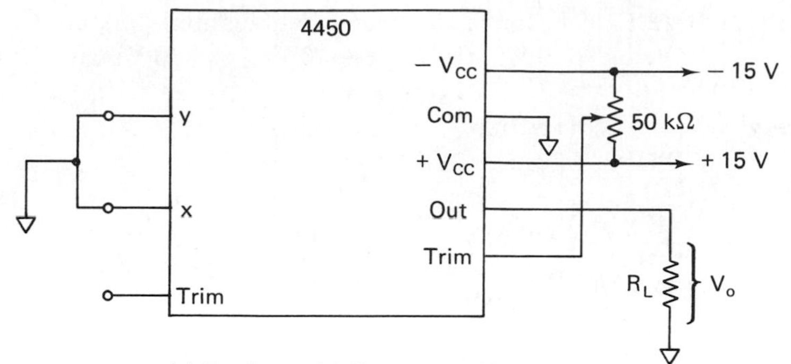

(a) Zeroing multiplier output offset voltage

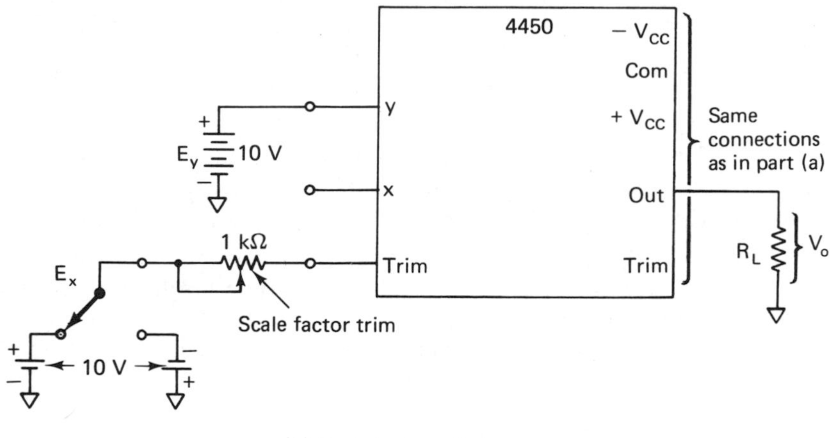

(b) Scale factor adjustment

Figure 11-3 Improving multiplier accuracy.

11-2 Squaring a Number or dc Voltage

Any positive or negative number can be squared by a multiplier, providing that the number can be represented by a voltage between 0 and 10 V. Simply connect the voltage E_i to *both* inputs as shown in Fig. 11-4. This type of connection is known as a *squaring circuit.*

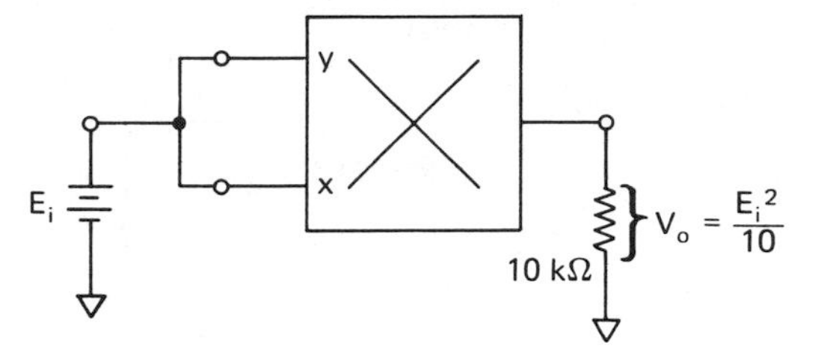

Figure 11-4 Squaring a dc voltage.

Example 11-2: Find V_o in Fig. 11-4 if (a) $E_i = +10$ V and (b) $E_i = -10$ V.
Solution: From Fig. 11-4,

(a) $V_o = \dfrac{10^2}{10} = 10 \text{ V}$

(b) $V_o = \dfrac{(-10)(-10)}{10} = 10 \text{ V}$

Example 11-2 shows that the output of the multiplier follows the rules of algebra; that is, when either a positive or negative number is squared, the result is a positive number.

11-3 Frequency Doubling

11-3.1 Principle of the Frequency Doubler. An ideal frequency doubler would give an output voltage whose frequency is twice the frequency of the input voltage. The doubler circuit should not incorporate a tuned circuit, since the tuned circuit can be tuned only to one frequency. A true doubler should double any frequency. The multiplier is very nearly an ideal doubler if only one frequency is applied to both inputs. The output voltage for a doubler circuit is given by the trigonometric identity

$$(\sin 2\pi ft)^2 = \frac{1}{2} - \frac{\cos 2\pi(2f)t}{2} \tag{11-2}$$

Equation (11-2) predicts that squaring a sine wave with a frequency of (for example) $f = 10$ kHz gives a negative cosine wave with a frequency of $2f$ or 20 kHz plus a dc term of $\frac{1}{2}$. Note that *any* input frequency f will be doubled when passed through a squaring circuit.

11-3.2 Squaring a Sinusoidal Voltage. In Fig. 11-5(a), sine wave voltage E_i is applied to both multiplier inputs. E_i has a peak value of 5 V and a frequency of 10 kHz. The output voltage V_o is predicted by the calculations shown in Example 11-3.

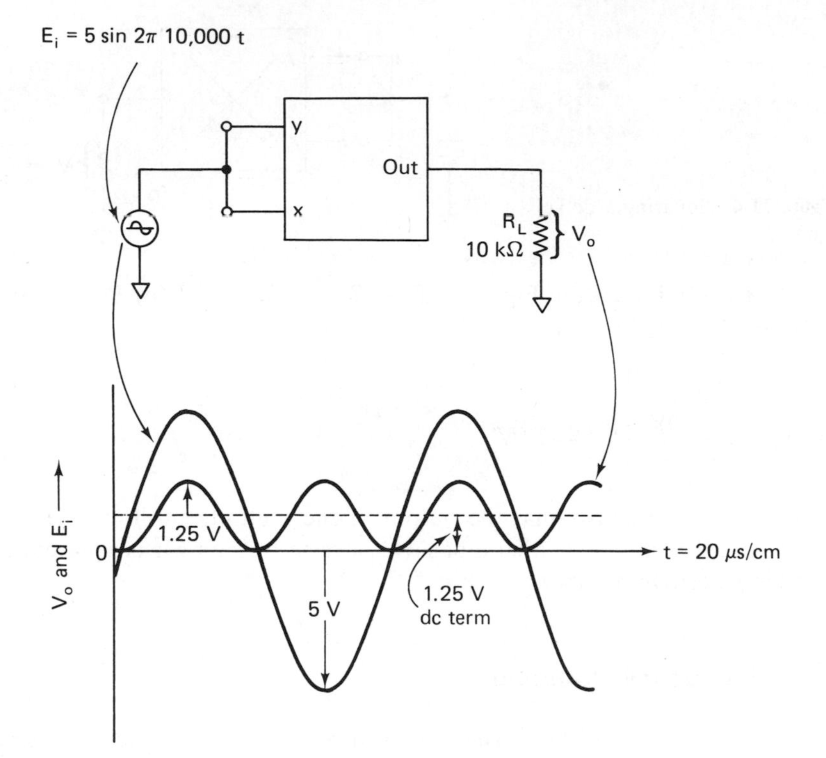

Figure 11-5 Squaring circuit as a frequency doubler.

Example 11-3: Calculate V_o in the squaring circuit or frequency doubler of Fig. 11-5.
Solution: Input $E_x = E_y = E_i$ and is expressed in volts by

$$E_i = E_x = E_y = 5 \sin 2\pi 10{,}000t$$

Substituting into Eq. (11-1b),

$$V_o = \frac{E_i^2}{10} = \frac{5^2}{10}(\sin 2\pi 10{,}000t)^2 \tag{11-3}$$

Applying Eq. (11-2),

$$V_o = 2.5\left[\frac{1}{2} - \frac{\cos 2\pi 20{,}000t}{2}\right] \text{V}$$
$$= \underbrace{1.25} - \underbrace{1.25 \cos 2\pi 20{,}000t}$$
$$= \text{dc term of 1.25 V} - \text{frequency doubled to 20,000 Hz, 1.25 V peak.}$$

Both E_i and V_o are shown in Fig. 11-5. If you want to remove the dc voltage simply install a 1 μF coupling capacitor between R_L and the output terminal. If you want to measure the dc voltage, simply connect a dc voltmeter to V_o.

Conclusion: V_o has two voltage components: (1) a dc voltage equal to $\frac{1}{20}(E_{i_p})^2$ and (2) an ac sinusoidal wave whose peak value is $\frac{1}{20}(E_{i_p})^2$ and whose frequency is double that of E_i.

Example 11-4: What are the dc and ac output voltage components of Fig. 11-5 if (a) $E_i = 10$ V peak at 1 kHz and (b) $E_i = 2$ V peak at 2.5 kHz?

Solution: (a) dc value $= (10)^2/20 = 5$ V; peak ac value $= (10)^2/20 = 5$ V at 2 kHz.

(b) dc value $= (2)^2/20 = 0.2$ V; peak ac value $= (2)^2/20 = 0.2$ V at 5 kHz.

11-4 Phase Angle Detection

11-4.1 Basic Theory. If two sine waves of the *same* frequency are applied to the multiplier inputs in Fig. 11-6(a), output voltage, V_o, has a dc voltage component and ac component whose frequency is twice that of the input frequency. This conclusion was developed in Section 11-3.2. The dc voltage is actually proportional to the difference in phase angle, θ, between E_x and E_y. For example in Fig. 11-5, $\theta = 0°$, because there was no phase difference between E_x and E_y. Fig. 11-6(b) shows a phase difference of 90°, therefore $\theta = 90°$.

If one input sine wave differs in phase angle from the other, it is possible to calculate or measure the phase angle difference from the dc voltage component in V_o. This dc component $V_{o\,dc}$ is given by

$$V_{o\,dc} = \frac{E_{x_p}E_{y_p}}{20}(\cos\theta) \tag{11-4a}$$

where E_{x_p} and E_{y_p} are peak amplitudes of E_x and E_y. For example, if $E_{x_p} = 10$ V, $E_{y_p} = 5$ V, and they are in phase, then $V_{o\,dc}$ would indicate 2.5 V on a dc voltmeter. This voltmeter point would be marked as a phase angle of 0° ($\cos 0° = 1$). If $\theta = 45°$ ($\cos 45° = 0.707$), the dc meter would read $0.707 \times 2.5\text{ V} \cong 1.75$ V. Our dc voltmeter can be calibrated as a phase angle meter 0° at 2.5 V, 45° at 1.75 V, and 90° at 0 V.

Equation (11-4a) may also be expressed by

$$\cos\theta = \frac{20V_{o\,dc}}{E_{x_p}E_{y_p}} \tag{11-4b}$$

If we could arrange for the product E_{x_p} E_{y_p} to equal 20 we could use a 0 V to 1 V dc voltmeter to read $\cos\theta$ directly from the meter face and calibrate the meter face in degrees from a cosine table. That is, Eq. (11-4b) reduces to

$$V_{o\,dc} = \cos\theta \qquad \text{for } E_{x_p} = E_{y_p} = 4.47 \text{ V} \tag{11-4c}$$

This point is explored further in Section 11.4.2.

Example 11-5: In Fig. 11-6, $E_{x_p} = E_{y_p} = 5$ V and the dc component of V_o is 1.25 V from Eq. (11-4b). Prove that there is 0° phase angle between E_x and E_y (since they are the same voltage).
Solution: From Eq. (11-4b),

$$\frac{20 \times 1.25}{5 \times 5} = \cos\theta = \frac{25}{25} = 1$$

Since $\cos 0° = 1$, $\theta = 0°$

11-4.2 A Phase Angle Meter. Equation (11-4b) points the way to making a phase angle meter. Assume that the peak values of E_x and E_y in Fig. 11-6(a) are scaled to 4.47 V by amplifiers or voltage dividers. Then a dc voltmeter is connected as shown in Fig. 4-6(a) to measure just the dc voltage component. The meter face then can be calibrated directly in degrees. The procedure is developed in Examples 11-6 and 11-7.

Example 11-6: The average value of V_o in Fig. 11-6(c) is 0, so the dc component of $V_o = 0$ V. Calculate θ.
Solution: From Eq. (11-4b), $\cos\theta = 0$, so $\theta = 90°$.

Example 11-7: Calculate $V_{o\,dc}$ for phase angles of (a) $\theta = 30°$, (b) $\theta = 45°$, and (c) $\theta = 60°$.
Solution: From a trig table, obtain the $\cos\theta$ and apply Eq. (11-4b).

θ	$\cos\theta$	$V_{o\,dc}$
30°	0.866	0.866 V
45°	0.707	0.707 V
60°	0.500	0.500 V
0°	1.000	1.000 V
90°	0.000	0.000 V

(The last two rows of this table come from Examples 11-5 and 11-6.)

The 0–1 V voltmeter scale can now be calibrated in degrees, 0 V for a 90° phase angle and 1.0 V for 0° phase angle. At 0.866 V, $\theta = 30°$, and so forth. The phase angle meter does not indicate whether θ is a leading or lagging phase angle but only the phase difference between E_x and E_y.

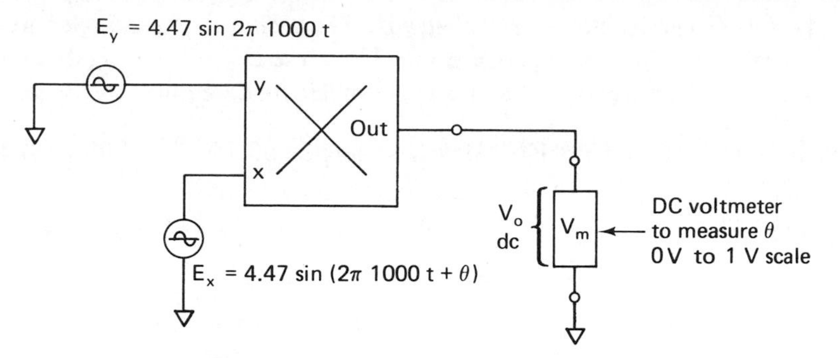

(a) Phase angle measurement

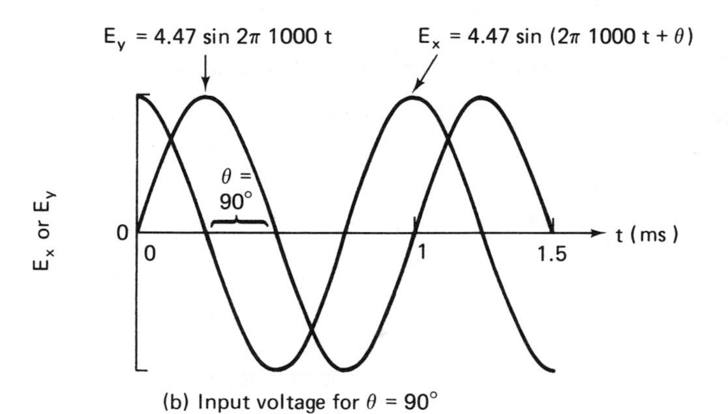

(b) Input voltage for $\theta = 90°$

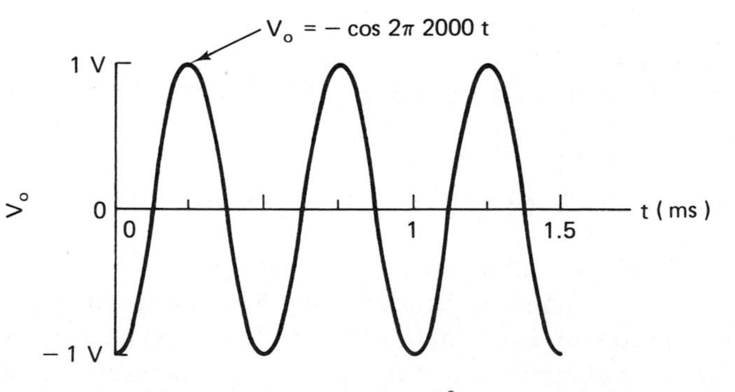

(c) Output voltage for $\theta = 90°$, dc term is 0 V

Figure 11-6 Multiplier used to measure phase angle difference between two equal frequencies.

11-4.3 Phase Angles Greater than 90°. The cosine of phase angles greater than $+90°$ or $-90°$ is a negative value. Therefore, V_o will be negative. This extends the capability of the phase angle meter in Example 11-7.

Example 11-8: Calculate $V_{o\,dc}$ for phase angles of (a) $\theta = \pm 90°$, (b) $\theta = \pm 120°$, (c) $\theta = \pm 135°$, (d) $\theta = \pm 150°$, and (e) $\theta = \pm 180°$.
Solution: Using Eq. (11-4b) and tabulating results, we have

θ	$\pm 90°$	$\pm 120°$	$\pm 135°$	$\pm 150°$	$\pm 180°$
$V_{o\,dc}$	0 V	−0.5 V	−0.70 V	−0.866 V	−1 V

From the results of Examples 11-7 and 11-8, a ± 1 V voltmeter can be calibrated to read from 0 to 180°.

11-5 Introduction to Amplitude Modulation

11-5.1 Need for Amplitude Modulation. Low-frequency audio or data signals cannot be transmitted from antennas of reasonable size. Audio signals can be transmitted by changing or *modulating* some characteristic of a higher frequency *carrier* wave. If the amplitude of the carrier wave is changed in proportion to the audio signal, the process is called *amplitude modulation* (AM). Changing the frequency or the phase angle of the carrier wave results in *frequency modulation* (FM) and *phase angle modulation* (PM), respectively.

Of course, the original audio signal must eventually be recovered by a process called *demodulation*. The remainder of this section concentrates on using the multiplier for amplitude modulation.

11-5.2 Defining Amplitude Modulation. The introduction to amplitude modulation begins with the amplifier in Fig. 11-7(a). The input voltage E_c is amplified by a constant gain A. Amplifier output V_o is the product of gain A and E_c. Now suppose that the amplifier's gain is varied. This concept is represented by an arrow through A in Fig. 11-7(b). Assume that A is varied from 0 to a maximum and back to 0 as shown in Fig. 11-7(b) by the plot of A vs t. This means that the amplifier multiplies the input voltage E_c by a different value (gain) over a period of time. V_o is now the amplitude of input E_c varied or multiplied by the amplitude of A. This process is an example of amplitude modulation, and the output voltage V_o is called the *amplitude modulated signal*. Therefore, to obtain an amplitude-modulated signal (V_o), the amplitude of a high-frequency carrier signal (E_c) is varied by an intelligence or data signal A.

11-5.3 The Multiplier Used as a Modulator. From Section 11-5.2 and Fig. 11-7(b), V_o equals E_c multiplied by A. Therefore, amplitude modulation

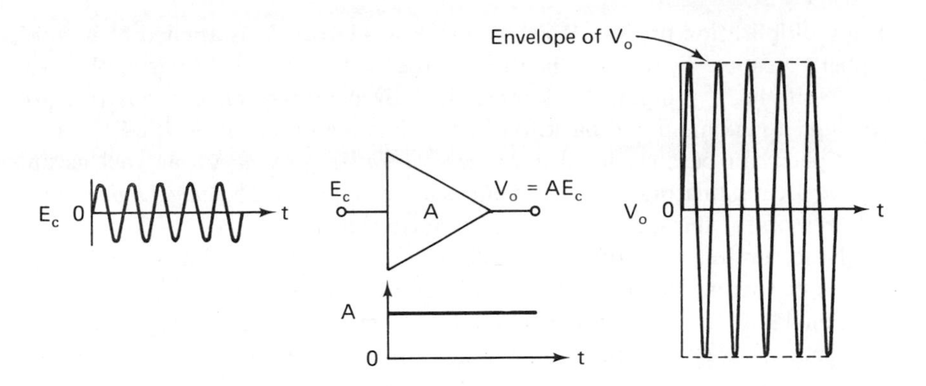

(a) Input E_c is amplified by constant gain A to give output $V_o = AE_c$

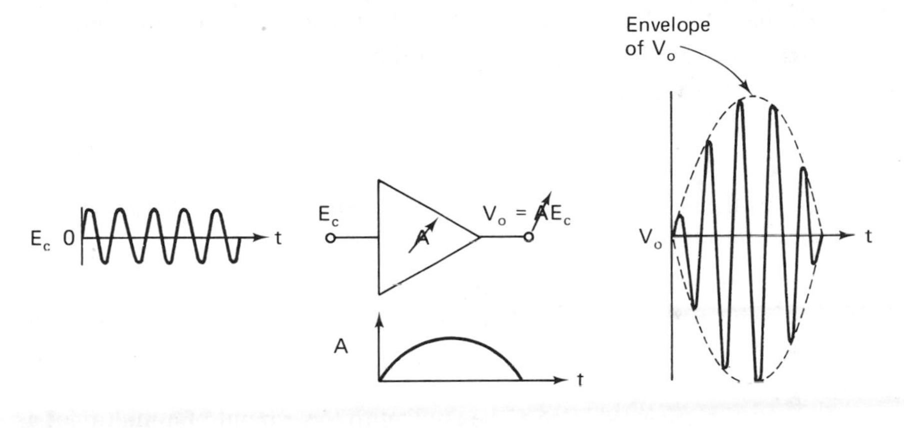

(b) If amplifier gain A is varied with time, the envelope of V_o is varied with time

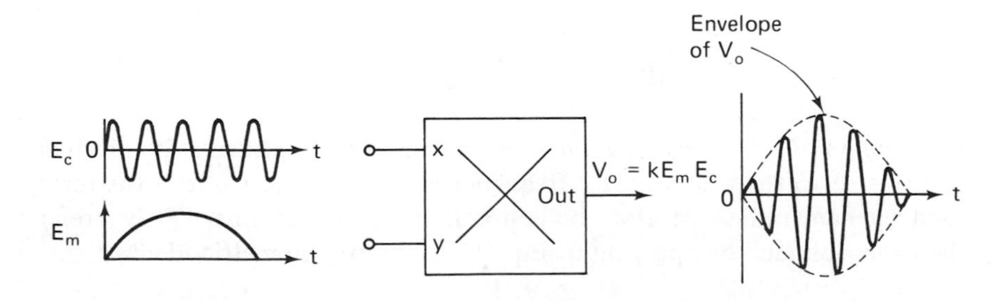

(c) If E_m varies as A in part (b), then V_o has the same general shape as in part (b)

Figure 11-7 Introduction to modulation.

is a multiplication process. As shown in Fig. 11-7(c), E_c is applied to a multiplier's x input. E_m (having the same shape as A in Fig. 11-7(b)) is applied to the multiplier's y input. E_c is multiplied by a voltage that varies from 0 through a maximum and back to 0. So V_o has the same envelope as E_m. The multiplier can be considered a *voltage-controlled gain device* as well as an amplitude modulator. The wave shape shown is that of a *balanced modulator.* The reason for this name will be given in Section 11-6.3.

Note carefully in Fig. 11-7(c) that E_c is a sine wave; that is, the peak values of successive half cycles are different. This principle will be used in Section 11-10 to show how a frequency-shifter (heterodyne) circuit works. But first, we examine amplitude modulation in greater detail.

11-5.4 Mathematics of the Balanced Modulator. A high-frequency sinusoidal *carrier wave*, E_c, is applied to one input of a multiplier. A lower-frequency audio or data signal is applied to the second input of a modulator and will be called the *modulating wave*, E_m. For test and analysis, both E_c and E_m will be sine waves described as follows.

Carrier wave, E_c

$$E_c = E_{c_p} \sin 2\pi f_c t \tag{11-5a}$$

where E_{c_p} is the peak value of the carrier wave and f_c is the carrier frequency.

Modulating Wave, E_m

$$E_m = E_{m_p} \sin 2\pi f_m t \tag{11-5b}$$

where E_{m_p} is the peak value of the modulating wave and f_m is the modulating frequency.

Now let the carrier voltage E_c be applied to the x input of a multiplier as E_x, and let the modulating voltage E_m be applied to the y input of a multiplier as E_y. The multiplier's output voltage V_o is expressed as a product term from Eq. (11-1b) as

$$V_o = \frac{E_m E_c}{10} = \frac{E_{m_p} E_{c_p}}{10}(\sin 2\pi f_m t)(\sin 2\pi f_c t) \tag{11-6}$$

Equation (11-6) is called the *product term*, because it represents the product of two sine waves with different frequencies. However, it is not in the form used by ham radio operators or communication's personnel. They prefer the form obtained by applying to Eq. (11-6) the trigonometric identity

$$(\sin A)(\sin B) = \frac{1}{2}[\cos (A - B) - \cos (A + B)] \tag{11-7}$$

Substituting Eq. (11-7) into Eq. (11-6), where $A = E_c$ and $B = E_m$, we have

$$V_o = \frac{E_{m_p}E_{c_p}}{20}\cos 2\pi(f_c - f_m)t - \frac{E_{m_p}E_{c_p}}{20}\cos 2\pi(f_c + f_m)t \qquad (11\text{-}8)$$

Equation (11-8) is analyzed in Section 11-5.5.

11-5.5 Sum and Difference Frequencies. Recall from Section 11-5.3 that E_c is a sine wave and E_m is a sine wave, but no part of V_o is a sine wave. V_o in Fig. 11-7(c) is expressed mathematically by either Eq. (11-6) or Eq. (11-8). But Eq. (11-8) shows that V_o is made up of two sine waves with frequencies different from either E_m or E_c. They are the *sum frequency* $f_c + f_m$ and the *difference frequency* $f_c - f_m$. The sum and difference frequencies are evaluated in Example 11-9.

Example 11-9: In Fig. 11-8, carrier signal E_c has a peak voltage of $E_{c_p} = 5$ V and a frequency of $f_c = 10{,}000$ Hz. The modulating signal E_m has a peak voltage of $E_{m_p} = 5$ V and a frequency of $f_m = 1{,}000$ Hz. Calculate the peak voltage and frequency of (a) the sum frequency and (b) the difference frequency.

Solution: From Eq. (11-8), the peak value of both sum and difference voltages is

$$\frac{E_{m_p}E_{c_p}}{20} = \frac{5\text{ V} \times 5\text{ V}}{20} = 1.25\text{ V}$$

The sum frequency is $f_c + f_m = 10{,}000\text{ Hz} + 1000\text{ Hz} = 11{,}000$ Hz; the difference frequency is $f_c - f_m = 10{,}000\text{ Hz} - 1000\text{ Hz} = 9000$ Hz. Thus V_o is made up of the difference of two cosine waves.

$$V_o = 1.25\cos 2\pi 9000t - 1.25\cos 2\pi 11{,}000t$$

This result can be verified by connecting a wave or spectrum analyzer to the multiplier's output; a 1.25 V deflection occurs at 11,000 Hz and at 9,000 Hz.

A cathode-ray oscilloscope can be used to show input and output voltages of the multiplier of Example 11-8. The product term for V_o is found from Eq. (11-6):

$$V_o = 2.5\text{ V}\underbrace{(\sin 2\pi 10{,}000t)}_{}\underbrace{(\sin 2\pi 1000t)}_{}$$
$$V_o = 2.5 \times \quad E_c \quad \times \quad E_m$$

V_o is shown with E_m in the top drawing and with E_c in the bottom drawing. Observe that E_m and E_c have peak voltages of 5 V. The peak value of V_o is 2.5 V.

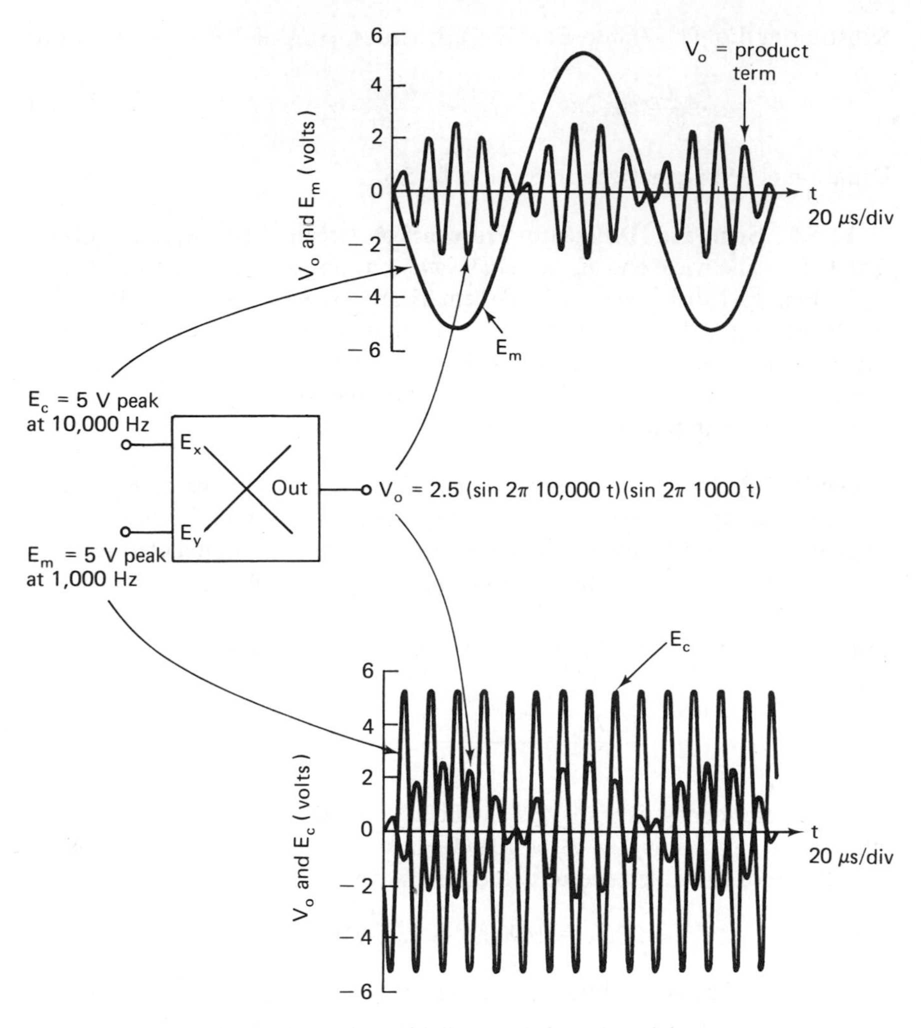

Figure 11-8 The multiplier as a balanced modulator.

11-5.6 Side Frequencies and Side Bands. Another way of displaying the output of a modulator is by a graph showing the peak amplitude as a vertical line for each frequency. The resulting *frequency spectrum* is shown in Fig. 11-9(a). The sum and difference frequencies in V_o are called *upper* and *lower side* frequencies because they are above and below the carrier frequency on the graph.

When more than one modulating signal is applied to the modulator (y input) input in Fig. 11-8, each generates a sum and difference frequency in the output. Thus, there will be two side frequencies for each y input frequency, placed symmetrically on either side of the carrier. If the expected range of

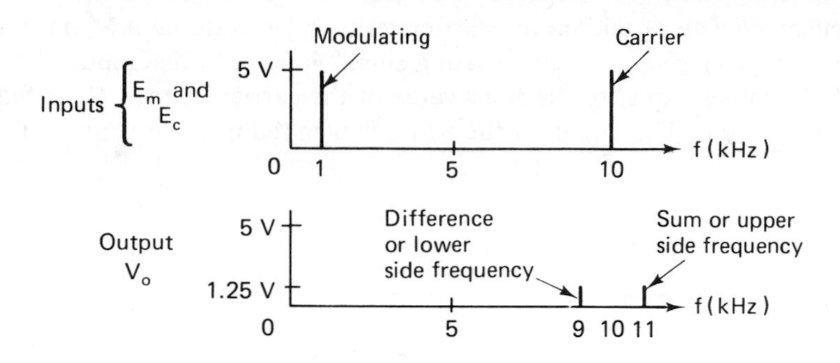

(a) Frequency spectrum for f_c = 10 kHz and f_m = 1 kHz in Example 11.8

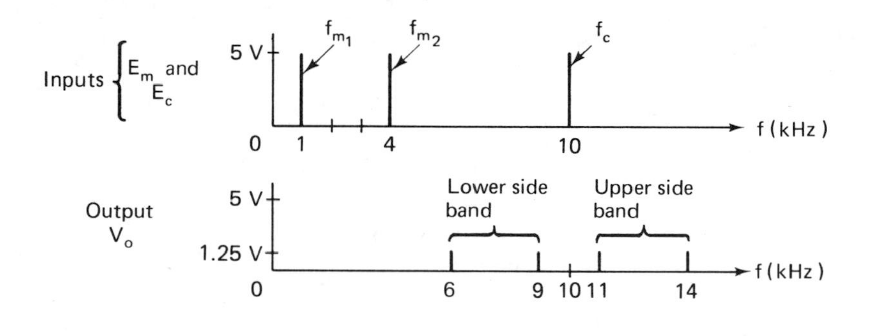

(b) Frequency spectrum for f_c = 10 kHz and f_{m_1} = 1 kHz, f_{m_2} = 4 kHz

Figure 11-9 Frequency spectrum for a balanced modulator.

modulating frequencies is known, then the resulting range of side frequencies can be predicted. For example, if the modulating frequencies range between 1 kHz and 4 kHz, the lower side frequencies fall in a band between (10 — 4) kHz = 6 kHz and (10 — 1) kHz = 9 kHz. The band between 6 kHz and 9 kHz is called the *lower side band.* For this same example, the *upper side band* ranges from (10 + 1) kHz = 11 kHz to (10 + 4) kHz = 14 kHz. Both upper and lower side bands are shown in Fig. 11-9(b).

11-6 Standard Amplitude Modulation

11-6.1 Amplitude Modulator Circuit. The circuits of Section 11-5 multiplied the carrier and modulating signals to generate a balanced output that is expressed either as (1) a product term or (2) a sum and difference frequency. The classical or standard amplitude modulator AM adds the carrier term to the output. One way of adding the carrier term to generate an AM output is shown in Fig. 11-10(a). The modulating signal is fed into one input of an adder. A dc voltage equal to the peak value of the carrier voltage E_{c_p} is fed into the other input. The output of the adder is then fed into the y input of a

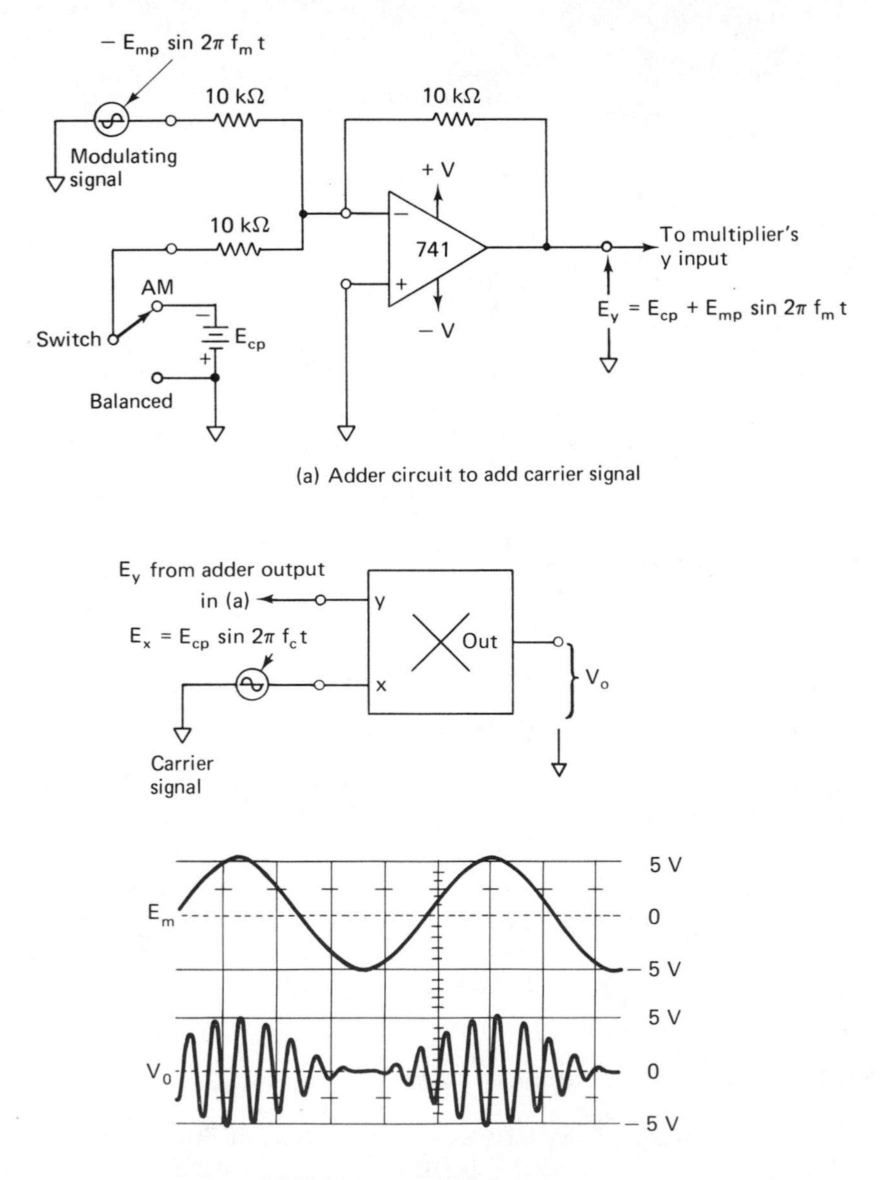

Figure 11-10 Circuit to demonstrate amplitude modulation or balanced modulation. See also Fig. 11-12.

multiplier, as shown in Fig. 11-10(b). The carrier signal is fed into the x input. The multiplier multiplies E_x by E_y, and its output voltage is the standard AM voltage given by either of the following equations.

$$V_o = \begin{cases} \dfrac{E_{c_p}^2}{10} \sin 2\pi f_c t \quad \text{(carrier term)} \\ \qquad + \\ \dfrac{E_{c_p} E_{m_p}}{10} (\sin 2\pi f_c t)(\sin 2\pi f_m t) \quad \text{(product term)} \end{cases} \tag{11-9}$$

or

$$V_o = \begin{cases} \dfrac{E_{c_p}^2}{10} \sin 2\pi f_c t \quad \text{(carrier term)} \\ \qquad + \\ \dfrac{E_{c_p} E_{m_p}}{20} \cos 2\pi (f_c - f_m) t \quad \text{(lower side frequency)} \\ \qquad - \\ \dfrac{E_{c_p} E_{m_p}}{20} \cos 2\pi (f_c + f_m) t \quad \text{(upper side frequency)} \end{cases} \tag{11-10}$$

The output voltage V_o is shown in Fig. 11-10(b). The voltage levels are worked out in the following example.

Example 11-10: In Fig. 11-10, $E_{c_p} = E_{m_p} = 5$ V. The carrier frequency $f_c =$ 10 kHz, and the modulating frequency is $f_m = 1$ kHz. Evaluate the peak amplitudes of the output carrier and product terms.
Solution: From Eq. (11-9), the carrier term peak voltage is

$$\frac{(5\text{ V})(5\text{ V})}{10} = 2.5\text{ V}$$

The product term peak voltage is

$$\frac{(5\text{ V})(5\text{ V})}{10} = 2.5\text{ V}$$

The wave shape of V_o is shown in Fig. 11-10(b). Observe that the envelope of V_o is the same shape as E_m. This is characteristic of a standard amplitude modulator, AM, *not* of the balanced modulator.

11-6.2 Frequency Spectrum of a Standard AM Modulator. The signal frequencies present in V_o for the standard AM output of Fig. 11-10 are found from Eq. (11-10). Using the voltage values in Example 11-10,

$$\begin{aligned} \text{Carrier term} &= 2.5\text{ V peak at 10,000 Hz} \\ \text{Lower side frequency} &= 1.25\text{ V peak at 9000 Hz} \\ \text{Upper side frequency} &= 1.25\text{ V peak at 11,000 Hz} \end{aligned}$$

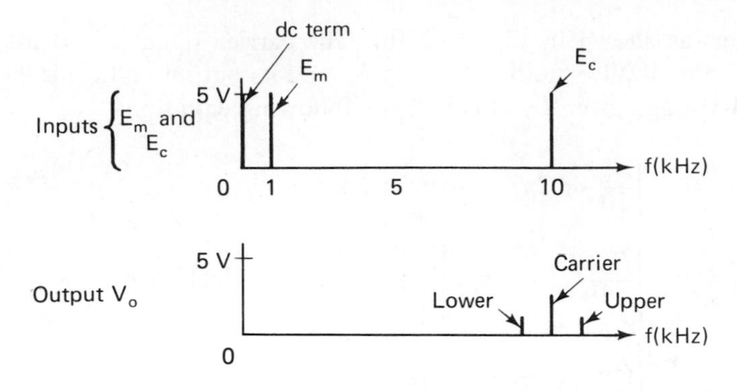

Figure 11-11 Frequency spectrum for a standard AM modulator, f_c = 10 kHz, f_m = 1 kHz.

These frequencies are plotted in Fig. 11-11 and should be compared with the balanced modulator of Fig. 11-9.

11-6.3 Comparison Between Standard AM Modulators and Balanced Modulators. If the switch in Fig. 11-10(a) is positioned to AM, V_o will contain three frequencies, f_c, $f_c + f_m$, and $f_c - f_m$. Observe that the envelope of V_o has the same shape as the intelligence signal E_m. This observation can be used to recover E_m from the AM signal.

If the switch in Fig. 11-10(a) is positioned to "Balanced," V_o will contain only the product term with only two frequencies $f_c + f_m$ and $f_c - f_m$. The envelope of V_o does *not* follow E_m. Since V_o does not contain f_c, this type of modulation is called *balanced modulation* in the sense that the carrier has been balanced out. It is also called *suppressed carrier modulation*, since the carrier is suppressed in the output.

For comparison of balanced and standard AM modulation, both outputs are shown together in Fig. 11-12.

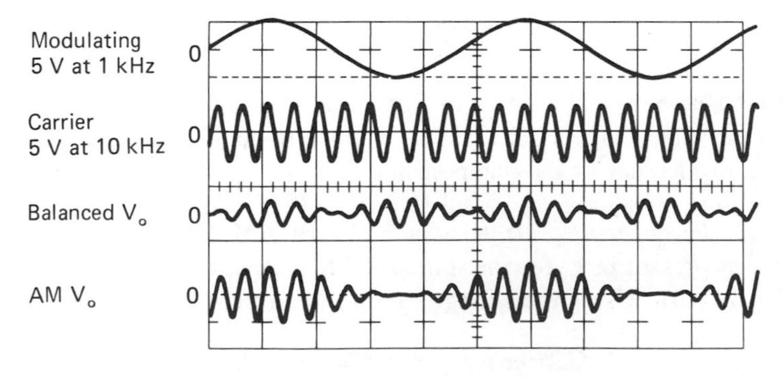

Figure 11-12 Comparison of balanced modulation and standard AM from Fig. 11-10.

11-7 Demodulating an AM Voltage

Demodulation, or *detection*, is the process of recovering a modulating signal E_m from the modulated output voltage V_o. To explain how this is accomplished, the AM modulated wave is applied to the y input of a multiplier as shown in Fig. 11-13. Each y input frequency is multiplied by the x input carrier frequency and generates a sum and difference frequency as shown in Fig. 11-13(b). Since only the 1-kHz frequency is the modulating signal, use a low-pass filter to extract E_m. Thus, the demodulator is simply a multiplier with the carrier frequency applied to one input, and the AM signal to be demodulated is fed into the other input. The multiplier's output is fed into a low pass filter whose output is the original modulating data signal E_m. Thus a multiplier plus a low-pass filter and carrier signal equals a demodulator.

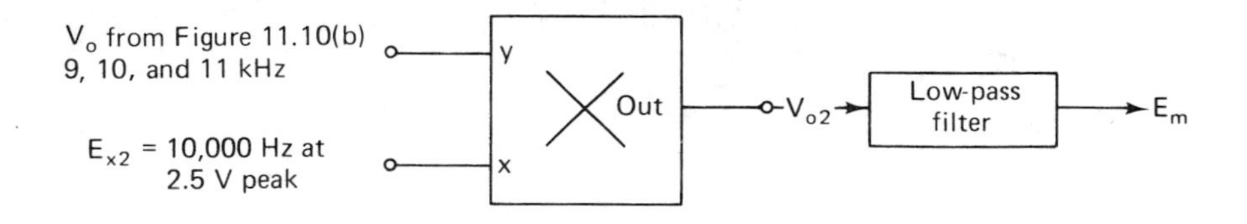

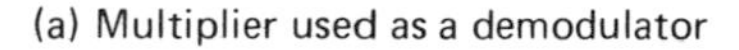

(a) Multiplier used as a demodulator

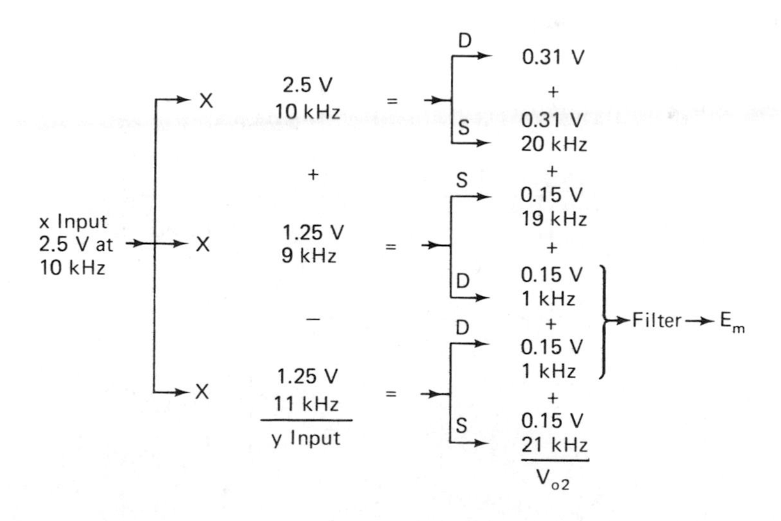

(b) Frequency and peak amplitude of signal components at x-input, y-input, multiplier output and filter

Figure 11-13 Demodulator is a multiplier plus low pass filter.

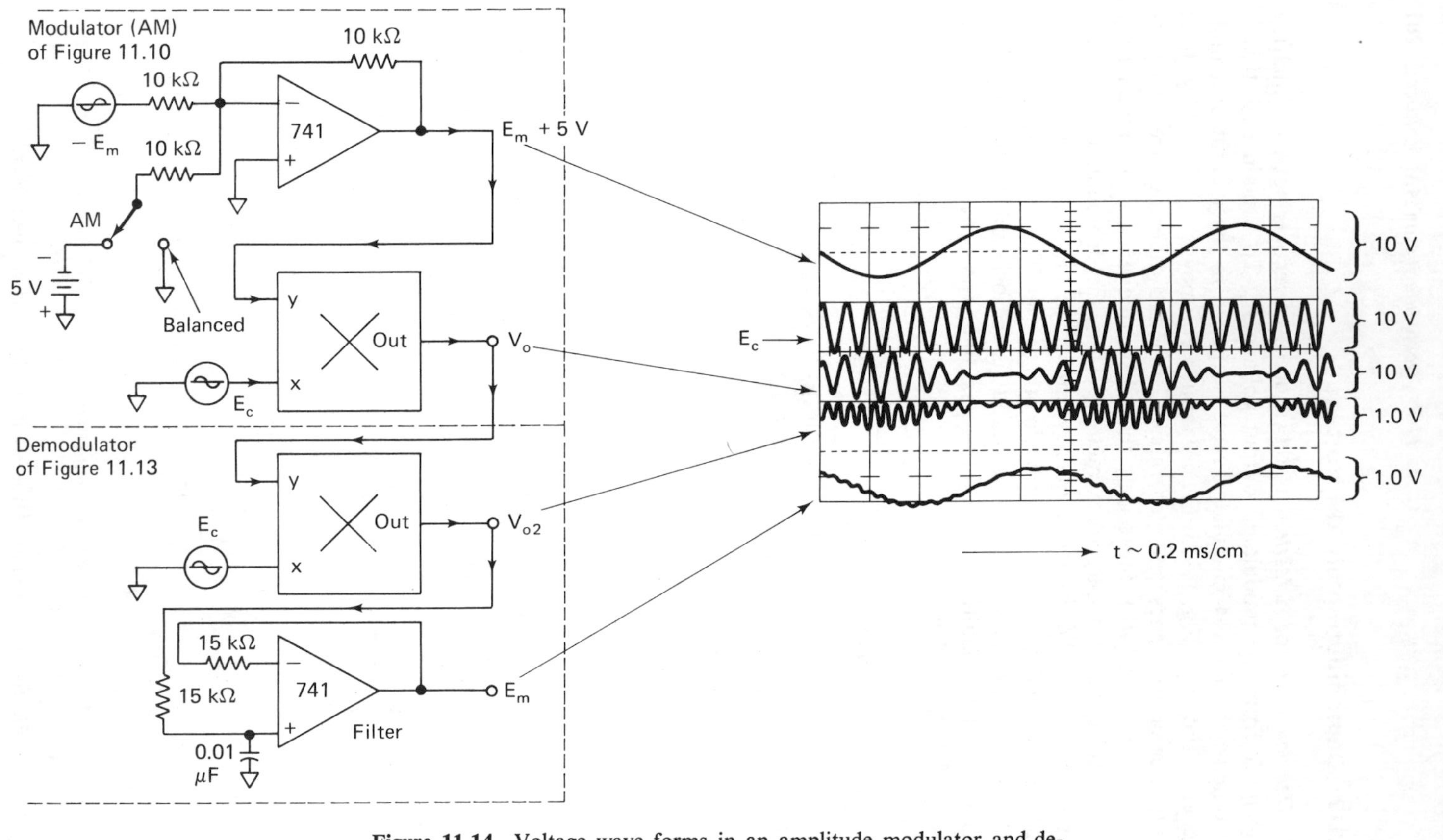

Figure 11-14 Voltage wave forms in an amplitude modulator and demodulator ($f_c = 10$ kHz, $f_m = 1$ kHz).

Wave shapes at inputs and outputs of both the AM modulator and demodulator are shown in Fig. 11-14. Note the unusual shape of V_{o2} because it contains six components (detailed in Fig. 11-13(b)).

11-8 Demodulating a Balanced Modulator Voltage

Modulating signal E_m is recovered from a balanced modulator by means of the same technique employed in Fig. 11-13 and Section 11-7. The only difference is due to the absent carrier frequency of 10 kHz at the demodulator's y input. This missing 10 kHz also eliminates both the dc and 20 kHz term in V_{o2}. The circuit arrangement of Fig. 11-15 was built to demonstrate the demodulating technique and show the resulting wave shapes. The demodulated E_m is not a pure sine wave, because only a simple filter was used. If f_c is increased to 100 kHz, E_m will be closer to being a pure sine wave.

The carrier's frequency fed into the demodulator should be *exactly* equal to the carrier's frequency driving the modulator.

11-9 Single Side Band Modulation and Demodulation

In the balanced modulator of Figs. 11-8 and 11-9, we could add a high-pass filter (see Chapter 12) to the modulator's output. If the filter removed all of the lower side frequencies, the output is *single side band* (SSB). If the filter only attenuated the lower side frequencies (to leave a *vestige* of the lower side band), we would have a *vestigial* side band modulator.

Assume that only one modulating frequency f_m is applied to our single side band modulator together with carrier f_c. Its output would be a single upper side frequency $f_c + f_m$. To demodulate this signal and recover f_m, all we have to do is connect the SSB signal $f_c + f_m$ to one multiplier input and f_c to the other input. According to the principles set forth in Section 11-5.4, the demodulators output would have a sum frequency of $(f_c + f_m) + f_c$ and a difference frequency of $(f_c + f_m) - f_c = f_m$. A low-pass filter would recover the modulating signal f_m and easily eliminate the high frequency signal, whose frequency is $2f_c + f_m$.

11-10 Frequency Shifting

In radio communication circuits, it is often necessary to shift a carrier frequency f_c with its accompanying side frequencies down to a lower intermediate frequency f_{IF}. This shift of each frequency is accomplished with the

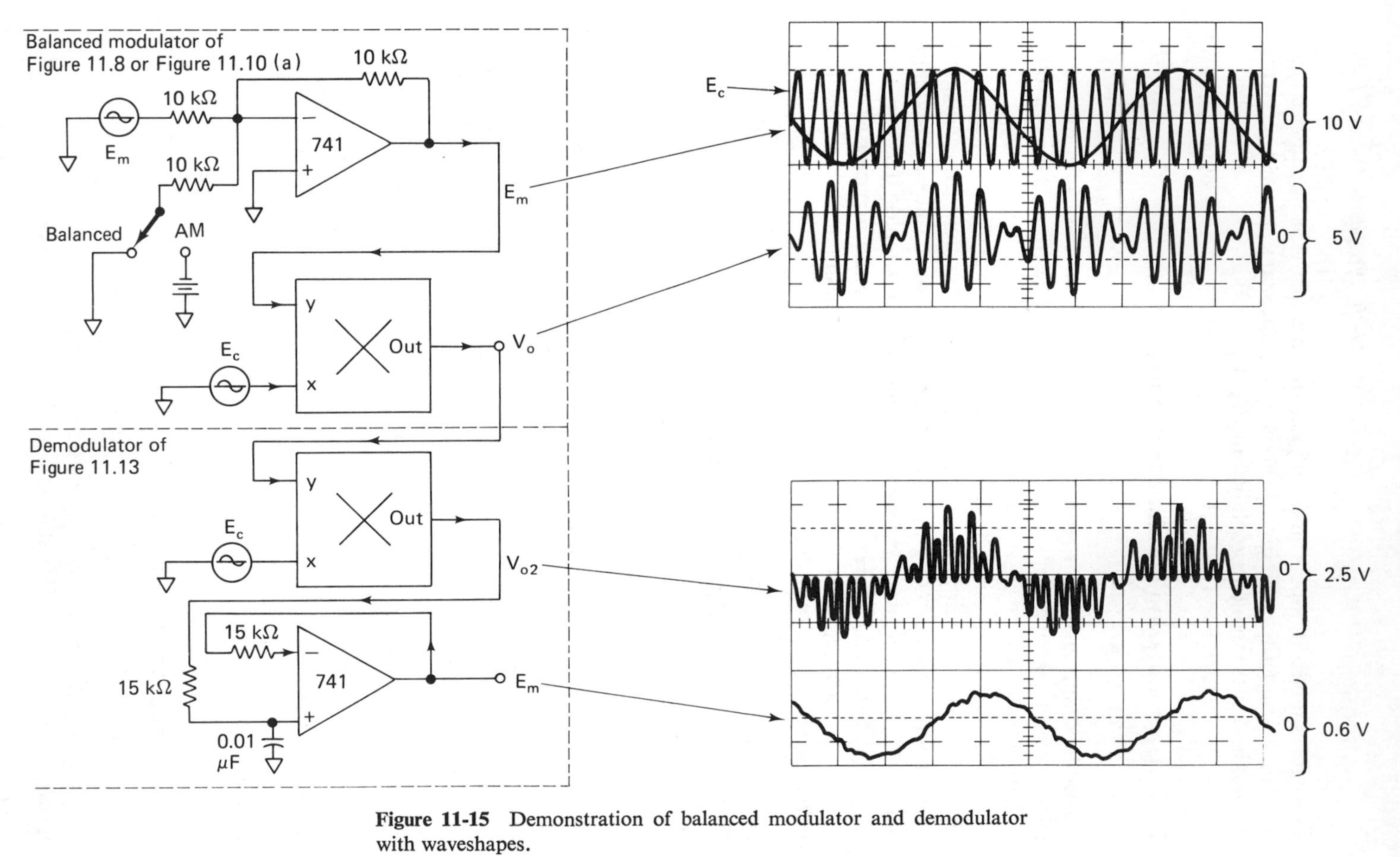

Figure 11-15 Demonstration of balanced modulator and demodulator with waveshapes.

multiplier connections of Fig. 11-16(a). The modulated carrier signals are applied to the y input. A local oscillator is adjusted to a frequency, f_o, equal to the sum of the carrier and desired intermediate frequency and applied to the y input. The frequencies present in the output of the multiplier are calculated in the following example.

Example 11-11: In Fig. 11-16(a), amplitudes and frequencies at each input are as follows:

y input:

Peak amplitude	*Frequency*
1 V	$(f_c + f_m) = 1005$ kHz
4 V	$f_c = 1000$ kHz
1 V	$(f_c - f_m) = 995$ kHz

where f_c is the broadcasting station's carrier frequency and $(f_c + f_m)$ and $(f_c - f_m)$ are the upper and lower side frequencies due to a 5-kHz modulating frequency.

x input:
The local oscillator is set for a 5-V peak sine wave at 1445 kHz, because the desired intermediate frequency is 455 kHz. Find the peak value and frequency of each signal component in the output of the multiplier.
Solution: From Eq. (11-10), the peak amplitude of each y input frequency is multiplied by the peak amplitude of the local oscillator frequency. This product is multiplied by $\frac{1}{20}$ ($\frac{1}{10}$ for the scale factor $\times \frac{1}{2}$ from the trigonometric identity) to obtain the peak amplitude of the resulting sum and difference frequencies at the multiplier's output. The results are tabulated in Fig. 11-16(a).

All frequencies present in the multiplier's output are plotted on the frequency spectrum of Fig. 11-16(c). A low-pass filter or band-pass filter is used to pass only the three lower intermediate frequencies of 450 kHz, 455 kHz, and 460 kHz. The upper intermediate frequencies of 2450, 2455, and 2460 may be used if desired, but they are usually filtered out.

We conclude from Example 11-11 that each frequency present at the y input is shifted down and up to new intermediate frequencies. The lower set of intermediate frequencies can be extracted by a filter. Thus, the information contained in the carrier f_c has been preserved and shifted to another subcarrier or intermediate frequency. The process of frequency shifting is also called *heterodyne*.

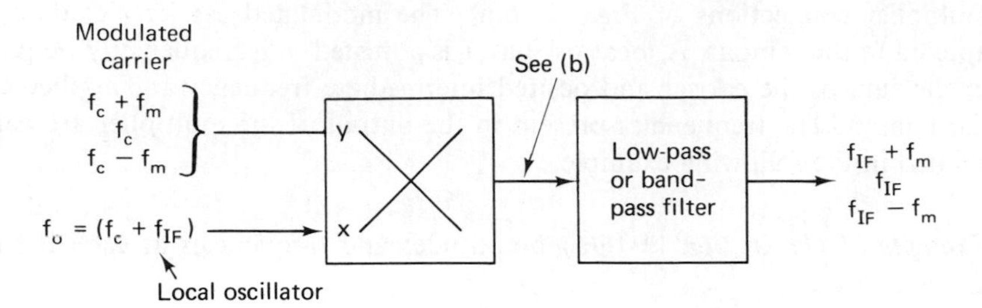

(a) Circuit for a frequency shifter

	Multiplier output	
Frequency at y input (kHz)	Peak (volts)	Frequency (kHz)
1005	$\frac{1 \times 5}{20} = 0.25$ V	1455 + 1005 = 2460 1455 − 1005 = 450
1000	$\frac{4 \times 5}{20} = 1.0$ V	1455 + 1000 = 2455 1455 − 1000 = 455
995	$\frac{1 \times 5}{20} = 0.25$ V	1455 + 995 = 2450 1455 − 995 = 460

(b) Frequencies present in multiplier output

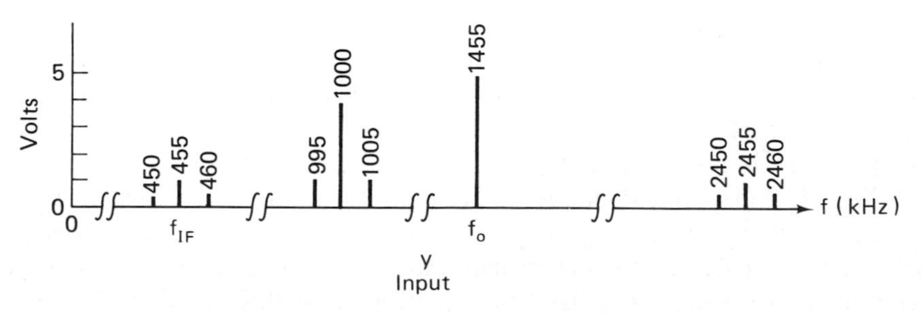

(c) y Input frequencies are shifted to the intermediate frequency

Figure 11-16 The multiplier as a frequency shifter.

11-11 Analog Divider

An analog divider will give the ratio of two signals or provide gain control. It is constructed as shown in Fig. 11-17 by inserting a multiplier in the feedback loop of an op amp. Since the op amp's (—) input draws negligible current, the current I is equal in the equal resistors R. Therefore, the output voltage of the multiplier V_m is equal in magnitude but opposite in polarity (with respect to ground) to E_Z or

$$E_Z = -V_m \tag{11-11a}$$

But V_m is also equal to one-tenth (scale factor) of the product of input E_x and output of the op amp V_o. Substituting for V_m

$$E_Z = -\frac{V_o E_x}{10} \tag{11-11b}$$

Solving for V_o,

$$V_o = -\frac{10 E_Z}{E_x} \tag{11-11c}$$

Equation (11-11c) shows that the divider's output V_o is proportional to the ratio of inputs E_Z and E_x. E_x should never be allowed to go to 0 V or

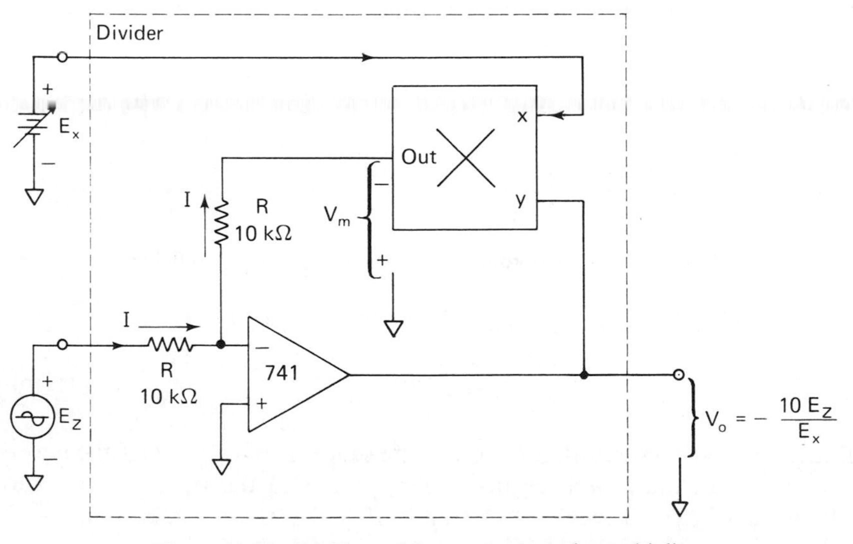

Figure 11-17 Division with an op amp and a multiplier.

to a negative voltage, because the op amp will saturate. E_Z can be either positive, negative, or 0 V. Note that the divider can be viewed as a voltage gain $10/E_x$ acting on E_Z. So if x is changed, the gain will change. This voltage control of the gain is useful in automatic gain-control circuits.

11-12 Finding Square Roots

A divider can be made to find square roots by connecting both inputs of the multiplier to output of the op amp (see Fig. 11-18). Equation (11-11a) also pertains to Fig. 11-18. But now V_m is one-tenth (scale factor) of $V_o \times V_o$ or

$$-E_Z = V_m = \frac{V_o^2}{10} \tag{11-12a}$$

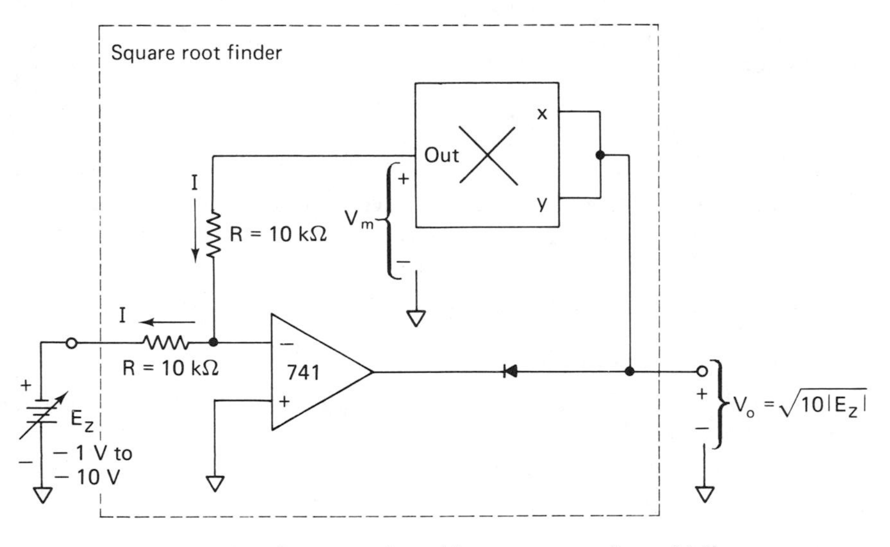

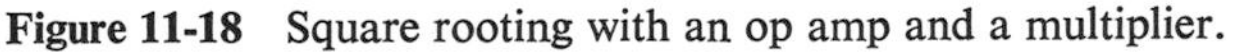
Figure 11-18 Square rooting with an op amp and a multiplier.

Solving for V_o (eliminate $\sqrt{-1}$),

$$V_o = \sqrt{10|E_Z|} \tag{11-12b}$$

Equation (11-12b) states that V_o equals the square root of 10 times the *magnitude* of E_Z. E_Z must be a negative voltage, or else the op amp saturates. The range of E_Z is between −1 V and −10 V. Voltages smaller than −1 V will cause inaccuracies.

Problems

11-1 Find V_o in Fig. 11-1 for the following combination of inputs: (a) $x = 5$ V, $y = 5$ V; (b) $x = -5$ V, $y = 5$ V; (c) $x = 5$ V, $y = -5$ V; (d) $x = -5$ V, $y = -5$ V.

11-2 State the operating point quadrant for each combination in Problem 11-1 (see Fig. 11-2(a)).

11-3 What is the name of the procedure used to make $V_o = 0$ V when both x and y inputs are at 0 V?

11-4 Find V_o in Fig. 11-4 if $E_i = -3$ V.

11-5 The peak value of E_i in Fig. 11-5 is 8 V, and its frequency is 400 Hz. Evaluate the outputs (a) dc terms and (b) ac term.

11-6 In Fig. 11-6, $E_{x_p} = 10$ V, $E_{y_p} = 10$ V, and $\theta = 30°$. Find V_o.

11-7 Repeat Problem 11-6 for $\theta = -30°$.

11-8 In the balanced modulator of Fig. 11-8, E_x is a 15 kHz sine wave at 8 V peak and E_y is a 3kHz sine wave at 5 V peak. Find the peak voltage of each frequency in the output.

11-9 In Fig. 11-8, the carrier frequency is 15 kHz. The modulating frequencies range between 1 kHz and 2 kHz. Find the upper and lower side bands.

11-10 The switch is on AM in Fig. 11-10. The modulating frequency is 10 kHz at 5 V peak. The carrier is 100 kHz at 8 V peak. Identify the peak value and each frequency contained in the output.

11-11 If the switch is thrown to "Balanced" in Problem 11-10, what changes result in the output?

11-12 The x input of Fig. 11-13 is three sine waves of 5 V at 20 kHz, 2.0 V at 21 kHz, and 2.0 V at 19 kHz. The y input is 5 V at 20 kHz. What are the output signal frequency components?

11-13 It is desired to shift a 550 kHz signal to a 455 kHz intermediate frequency. What frequency should be generated by the local oscillator?

11-14 $E_x = 10$ V and $E_Z = -1$ V in Fig. 11-17. Find V_o.

12

Active Filters

12-0 Introduction

A *filter* is a circuit that is designed to pass a specified band of frequencies while attenuating all signals outside this band. Filter networks may be either active or passive. *Passive filter networks* contain only resistors, inductors, and capacitors. *Active filters*, which are the only type covered in this text, employ transistors or op amps plus resistors, inductors, and capacitors. Inductors are not often used in active filters, because they are bulky and costly and may have large internal resistive components.

There are four types of filters.: *low-pass*, *high-pass*, *band-pass*, and *band-elimination* (also referred to as *band- reject* or *notch*) filters. Figure 12-1 illustrates frequency response plots for the four types of filters. A low-pass filter is a circuit that has a constant output voltage from dc up to a *cutoff frequency*, f_c. As the frequency increases above f_c, the output voltage is attenuated (decreases). Figure 12-1(a) is a plot of the magnitude of the output voltage of a low-pass filter versus frequency. The solid line is a plot for the ideal low-pass filter, while the dashed lines indicate the curves for practical low-pass filters. The range of frequencies that are *transmitted* is known as the *pass band.* The range of frequencies that are *attenuated* is known as the *stop band.* The cutoff frequency, f_c, is also called the 0.707 frequency, the 3-dB frequency, the corner frequency, or the break frequency.

High-pass filters attenuate the output voltage for all frequencies below the cutoff frequency f_c. Above f_c, the magnitude of the output voltage is constant. Figure 12-1(b) is the plot for ideal and practical high-pass filters.

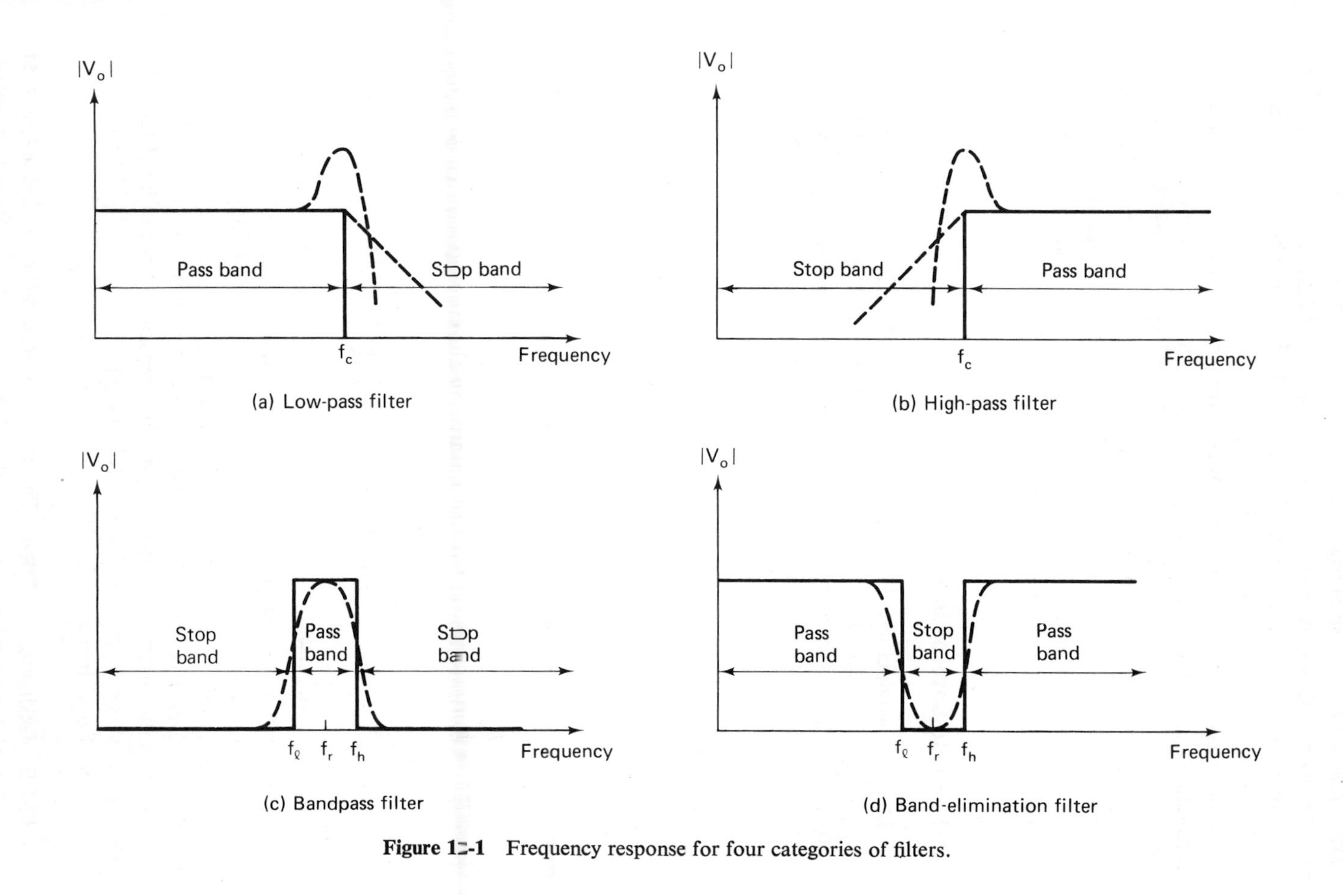

Figure 11-1 Frequency response for four categories of filters.

The solid line is the ideal curve, while the dashed curves show how practical high-pass filters deviate from the ideal.

Band-pass filters pass only a band of frequencies while attenuating all frequencies outside the band. Band-elimination filters perform in an exactly opposite way; that is, band-elimination filters reject a specified band of frequencies while passing all frequencies outside the band. Typical frequency-response plots for bandpass and band-climination filters are shown in Figs. 12-1(c) and 12-1(d). Once again, the solid line represents the ideal plot, while dashed lines show the practical curves.

12-1 Basic Low-pass Filter

12-1.1 Introduction. The circuit of Fig. 12-2(a) is a commonly used low-pass active filter. The filtering is done by the RC network, and the op amp is used as a unity-gain amplifier. The resistor R_f is equal to R and is included for dc offset. (At dc, the capacitive reactance is infinite and the dc resistance path to ground for both input terminals should be equal. See Section 9-4).)

The differential voltage between pins 2 and 3 is essentially 0 V. Therefore, the voltage across capacitor C equals output voltage V_o, because this circuit is a voltage follower. E_i divides between R and C. The capacitor voltage equals V_o and is

$$V_o = \frac{1/j\omega C}{R + 1/j\omega C} \times E_i \tag{12-1a}$$

where ω is the frequency of E_i in radians per second ($\omega = 2\pi f$) and j is equal to $\sqrt{-1}$. Rewriting Eq. (12-1a) to obtain the closed-loop voltage gain A_{CL}, we have

$$A_{CL} = \frac{V_o}{E_i} = \frac{1}{1 + j\omega RC} \tag{12-1b}$$

To show that the circuit of Fig. 12-2(a) is a low-pass filter, consider how A_{CL} in Eq. (12-1b) varies as frequency is varied. At very low frequencies, that is, as ω approaches 0, $|A_{CL}| = 1$, and at very high frequencies, as ω approaches infinity, $|A_{CL}| = 0$. ($|\quad|$ means magnitude.)

Figure 12-2b is a plot of $|A_{CL}|$ versus ω and shows that for frequencies *greater* than the cutoff frequency ω_c, $|A_{CL}|$ decreases at a rate of 20 db/decade. This is the same as saying that the voltage gain is divided by 10 when the frequency of ω is increased by 10.

12-1.2 Designing the Filter. The cutoff frequency ω_c is defined as that frequency of E_i where $|A_{CL}|$ is reduced to 0.707 times its low frequency value.

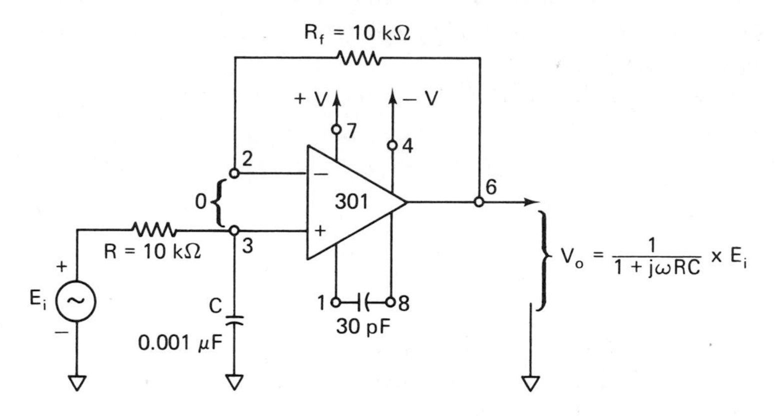

(a) Low-pass filter for a roll-off of − 20 db/decade

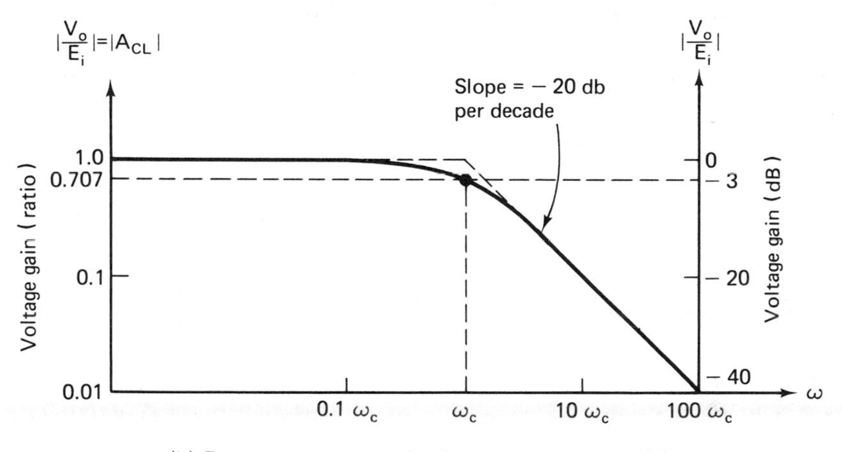

(b) Frequency-response plot for the circuit of part (a)

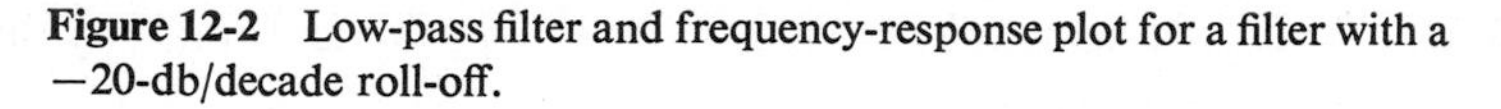

Figure 12-2 Low-pass filter and frequency-response plot for a filter with a −20-db/decade roll-off.

This important point will be discussed farther in Section 12-1.3. The cutoff frequency is evaluated from

$$\omega_c = \frac{1}{RC} = 2\pi f_c \tag{12-2a}$$

where ω_c is the cutoff frequency in radians per second, f_c is the cutoff frequency in Hertz, R is in ohms, and C is in farads. Equation (12-2a) may be rearranged to solve for C:

$$C = \frac{1}{\omega_c R} = \frac{1}{2\pi f_c R} \tag{12-2b}$$

Example 12-1: Let $R = 10\ \text{k}\Omega$ and $C = 0.001\ \mu\text{F}$ in Fig. 12-2(a); what is the cutoff frequency?
Solution: By Eq. (12-2a),

$$\omega_c = \frac{1}{(10 \times 10^3)(0.001 \times 10^{-6})} = 100\ \text{k rad/s}$$

or

$$f_c = \frac{\omega_c}{6.28} = \frac{100 \times 10^3}{6.28} = 15.9\ \text{kHz}$$

Example 12-2: For the low-pass filter of Fig. 12-2a, calculate C for a cutoff frequency of 2kHz and $R = 10\ \text{k}\Omega$.
Solution: From Eq. (12-2b),

$$C = \frac{1}{\omega_c R} = \frac{1}{(6.28)(2 \times 10^3)(10 \times 10^3)} \simeq 0.008\ \mu\text{F}$$

Example 12-3: Calculate C for Fig. 12-2(a) for a cutoff frequency of 30 k rad/s and $R = 10\ \text{k}\Omega$,
Solution: From Eq. (12-2b),

$$C = \frac{1}{\omega_c R} = \frac{1}{(30 \times 10^3)(10 \times 10^3)} = 0.0033\ \mu\text{F}$$

Conclusion: The design of a low-pass filter similar to Fig. 12-2(a) is accomplished in three steps:

1. Choose the cutoff frequency either ω_c or f_c
2. Choose the input resistance R, usually between $10\ \text{k}\Omega$ and $100\ \text{k}\Omega$
3. Calculate C from Eq. (12-2b)

12-1.3 Filter Response. The value of A_{CL} at ω_c is found by letting $\omega RC = 1$ in Eq. (12-1b).

$$A_{CL} = \frac{1}{1 + j1} = \frac{1}{\sqrt{2}\angle 45^\circ} = 0.707\angle -45^\circ$$

Therefore, the magnitude of A_{CL} at ω_c is

$$|A_{CL}| = \frac{1}{\sqrt{2}} = 0.707 = -3\ \text{db}$$

and the phase angle is -45°.

The solid curve in Fig. 12-2(b) shows how the magnitude of the actual frequency response deviates from the straight dashed line approximation in the vicinity of ω_c. At $0.1\omega_c$, $|A_{CL}| = 1$ (0 db), and at $10\ \omega_c$, $|A_{CL}| = 0.1$

(−20 db). Table 12-1 gives both the magnitude and the phase angle for different values of ω between 0.1 ω_c and 10 ω_c.

Table 12-1 MAGNITUDE AND PHASE ANGLE FOR THE LOW-PASS FILTER OF FIG. 12-2a.

ω	$\lvert A_{CL} \rvert$	*Phase Angle*
$0.1\omega_c$	1.0	−6°
$0.25\omega_c$	0.97	−14°
$0.5\omega_c$	0.89	−27°
ω_c	0.707	−45°
$2\omega_c$	0.445	−63°
$4\omega_c$	0.25	−76°
$10\omega_c$	0.1	−84°

Many applications require steeper roll-offs after the cutoff frequency. One common filter configuration that gives steeper roll-offs is the *Butterworth filter*.

12-2 Introduction to the Butterworth Filter

In many low-pass filter applications, it is necessary for the closed-loop gain to be as close to 1 as possible within the pass band. The *Butterworth filter* is best suited for this type of application. The Butterworth filter is also called a *maximally flat* or *flat-flat* filter, and all filters in this chapter will be of the Butterworth type. Figure 12-3 shows the ideal (solid line) and the

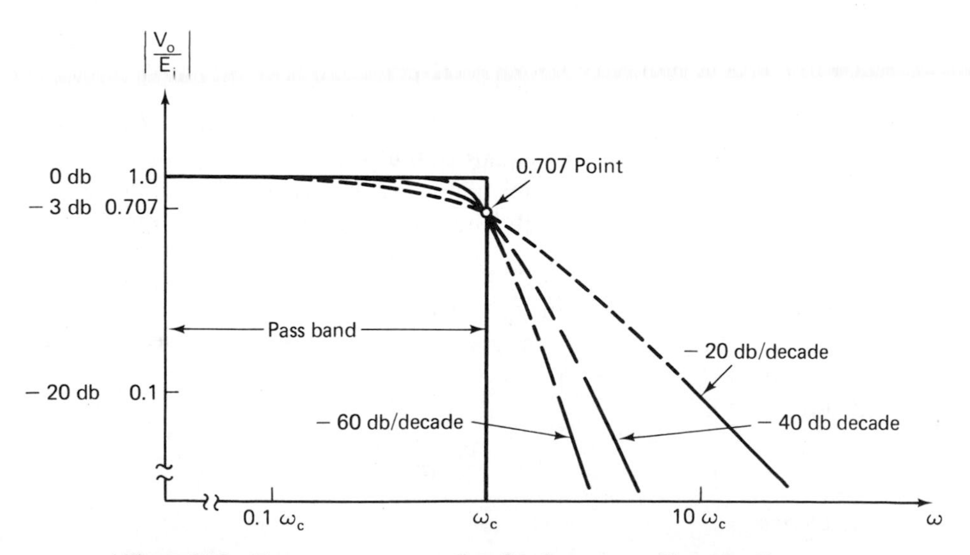

Figure 12-3 Frequency-response plots for three types of low-pass Butterworth filters.

practical (dashed lines) frequency response for three types of Butterworth filters. As the roll-offs become steeper, they approach the ideal filter more closely.

Two active filters similar to Fig. 12-2(a) could be coupled together to give a roll-off of —40 db/decade. This would not be the most economical design, because it would require two op amps. In Section 12-3.1, it is shown how one op amp can be used to build a Butterworth filter with a single op amp to give a —40-db/decade roll-off. Then in Section 12-4, a —40-db/decade filter will be cascaded with a —20-db/decade filter to produce a —60-db/decade filter.

Butterworth filters are not designed to keep a constant phase angle at the cutoff frequency. A basic low-pass filter of —20-db/decade has a phase angle of —45° at ω_c. A —40-db/decade Butterworth filter has a phase angle of —90° at ω_c and a —60-db/decade filter has a phase angle of —135° at ω_c. Therefore, for each increase of —20-db/decade, the phase angle will increase by —45°. We now proceed to a Butterworth filter that has a roll-off steeper than —20 db/decade.

12-3 −40-db/decade Low-pass Butterworth Filter

12-3.1 Simplified Design Procedure. The circuit of Fig. 12-4(a) is one of the most commonly used low-pass filters. It produces a roll-off of —40 db/decade; that is, after the cut-off frequency, the magnitude of A_{CL} decreases by 40 db as ω increases to $10\omega_c$. The solid line in Fig. 12-4(b) shows the actual frequency-response plot, which is explained in more detail in Section 12-3.2.

The op amp is connected for dc unity gain. Resistor R_f is included for dc offset, as explained in Section 9-4.

Since the op amp circuit is basically a voltage follower (unity-gain amplifier), the voltage across C_1 equals output voltage, V_o.

The design of the low-pass filter of Fig. 12-4(a) is greatly simplified by making resistors R_1 and R_2 equal. There are only four steps in the design procedure:

1. Choose the cutoff frequency ω_c or f_c
2. Let $R_1 = R_2 = R$, and choose a convenient value between 10 kΩ and 100 kΩ. Choose $R_f = 2R$.
3. Calculate C_1 from

$$C_1 = \frac{0.707}{\omega_c R} \tag{12-3}$$

4. Choose

$$C_2 = 2C_1 \tag{12-4}$$

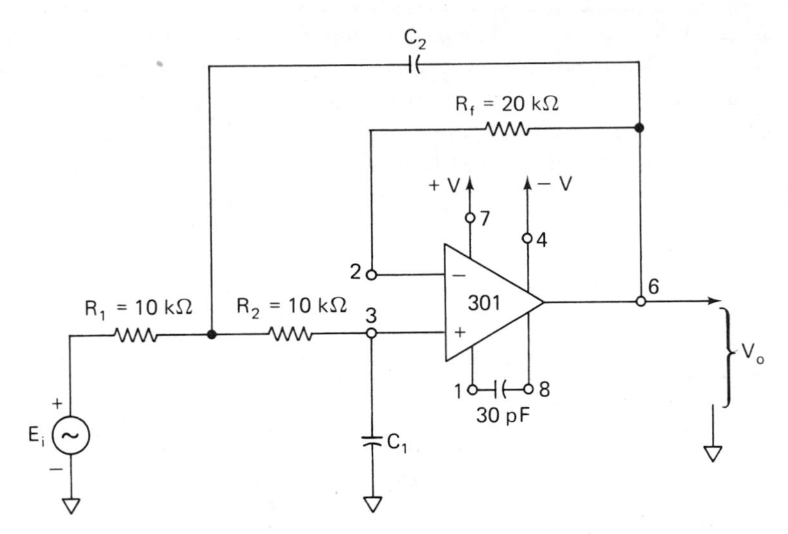

(a) Low-pass filter for a roll-off of – 40 db/decade

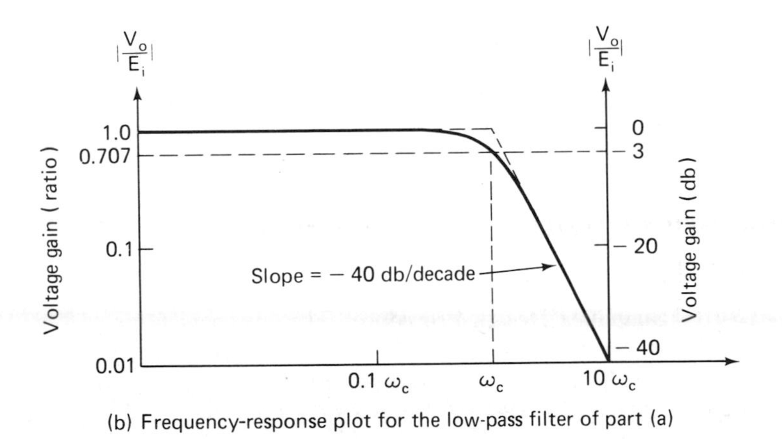

(b) Frequency-response plot for the low-pass filter of part (a)

Figure 12-4 Circuit and frequency plot for a low-pass filter of –40-db/decade.

Example 12-4: Determine (a) C_1 and (b) C_2 in Fig. 12-4(a) for a cutoff frequency of 30 k rad/s. Let $R_1 = R_2 = R = 10\ \text{k}\Omega$.

Solution: (a) From Eq. (12-3),

$$C_1 = \frac{0.707}{(30 \times 10^3)(10 \times 10^3)} = 0.0024\ \mu\text{F}$$

(b) $C_2 = 2C_1 = 2(0.0024\ \mu\text{F}) = 0.0048\ \mu\text{F}$.

12-3.2 Filter Response. The dashed curve in Fig. 12-4(b) shows that the filter of Fig. 12-4(a) not only has a steeper roll-off after ω_c than does Fig. 12-2(a) but also remains at 0 db almost up to about $0.25\omega_c$. The phase angles for the circuit of Fig. 12-4(a) range from 0° at $\omega = 0$ rad/s (dc condition) to $-180°$ as ω approaches ∞ (infinity). Table 12-2 compares magnitude and phase angle for the low-pass filters of Figs. 12-2(a) and 12-4(a) from $0.1\omega_c$ to $10\omega_c$.

Table 12-2 MAGNITUDE AND PHASE ANGLE FOR FIGS. 12-2(a) AND 12-4(a)

	$\lvert A_{CL}\rvert$		*Phase Angle*	
ω	*Fig. 12-2(a)*	*Fig. 12-4(a)*	*Fig. 12-2(a)*	*Fig. 12-4(a)*
$0.1\omega_c$	1.0	1.0	−6°	−8°
$0.25\omega_c$	0.97	0.998	−14°	−21°
$0.5\omega_c$	0.89	0.97	−27°	−43°
ω_c	0.707	0.707	−45°	−90°
$2\omega_c$	0.445	0.24	−63°	−137°
$4\omega_c$	0.25	0.053	−76°	−143°
$10\omega_c$	0.1	0.01	−84°	−172°

The next low-pass filter cascades the filter of Fig. 12-2(a) with the filter of Fig. 12-4(a) to form a roll-off of −60 db/decade.

As will be shown, the capacitiors are the only values that have to be calculated.

12-4 −60 db/decade Low-pass Butterworth Filter

12-4.1 Simplified Design Procedure. The low-pass filter of Fig. 12-5(a) is built using one low-pass filter of −40 db/decade cascaded with another of −20 db/decade to give an overall roll-off of −60 db/decade. The overall closed-loop gain A_{CL} is the gain of the first filter times the gain of second filter, or

$$A_{CL} = \frac{V_o}{E_i} = \frac{V_{o1}}{E_i} \times \frac{V_o}{V_{o1}} \tag{12-5}$$

For a Butterworth filter, the magnitude of A_{CL} must be 0.707 at ω_c; to guarantee that the frequency response is flat in the pass band, use the following design steps:

1. Choose the cutoff frequency ω_c or f_c
2. Choose the input resistors to be equal ($R_1 = R_2 = R_3 = R$); values between 10 kΩ and 100 kΩ are typical

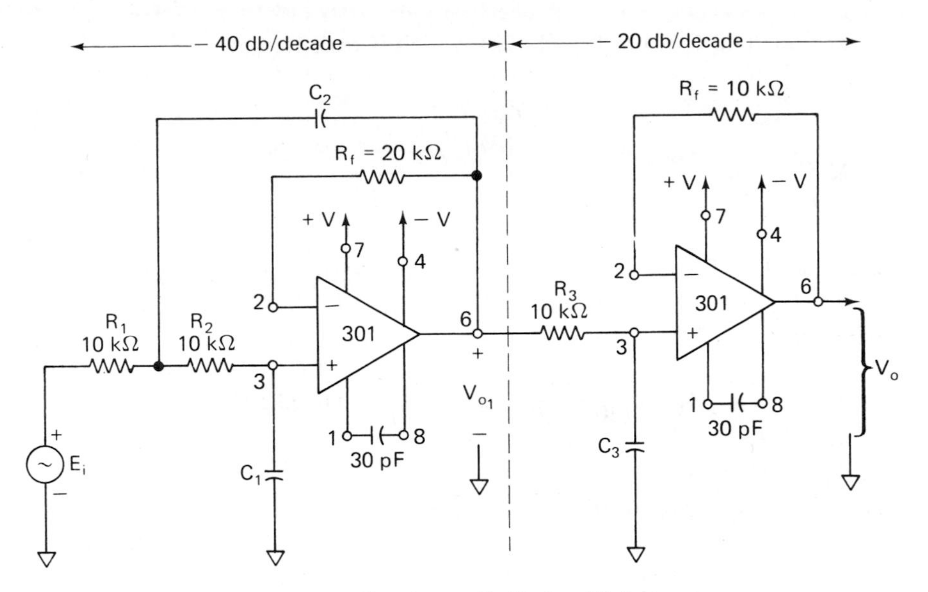

(a) Low-pass filter for a roll off of – 60 db/decade

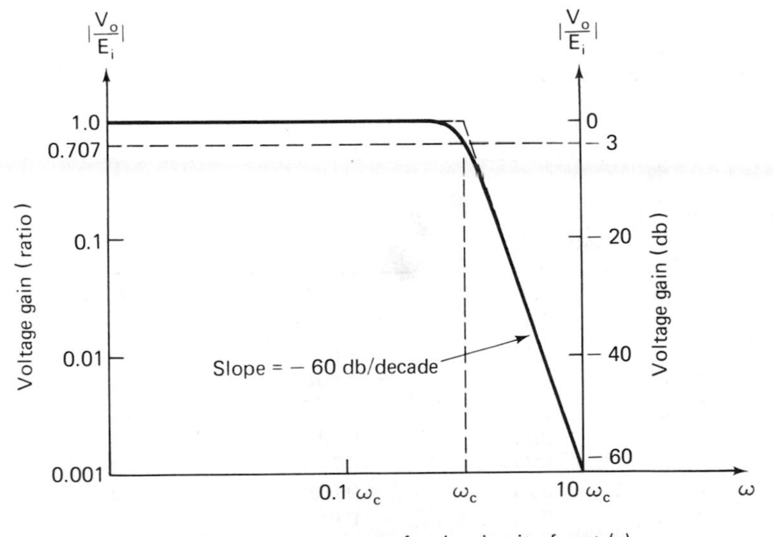

(b) Plot of frequency response for the circuit of part (a)

Figure 12-5 Low-pass filter designed for a roll-off of –60-db/decade and corresponding frequency-response plot.

3. Calculate C_3 from Eq. (12-2b), which is rewritten as

$$C_3 = \frac{1}{\omega_c R} \tag{12-6}$$

4. $C_1 = \frac{1}{2}C_3$ (12-7)
5. $C_2 = 2C_3$ (12-8)

Example 12-5: For the low-pass filter of Fig. 12-5(a), calculate (a) C_3, (b) C_1, and (3) C_2 for a cutoff frequency of 30 k rad/s. $R_1 = R_2 = R_3 = 10\ \text{k}\Omega$.
Solution: From Eq. (12-6),

$$C_3 = \frac{1}{(30 \times 10^3)(10 \times 10^3)} = 0.0033\ \mu F$$

(b) $C_1 = \frac{1}{2}C_3 = 0.0017\ \mu\text{F}$.
(c) $C_2 = 2C_3 = 0.0066\ \mu\text{F}$.

Example 12-5 shows that the capacitors C_1 and C_2 of Fig. 12-5(a) are different from those of Fig. 12-4(a), although the cutoff frequency is the same. This is necessary so that $|A_{CL}|$ remains at 0 db in the pass band until the cutoff frequency is nearly reached; then $|A_{CL}| = 0.707$ at ω_c.

12-4.2 Filter Response. The solid line in Fig. 12-5(b) is the actual plot of the frequency response for Fig. 12-5(a). The dashed curve in the vicinity shows the straight line approximation. Table 12-3 compares the magnitudes of A_{CL} for the three low-pass filters presented in this chapter. Note that the $|A_{CL}|$ for Fig. 12-5(a) remains quite close to 1 (0 db) until the cutoff frequency, ω_c; then the steep roll-off occurs.

Table 12-3 $|A_{CL}|$ FOR THE LOW-PASS FILTERS OF FIGS. 12-2(a), 12-4(a), AND 12-5(a)

ω	*−20 db/decade; Fig. 12-2(a)*	*−40 db/decade; Fig. 12-4(a)*	*−60 db/decade; Fig. 12-5(a)*
$0.1\omega_c$	1.0	1.0	1.0
$0.25\omega_c$	0.97	0.998	0.999
$0.5\omega_c$	0.89	0.97	0.992
ω_c	0.707	0.707	0.707
$2\omega_c$	0.445	0.24	0.124
$4\omega_c$	0.25	0.053	0.022
$10\omega_c$	0.1	0.01	0.001

The phase angles for the low-pass filter of Fig. 12-5(a) range from 0° at $\omega = 0$ (db condition) to −270° as ω approaches ∞. Table 12-4 compares the phase angles for the three low-pass filters.

Table 12-4 PHASE ANGLES FOR THE LOW-PASS FILTERS OF FIGS. 12-2(a), 12-4(a), AND 12-5(a)

ω	-20 *db/decade; Fig. 12-2(a)*	-40 *db/decade; Fig. 12-4(a)*	-60 *db/decade; Fig. 12-5(a)*
$0.1\omega_c$	$-6°$	$-8°$	$-12°$
$0.25\omega_c$	$-14°$	$-21°$	$-29°$
$0.5\omega_c$	$-27°$	$-43°$	$-60°$
ω_c	$-45°$	$-90°$	$-135°$
$2\omega_c$	$-63°$	$-137°$	$-210°$
$4\omega_c$	$-76°$	$-143°$	$-226°$
$10\omega_c$	$-84°$	$-172°$	$-256°$

12-5 High-pass Butterworth Filters

12-5.1 Introduction. A high-pass filter is a circuit that attenuates all signals below a specified cutoff frequency ω_c and passes all signals whose frequency is above the cutoff frequency. Thus, a high-pass filter performs the opposite function of the low-pass filter.

Figure 12-6 is a plot of the magnitude of the closed-loop gain versus ω for three types of Butterworth filters. The phase angle for a circuit of 20 db/decade is $+45°$ at ω_c. Phase angles at ω_c increase by $+45°$ for each increase of 20 db/decade. Comparison of phase angles for these three types of high-pass filters will be shown in Section 12-5.5.

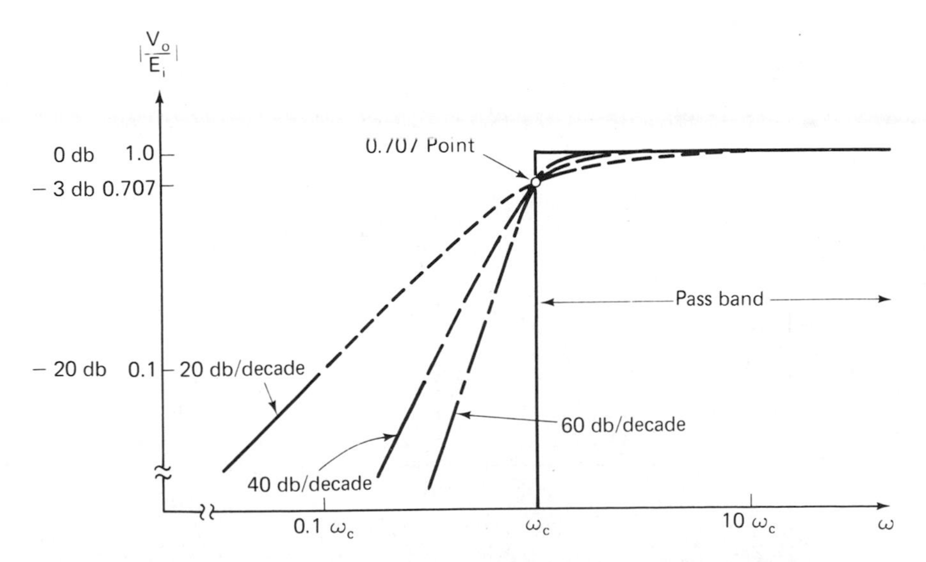

Figure 12-6 Comparison of frequency response for three high-pass Butterworth filters.

In this text, the design of high-pass filters will be similar to that of the low-pass filters. In fact, the only difference will be the position of the filtering capacitors and resistors.

12-5.2 20-db/decade Filter. Compare the high-pass filter of Fig. 12-7(a) with the low-pass filter of Fig. 12-2(a) and note that C and R are interchanged. The feedback resistor R_f is included to minimize dc offset. Since the op amp is connected as a unity-gain follower in Fig. 12-7(a), the output voltage V_o

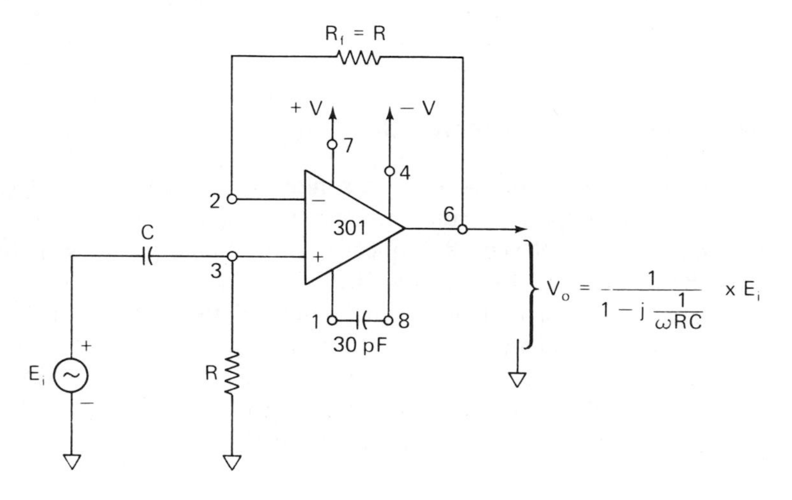

(a) High-pass filter with a roll-off of 20 db/decade

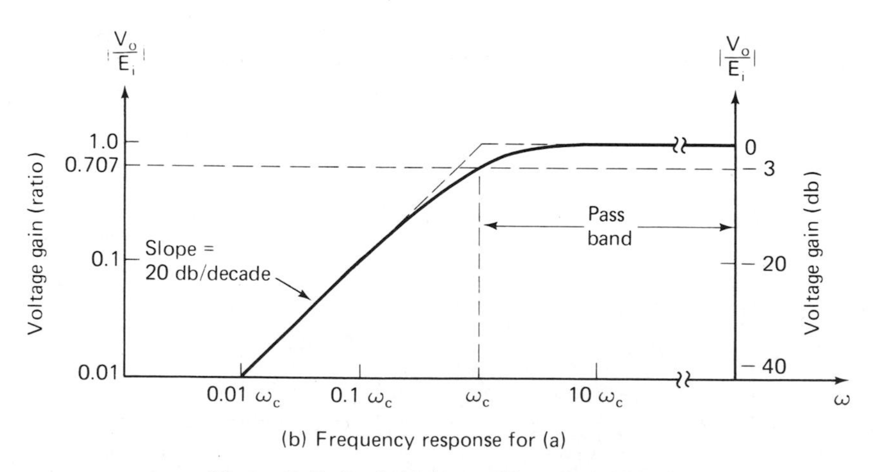

(b) Frequency response for (a)

Figure 12-7 Basic high-pass filter, 20-db/decade.

equals the voltage across R and is expressed by

$$V_o = \frac{1}{1 - j\frac{1}{\omega RC}} \times E_i \qquad (12\text{-}9)$$

When ω approaches 0 rad/s in Eq. (12-9), V_o approaches 0 V. At high frequencies, as ω approaches infinity, V_o equals E_i. Since the circuit is not an ideal filter, the frequency response is not ideal as shown by Fig. 12-7(b). The solid line is the actual response, while the dashed lines show the straight line approximation. The magnitude of the closed-loop gain equals 0.707 when $\omega RC = 1$. Therefore, the cutoff frequency, ω_c, is given by

$$\omega_c = \frac{1}{RC} = 2\pi f_c \qquad (12\text{-}10a)$$

or

$$R = \frac{1}{\omega_c C} = \frac{1}{2\pi f_c C} \qquad (12\text{-}10b)$$

The reason for solving for R and not C in Eq. (12-10b) is that for high-pass filters, usually C is chosen along with ω_c and R is calculated. The steps needed in designing Fig. 12-7(a) are

1. Choose the cutoff frequency ω_c or f_c
2. Choose a convenienent value of C
3. Calculate R from Eq. (12-10b)
4. Choose $R_f = R$

Example 12-6: Calculate R in Fig. 12-7(a) if $C = 0.002\ \mu$F and $f_c = 10$ kHz.
Solution: From Eq. (12-10b),

$$R = \frac{1}{(6.28)(10 \times 10^3)(0.002 \times 10^{-6})} = 8\ \text{k}\Omega$$

Example 12-7: In Fig. 12-7(a) if $R = 22\ \text{k}\Omega$ and $C = 0.01\ \mu$F, calculate (a) ω_c and (b) f_c.
Solution: (a) From Eq. (12-10a),

$$\omega_c = \frac{1}{(22 \times 10^3)(0.01 \times 10^{-6})} = 4.54\ \text{k rad/s}$$

(b)

$$f_c = \frac{\omega_c}{2\pi} = \frac{4.54 \times 10^3}{6.28} = 724\ \text{Hz}$$

12-5.3 40-db-decade Filter. The circuit of Fig. 12-8(a) is to be designed as a high-pass Butterworth filter with a roll-off of 40 db/decade below the

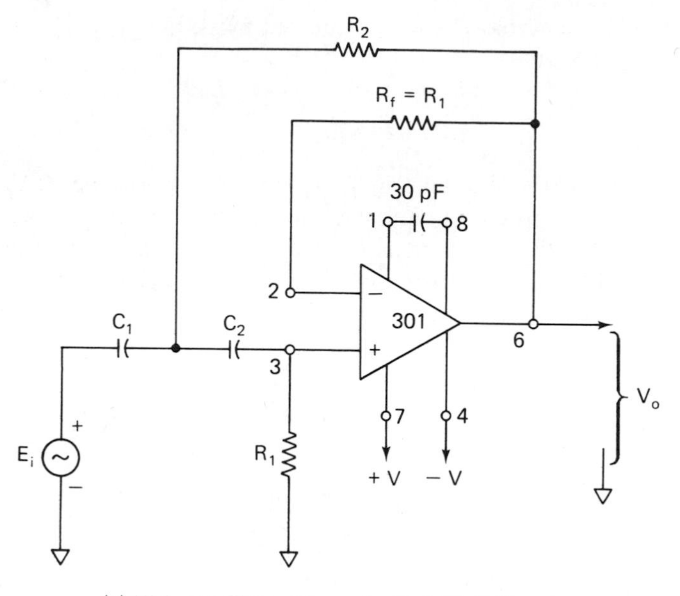

(a) High-pass filter with a roll-off of 40 db/decade

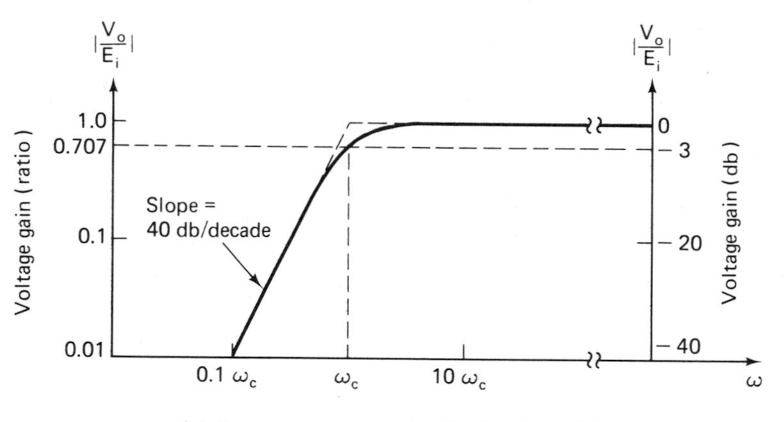

(b) Frequency response for circuit of part (a)

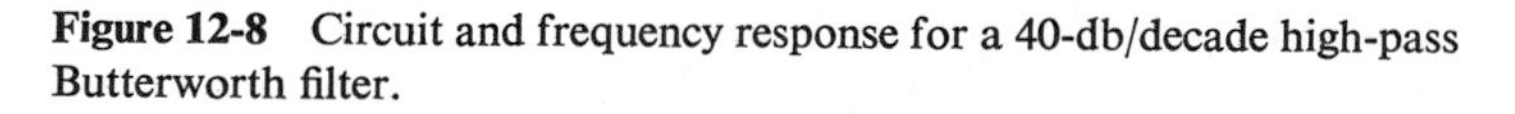

Figure 12-8 Circuit and frequency response for a 40-db/decade high-pass Butterworth filter.

cutoff frequency, ω_c. To satisfy the Butterworth criteria, the frequency response must be 0.707 at ω_c and be 0 db in the pass band. These conditions will be met if the following design procedure is followed:

1. Choose a cutoff frequency ω_c or f_c
2. Let $C_1 = C_2 = C$ and choose a convenient value

3. Calculate R_1 from

$$R_1 = \frac{1.414}{\omega_c C} \tag{12-11}$$

4. $R_2 = \frac{1}{2}R_1$ (12-12)
5. To minimize dc offset, let $R_f = R_1$

Example 12-8: In Fig. 12-8(a), let $C_1 = C_2 = 0.01\ \mu F$. Calculate (a) R_1 and (b) R_2 for a cutoff frequency of 1 kHz.
Solution: (a) From Eq. (12-11),

$$R_1 = \frac{1.414}{(6.28)(1 \times 10^3)(0.01 \times 10^{-6})} = 22.5\ k\Omega$$

(b) $R_2 = \frac{1}{2}(22.5\ k\Omega) = 11.3\ k\Omega$.

Example 12-9: Calculate (a) R_1 and (b) R_2 in Fig. 12-8(a) for a cutoff frequency of 80 k rad/s. $C_1 = C_2 = 125$ pF.
Solution: (a) From (12-11),

$$R_1 = \frac{1.414}{(80 \times 10^3)(125 \times 10^{-12})} = 140\ k\Omega$$

(b) $R_2 = \frac{1}{2}(140\ k\Omega) = 70\ k\Omega$

12-5.4 60-db/decade Filter. As with the low-pass filter of Fig. 12-5, a high-pass filter of 60 db/decade can be constructed by cascading a 40-db/decade filter with a 20-db/decade filter. This circuit (like the other high- and low-pass filters) is designed as a Butterworth filter to have the frequency response in Fig. 12-9(b). The design steps for Fig. 12-9(a) are

1. Choose the cutoff frequency ω_c or f_c
2. Let $C_1 = C_2 = C_3 = C$ and choose a convenient value
3. Calculate R_3 from

$$R_3 = \frac{1}{\omega_c C} \tag{12-13}$$

4. Let $R_1 = 2R_3$ (12-14)
5. Let $R_2 = \frac{1}{2}R_3$ (12-15)
6. To minimize dc offset current, let $R_{f1} = R_1$ and $R_{f2} = R_3$

Example 12-10: For Fig. 12-9(a), let $C_1 = C_2 = C_3 = C = 0.1\ \mu F$. Determine (a) R_3, (b) R_1, and (c) R_2 for a cutoff frequency of 1 k rad/s.
Solution: (a) By Eq. (12-13),

$$R_3 = \frac{1}{(1 \times 10^3)(0.1 \times 10^{-6})} = 10\ k\Omega$$

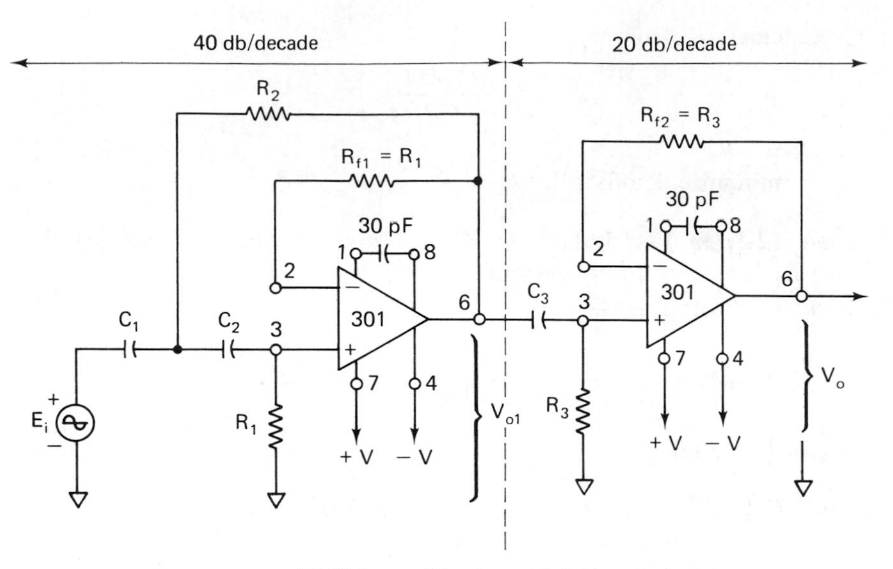

(a) High-pass filter for a 60-db/decade slope

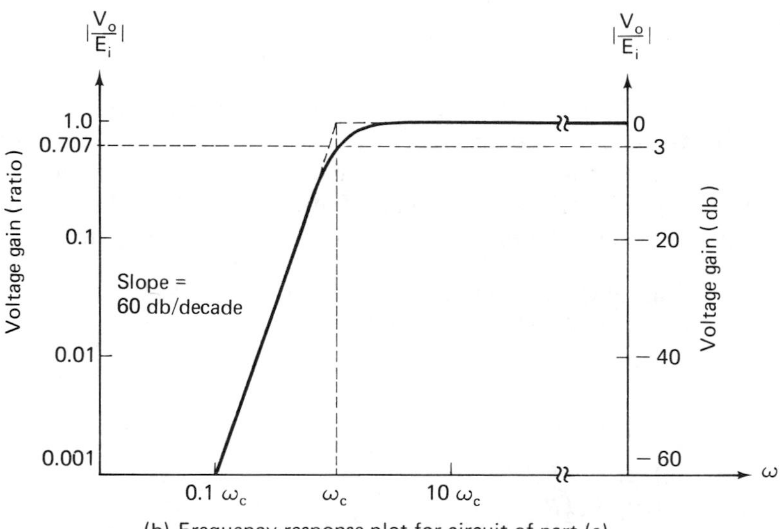

(b) Frequency-response plot for circuit of part (a)

Figure 12-9 Circuit and frequency response for a 60-db/decade Butterworth high-pass filter.

(b) $R_1 = 2R_3 = 2(10\ \text{k}\Omega) = 20\ \text{k}\Omega$
(c) $R_2 = \frac{1}{2}R_3 = \frac{1}{2}(10\ \text{k}\Omega) = 5\ \text{k}\Omega$

Example 12-11: Determine (a) R_3, (b) R_1 and (c) R_2 in Fig. 12-9(a) for a cutoff frequency of 60 kHz. Let $C_1 = C_2 = C_3 = C = 220$ pF.

Solution: (a) From Eq. (12-13),

$$R_3 = \frac{1}{(6.28)(60 \times 10^3)(220 \times 10^{-12})} = 12\text{ k}\Omega$$

(b) $R_1 = 2R_3 = 2(12\text{ k}\Omega) = 24\text{ k}\Omega$
(c) $R_2 = \frac{1}{2}R_3 = \frac{1}{2}(12\text{ k}\Omega) = 6\text{ k}\Omega$

If desired, the 20-db/decade section can come before the 40-db/decade section, because the op amps provide isolation and do not load one another.

12-5.5 Comparison of Magnitudes and Phase Angles. Table 12-5 compares the magnitudes of the closed-loop gain for the three high-pass filters. For each increase in 20 db/decade, the circuit not only has a steeper roll-off below ω_c but also remains closer to 0 db above ω_c.

Table 12-5 COMPARISON OF $|A_{CL}|$ FOR FIGS. 12-7(a), 12-8(a), AND 12-9(a)

ω	*20 db/decade; Fig. 12-7 (a)*	*40 db/decade; Fig. 12-8 (a)*	*60 db/decade; Fig. 12-9 (a)*
$0.1\omega_c$	0.1	0.01	0.001
$0.25\omega_c$	0.25	0.053	0.022
$0.5\omega_c$	0.445	0.24	0.124
ω_c	0.707	0.707	0.707
$2\omega_c$	0.89	0.97	0.992
$4\omega_c$	0.97	0.998	0.999
$10\omega_c$	1.0	1.0	1.0

The phase angle for a 20-db/decade Butterworth high-pass filter is 45° at ω_c. For a 40 db/decade filter it is 90°, and for a 60-db/decade filter it is 135°. Other phase angles in the vicinity of ω_c for the three filters are given in Table 12-6.

Table 12-6 COMPARISON OF PHASE ANGLES FOR FIGS. 12-7(a), 12-8(a), AND 12-9(a)

ω	*20 db/decade; Fig. 12-7(a)*	*40 db/decade; Fig. 12-8(a)*	*60 db/decade; Fig. 12-9(a)*
$0.1\omega_c$	84°	172°	256°
$0.25\omega_c$	76°	143°	226°
$0.5\omega_c$	63°	137°	210°
ω_c	45°	90°	135°
$2\omega_c$	27°	43°	60°
$4\omega_c$	14°	21°	29°
$10\omega_c$	6°	8°	12°

12-6 Band-pass Filters

12-6.1 Introduction. A *band-pass* filter is a circuit designed to pass signals only in a certain band of frequencies while rejecting all signals outside this band. Figures 12-1(c) and 12-10(a) show the frequency response of a band-pass filter. This type of filter has a maximum output voltage, V_{max}, or maximum voltage gain, A_r, at one frequency called the *resonant frequency*,

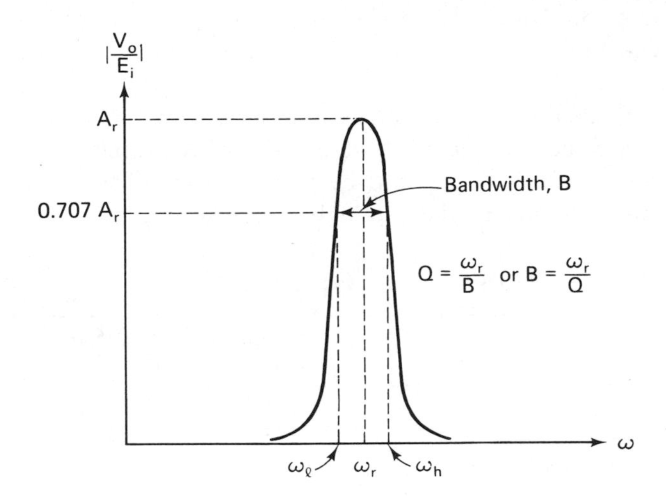

(a) Frequency response of a bandpass filter

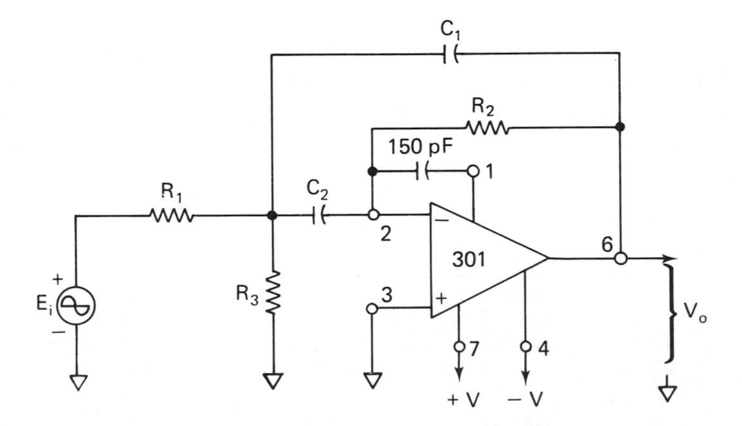

(b) Bandpass filter

Figure 12-10 Band-pass filter and frequency response.

ω_r. If the frequency varies from resonance, the output voltage decreases. There is one frequency above ω_r and one below ω_r at which the voltage gain is $0.707A_r$. These frequencies are designated by ω_h, the *high cutoff frequency*, and ω_l, the *low cutoff frequency*. The band of frequencies between ω_h and ω_l is the *bandwidth*, B

$$B = \omega_h - \omega_l \tag{12-16}$$

Band-pass filters are classified as either narrow-band or wide-band. A narrow-band filter is one that has a bandwidth of less than one-tenth the resonant frequency ($B < 0.1\,\omega_r$). If the bandwidth is greater than one-tenth the resonant frequency ($B > 0.1\,\omega_r$), the filter is a wide-band filter. The ratio of resonant frequency to band width is known as the *quality factor*, Q, of the circuit. Q indicates the selectivity of the circuit. The higher the value of Q, the more selective the circuit. In equation form,

$$Q = \frac{\omega_r}{B} \tag{12-17a}$$

or

$$B = \frac{\omega_r}{Q} \tag{12-17b}$$

For narrow-band filters, the Q of the circuit is greater than 10, and for wide-band filters, Q is less than 10.

12-6.2 Narrow-band Band-pass Filters. The circuit of Fig. 12-10(b) can be designed as either a wide-band filter ($Q < 10$) or as a narrow-band filter ($Q > 10$). Unlike either the low-pass or the high-pass filters of Sections 12-1 thru 12-5, the filter of Fig. 12-10(b) can be designed for a closed-loop gain greater than 1. The maximum gain, A_r, occurs at the resonant frequency as shown in Fig. 10-12(a). Normally, the designer of a band-pass filter first chooses the resonant frequency ω_r and the bandwidth B and calculates Q from Eq. (12-17a). For some designs, ω_r and Q are chosen and the bandwidth B is calculated from Eq. (12-17b). To simplify the design and reduce the number of calculations, choose $C_1 = C_2 = C$ and solve for R_1, R_2, and R_3 from the following equations:

$$R_2 = \frac{2}{BC} \tag{12-18}$$

$$R_1 = \frac{R_2}{2A_r} \tag{12-19}$$

$$R_3 = \frac{R_2}{4Q^2 - 2A_r} \tag{12-20}$$

To guarantee that R_3 is a positive value, be sure that $4Q^2 > 2A_r$. B in Eq. (12-18) is in radians per second.

Example 12-12: Design the band-pass filter of Fig. 12-10(b) to have $\omega_r = 10$ k rad/s, $A_r = 40$, $Q = 20$, and $C_1 = C_2 = C = 0.01\ \mu$F.
Solution: By Eq. (12-17b),

$$B = \frac{10 \times 10^3}{20} = 0.5 \text{ k rad/s}$$

From Eqs. (12-18) to (12-20),

$$R_2 = \frac{2}{(0.5 \times 10^3)(0.01 \times 10^{-6})} = 400 \text{ k}\Omega$$

$$R_1 = \frac{400 \times 10^3}{2(40)} = 5\text{k}\Omega$$

$$R_3 = \frac{400 \times 10^3}{4(400) - 2(40)} = 263\ \Omega$$

Example 12-13: If the band-width of Example 12-12 is to be increased to 1 k rad/sec, calculate (a) Q, (b) R_2, (c) R_1, and (d) R_3.
Solution: (a) By Eq. (12-17a),

$$Q = \frac{10 \times 10^3}{1 \times 10^3} = 10$$

(b) By Eq. (12-18),

$$R_2 = \frac{2}{(1 \times 10^3)(0.01 \times 10^{-6})} = 200 \text{ k}\Omega$$

(c) By Eq. (12-19),

$$R_1 = \frac{200 \times 10^3}{2(40)} = 2.5 \text{ k}\Omega$$

(d) By Eq. (12-20),

$$R_3 = \frac{200 \times 10^3}{4(100) - 2(40)} = 625\ \Omega$$

12-6.3 Wide-band Filters. As stated previously, a wide-band band-pass filter is a circuit in which $Q < 10$. The circuit of Fig. 12-10(b) can be designed as a wide-band filter, and Eqs. (12-18), (12-19), and (12-20) can be used, provided that $4Q^2 > 2A_r$.

Example 12-14: Design Fig. 12-10(b) to have $\omega_r = 20$ k rad/s, $A_r = 10$, $Q = 5$, and $C_1 = C_2 = C = 0.01\ \mu$F.

Solution: By Eq. (12-17b),

$$B = \frac{20 \times 10^3}{5} = 4 \text{ k rad/s}$$

From Eqs. (12-18) to (12-20),

$$R_2 = \frac{2}{(4 \times 10^3)(0.01 \times 10^{-6})} = 50 \text{ k}\Omega$$

$$R_1 = \frac{50 \times 10^3}{2(10)} = 2.5 \text{ k}\Omega$$

$$R_3 = \frac{50 \times 10^3}{4(25) - 2(10)} = 625 \ \Omega$$

Another idea for a wide-band filter is to connect a low-pass filter to a high-pass filter. For example, the low-pass filter of Example 12-5 connected to the high-pass filter of Example 12-10 gives the frequency response shown in Fig. 12-11. Although this wide-band filter uses four op amps, the roll-off is 60 db/decade at both the high and low cutoff frequencies. The gain in the pass band is 1, because the gain for both the low- and high-pass filter is 1. When one is building this type of wide-band filter, it makes no difference which filter comes first.

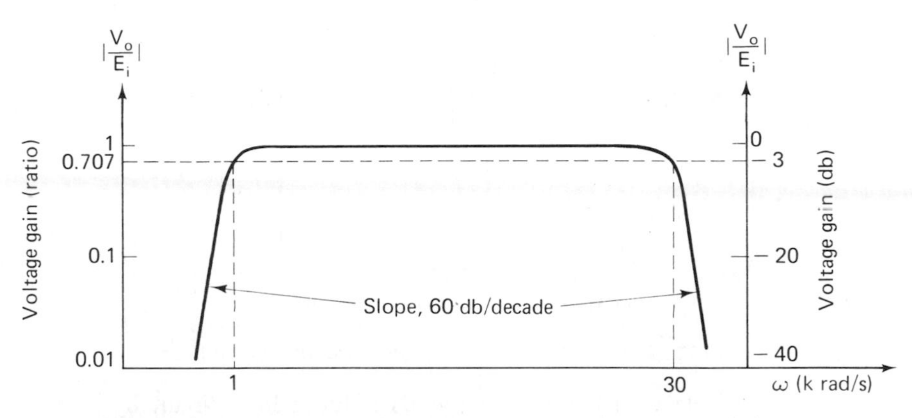

Figure 12-11 Wide-band filter response obtained by connecting Fig. 12-5(a) to Fig. 12-9(a).

12-7 Notch Filters

The circuit of Fig. 12-12(a) is a *notch* or *band-elimination filter*. Its frequency-response curve is shown in Fig. 12-1(d) and 12-12(b). Undesired frequencies are attenuated in the stop band. For example, it may be necessary

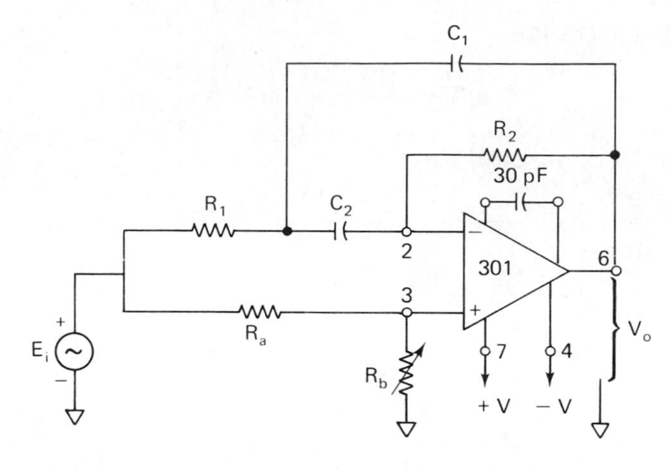

(a) Notch filter

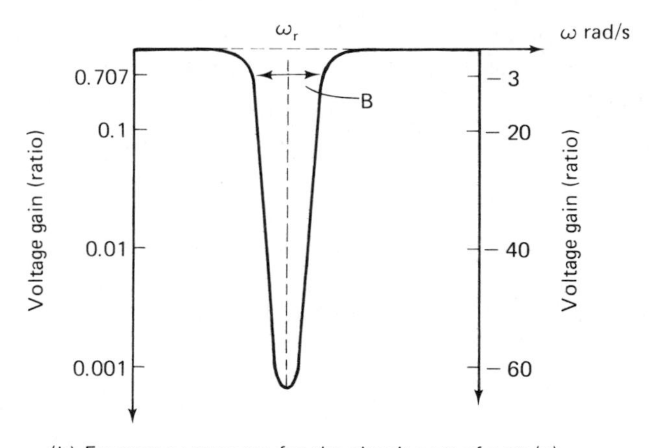

(b) Frequency response for the circuit part of part (a)

Figure 12-12 Circuit and frequency response for a notch filter.

to attenuate 60-Hz or 400-Hz noise signals induced in a circuit by motor generators. Design of the notch filter is carried out in 5 steps. You usually know or are designing for a required bandwidth B or Q and resonant frequency ω_r. Then proceed as follows:

1. Choose $C_1 = C_2 = C$ (some convenient value)
2. Calculate R_2 from

$$R_2 = \frac{2}{BC} \tag{12-21}$$

B is in radians per second

3. Calculate R_1 from

$$R_1 = \frac{R_2}{4Q^2} \tag{12-22}$$

4. Choose R_a, a convenient value such as 1 kΩ
5. Calculate R_b from

$$R_b = 2Q^2 R_a \tag{12-23}$$

This procedure is illustrated by an example.

Example 12-15: Design a notch filter from Fig. 12-12(a) for $f_r = 400$ Hz and $Q = 5$. Let $C_1 = C_2 = C = 0.01\ \mu$F.
Solution: $\omega_r = 2\pi f_r = (6.28)(400) = 2.51$ k rad/s. From Eq. (12-17a),

$$B = \frac{2.51 \times 10^3}{5} \cong 500 \text{ rad/s}$$

From Eq. (12-21),

$$R_2 = \frac{2}{(500)(0.01 \times 10^{-6})} = 400 \text{ k}\Omega$$

From Eq. (12-22),

$$R_1 = \frac{400 \text{ k}\Omega}{4(25)} = 4 \text{ k}\Omega$$

Choose $R_a = 1$ kΩ and from Eq. (12-23); $R_b = 2(25)\ 1$ kΩ $= 50$ kΩ.

When building the notch filter of Fig. 12-12, the following procedure should be used:

1. Ground the (+) terminal of the op amp. The resulting network is a band-pass filter similar to Fig. 12-10(b) but without R_3. The gain for this band-pass filter at ω_r is $2Q^2$. (For Example 12-15, the gain is 50). Adjust R_1 and R_2 to fine-tune ω_r and B.
2. Remove the ground at the (+) input and adjust R_b to the value obtained from Eq. (12-23).

The frequency response of a notch filter is shown in Fig. 12-12(b). Note that the band-width is still that band of frequencies at −3 db from the maximum value.

Problems

12-1 List the four types of filters.

12-2 What type of filter has a constant output voltage from dc up to the cutoff frequency?

12-3 What is a filter that passes a band of frequencies while attenuating all frequencies outside the band called?

12-4 In Fig. 12-2(a), if $R = 100\ \text{k}\Omega$ and $C = 0.02\ \mu\text{F}$, what is the cutoff frequency?

12-5 The low-pass filter of Fig. 12-2(a) is to be designed for a cutoff frequency of 4.5 kHz. If $C = 0.005\ \mu\text{F}$, calculate R.

12-6 If the cutoff frequency in Fig. 12-2(a) is 50 k rad/s and $R = 20\ \text{k}\Omega$, determine C.

12-7 What are the two characteristics of a Butterworth filter?

12-8 Design a -40-db/decade low-pass filter at a cutoff frequency of 10 k rad/s. Let $R_1 = R_2 = 50\ \text{k}\Omega$.

12-9 In Fig. 12-4(a), if $R_1 = R_2 = 10\ \text{k}\Omega$, $C_1 = 0.001\ \mu\text{F}$, and $C_2 = 0.002\ \mu\text{F}$, calculate the cutoff frequency f_c.

12-10 Calculate (a) C_3, (b) C_1 and (c) C_2 in Fig. 12-5(a) for a cutoff frequency of 10 k rad/s. $R_1 = R_2 = R_3 = 10\ \text{k}\Omega$.

12-11 If $R_1 = R_2 = R_3 = 20\ \text{k}\Omega$, $C_1 = 0.002\ \mu\text{F}$, $C_2 = 0.008\ \mu\text{F}$, and $C_3 = 0.004\ \mu\text{F}$ in Fig. 12-5(a), determine the cutoff frequency ω_c.

12-12 In Fig. 12-5(a), $C_1 = 0.01\ \mu\text{F}$, $C_2 = 0.04\ \mu\text{F}$, and $C_3 = 0.02\ \mu\text{F}$. Calculate R for a cutoff frequency of 1 kHz.

12-13 Calculate R in Fig. 12-7(a) if $C = 0.04\ \mu\text{F}$ and $f_c = 500$ Hz.

12-14 In Fig. 12-7(a) calculate (a) ω_c and (b) f_c if $R = 10\ \text{k}\Omega$ and $C = 0.001\ \mu\text{F}$.

12-15 Design a 40-db/decade high-pass filter for $\omega_c = 5$ k rad/s. $C_1 = C_2 = 0.02\ \mu\text{F}$.

12-16 Calculate (a) R_1 and (b) R_2 in Fig. 12-8(a) for a cutoff frequency of 40 k rad/s. $C_1 = C_2 = 250$ pF.

12-17 For Fig. 12-9(a), let $C_1 = C_2 = C_3 = 0.05\ \mu\text{F}$. Determine (a) R_3, (b) R_1, and (c) R_2 for a cutoff frequency of 500 Hz.

12-18 The circuit of Fig. 12-9(a) is designed with the values $C_1 = C_2 = C_3 = 400$ pF, $R_1 = 100\ \text{k}\Omega$, $R_2 = 25\ \text{k}\Omega$, and $R_3 = 50\ \text{k}\Omega$. Calculate the cutoff frequency, f_c.

12-19 If $\omega_h = 22.5$ k rad/s and $\omega_l = 22.1$ k rad/s, what is the bandwidth (a) in rad/s and (b) in hertz?

12-20 For the values given in Problem 12-19, determine the quality factor Q.

12-21 Design the band-pass filter of Fig. 12-10(b) to have $\omega_r = 10$ k rad/s, $A_r = 5$, $Q = 10$, and $C_1 = C_2 = 0.001\ \mu\text{F}$.

12-22 If the gain at the resonant frequency in Problem 12-21 is increased to 10, what are the values for R_1, R_2, and R_3?

12-23 In the band-pass filter of Fig. 12-10(b), $C_1 = C_2 = 0.01\ \mu\text{F}$, $R_1 = 40\ \text{k}\Omega$, $R_2 = 400\ \text{k}\Omega$, and $R_3 = 252\ \Omega$; determine (a) bandwidth (rad/s), (b) the gain at the resonant frequency, (c) Q, and (d) the resonant frequency in hertz.

12-24 Design the notch filter of Fig. 12-12(a) to have $\omega_r = 2$ k rad/s, $Q = 10$, and $C_1 = C_2 = 0.1\ \mu\text{F}$. Let $R_a = 1\ \text{k}\Omega$.

13

Integrated-Circuit Timers

13-0 Introduction

Applications such as oscillators, pulse generators, ramp or square wave generators, one-shot multivibrators, burglar alarms and voltage monitors all require a circuit capable of producing timing intervals. The most popular integrated-circuit timer is the 555, first introduced by Signetics Corporation (see Appendix 4). Similar to general-purpose op amps, the 555 is reliable, easy to use in a variety of applications, and low in cost. The 555 can also operate from supply voltages of 5 V to +18 V, making it compatible with both TTL (transistor–transistor logic) circuits and op amp circuits. The 555 timer can be considered a functional block that contains two comparators, two transistors, three equal resistors, a flip-flop, and an output stage. These are shown in Fig. 13-1.

Besides the 555 timer, there are also available counter timers such as Exar's XR—2240 (see Appendix 5). The 2240 contains a 555 timer plus a programmable binary counter in a single 16-pin package. A single 555 has a maximum timing range of approximately 15 minutes. Counter timers have a maximum timing range of days. The timing range of both can be extended to months or even years by cascading. Our study of timers will begin with the 555 and its applications and then proceed to the counter timers.

13-1 Operating Modes of the 555 Timer

The 555 IC timer has two modes of operation, either as an astable (free-running) multivibrator or as a monostable (one-shot) multivibrator. Free-running operation of the 555 is shown in Fig. 13-2(a). The output voltage

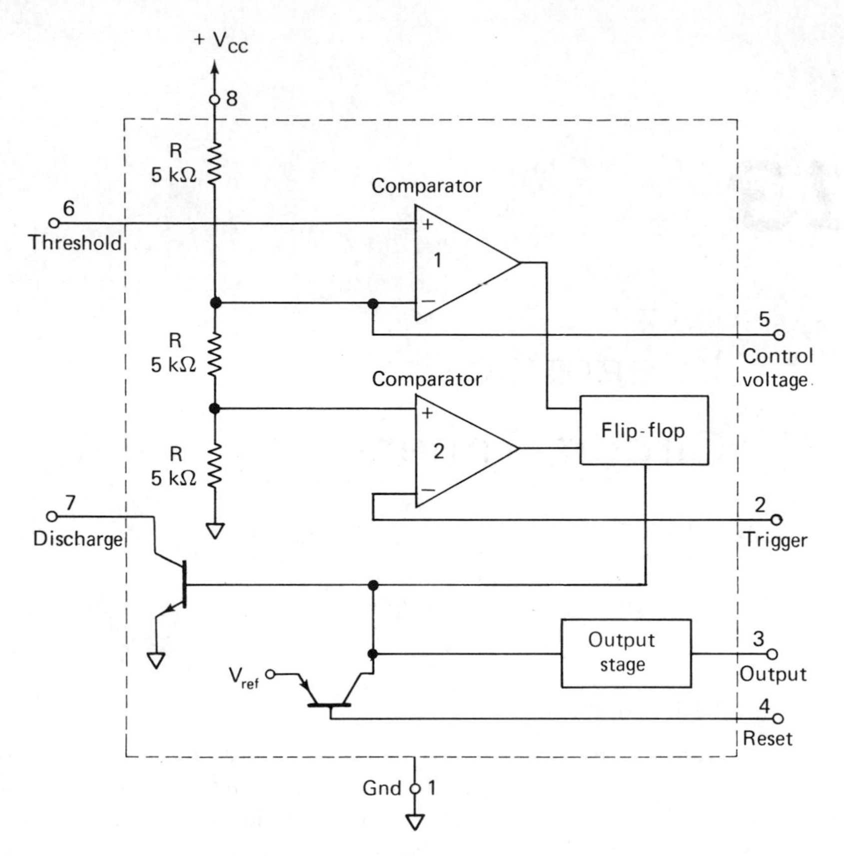

Figure 13-1 555 integrated-circuit timer.

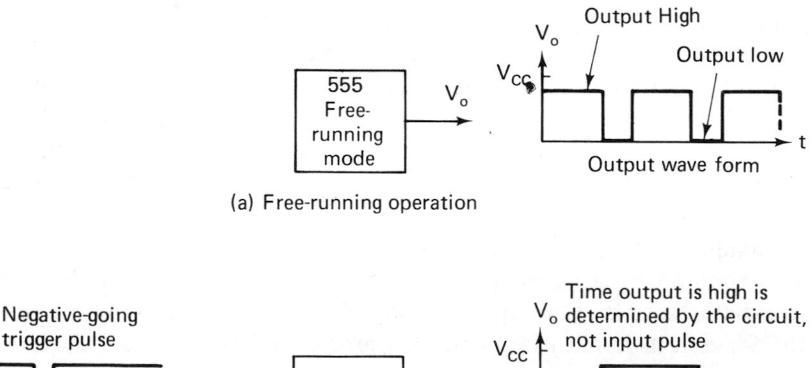

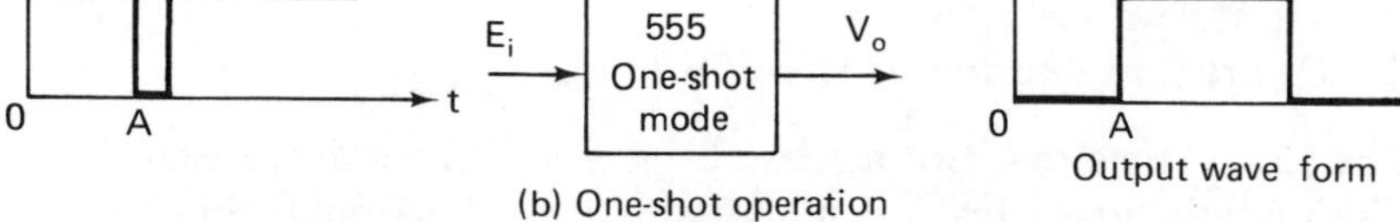

Figure 13-2 Operating modes of a 555 timer.

switches from a high to a low state and back again. The time the output is either high or low is determined by a resistor-capacitor network connected externally to the 555 timer (see Section 13-2). The value of the high output voltage is slightly less than V_{CC}. The value of the output voltage in the low state is approximately 0.1 V.

When the timer is operated as a one-shot multivibrator, the output voltage is low until a negative-going trigger pulse is applied to the timer; then the output switches high. The time the output is high is determined by a resistor and capacitor connected to the IC timer. At the end of the timing interval, the output returns to the low state. Monostable operation is examined further in Sections 13-5 and 13-6.

To understand how a 555 timer operates, a brief description of each terminal is given in Section 13-2.

13-2 Terminals of the 555

13-2.1 Packaging and Power Supply Terminals. The 555 timer is available in two package styles, TO 99 and DIP, as shown in Fig. 13-3 and Appendix 4. Pin 1 is the common, or ground, terminal, and pin 8 is the positive voltage supply terminal V_{CC}. V_{CC} can be any voltage between +5 V and +18 V. Thus, the 555 can be powered by existing digital logic supplies (+5 V), linear IC supplies (+15 V), and automobile or dry cell batteries. Internal circuitry requires about 0.7 mA per supply volt (10 mA for V_{CC} = +15) to set up internal bias currents. Maximum power dissipation for the package is 600 mW.

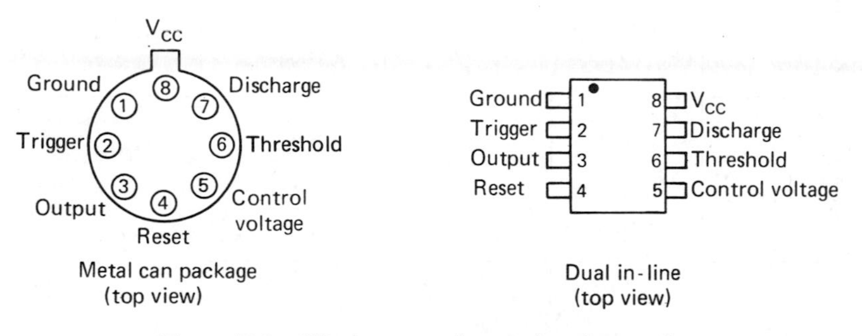

Figure 13-3 555 pin connections and package styles.

13-2.2 Output Terminal. The output terminal of a 555 is pin 3, and its operation can be explained by the models in Fig. 13-4. The output has two output states, a low state and a high state. In the low state the output of the 555 acts as a low resistance ($\cong$ 10 Ω) to ground as shown in Fig. 13-4a and c. When the output of the 555 is high, there appears to be an equivalent 10 Ω resistor between V_{CC} and pin 3 as shown in Fig. 13-4(b) and (d).

There are two ways of connecting a load to the output terminal; (1)

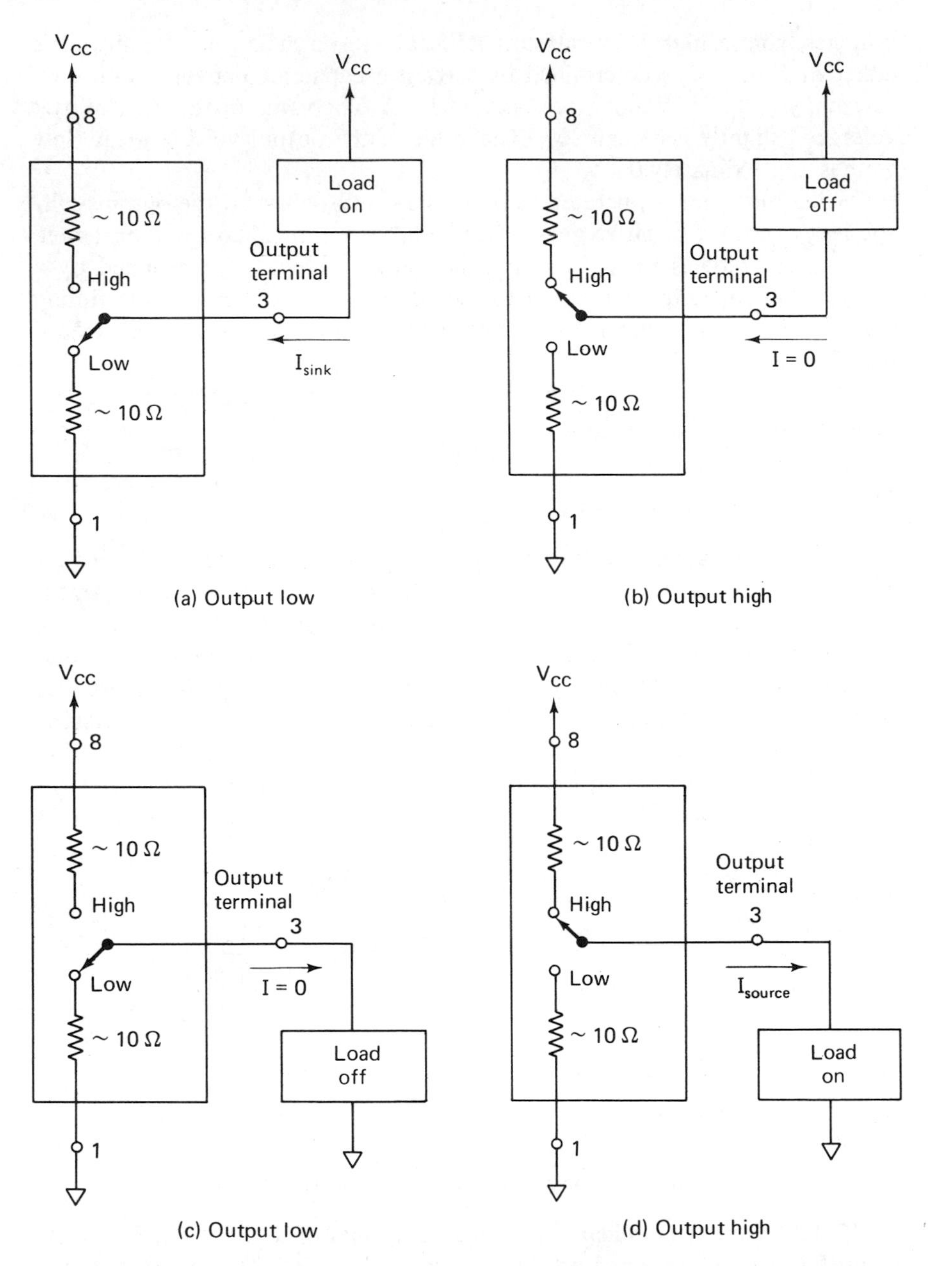

Figure 13-4 555 timer-output terminal operation.

between pin 3 and V_{CC} and (2) between pin 3 and ground. In Fig. 13-4(a), the output voltage of the 555 is low and the load conducts current. Load current flows into the output terminal and is called *sink* current. For sink currents less than 10 mA, the low output voltage is approximately 0.1 V.

For this type of load connection the load turns off when the output goes high (no current flow), as shown in Fig. 13-4(b).

Figure 13-4(c) shows a grounded load. The load is off when the output of the 555 is low. When the output goes high, the output terminal supplies or *sources* current to a grounded load (see Fig. 13-4(d)). The maximum sink and source current is 200 mA. The high output voltage is approximately 0.5 V below V_{CC}.

13-2.3 Trigger Terminal. Pin 2 of a 555 timer is the trigger terminal. If the voltage at this terminal is held greater than $\frac{2}{3}V_{CC}$, the output remains low. A negative-going pulse of sufficient amplitude on pin 2 triggers the output to go high. The width of the trigger pulse must be less than the width of the expected output pulse. If the trigger terminal is held at a low voltage, the output remains in its high state. The trigger terminal should not be grounded for long time intervals. These principles are discussed in more detail in Section 13-3.

13-2.4 Reset Terminal. The reset terminal, pin 4, allows the 555 to be disabled and overide command signals on the trigger input. When not used, the reset terminal should be wired to $+V_{CC}$. If the reset terminal is grounded or its potential reduced below 0.4 V, both the output terminal, pin 3, and the discharge terminal, pin 7, are at approximately ground potential. In other words, the output is held low. If the output was high, a ground on the reset terminal immediately forces the output low.

13-2.5 Discharge Terminal. Discharge terminal, pin 7, is used to discharge an external timing capacitor during the time the output is low. When the output is high, pin 7 acts as an open circuit and allows the capacitor to charge at a rate determined by an external resistor or resistors and capacitor. Figure 13-5 shows a model of the discharge terminal for when C is discharging and for when C is charging.

13-2.6 Threshold Terminal. The threshold terminal, pin 6, monitors an external capacitor's voltage. For example, once the 555 is triggered into its high state, the threshold terminal watches the rising capacitor voltage V_C as shown in Fig. 13-5(b). When V_C increases to the *threshold voltage*, approximately $\frac{2}{3}V_{CC}$, the output of the 555 switches to its low state.

13-2.7 Control Voltage Terminal. A 0.01-μF filter capacitor is usually connected from the control voltage terminal, pin 5, to ground. The capacitor by-passes noise and/or ripple voltages from the power supply to minimize their effect on threshold voltage. The control voltage terminal may also be used to change both the threshold and trigger voltage levels. For example, connecting a 5-kΩ resistor between pins 5 and 8 changes threshold voltage to $0.8V_{CC}$ and the trigger voltage to $0.2\ V_{CC}$. An external voltage applied to

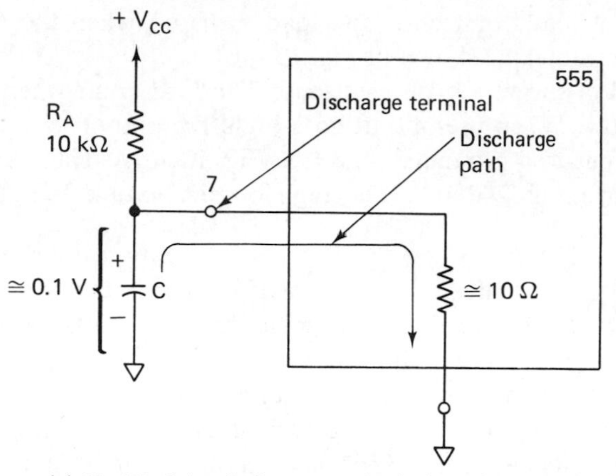

(a) Model of the discharge terminal when the output is low, and capacitor is discharging

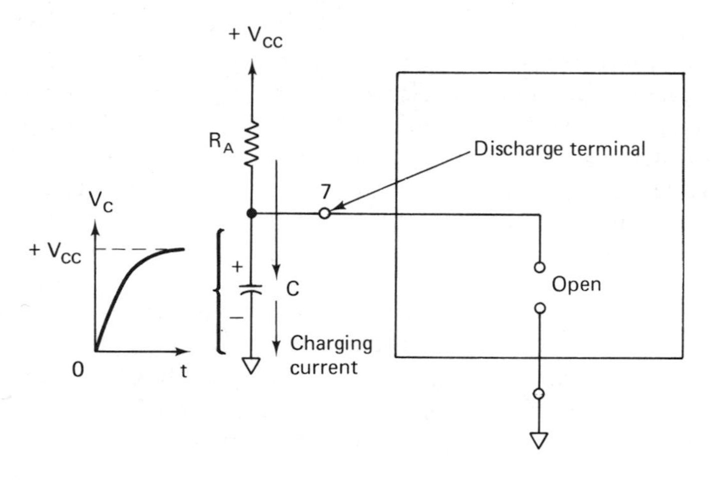

(b) Model of the discharge terminal when the output is high, and capacitor is charging

Figure 13-5 Operation of discharge terminal.

pin 5 will change both threshold and trigger voltages and can also be used to modulate the output waveform.

13-3 Free-Running or Astable Operation

13-3.1 Circuit Operation. The 555 is connected in Fig. 13-6(a) as a free-running multivibrator. Capacitor C charges through R_A and R_B from V_{CC}. R_B and pin 7 discharge the capacitor. When the capacitor is charging, the out-

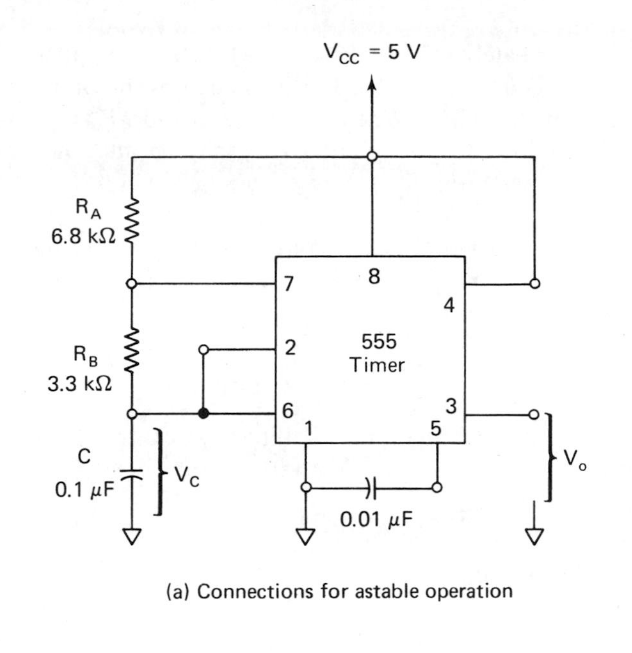

(a) Connections for astable operation

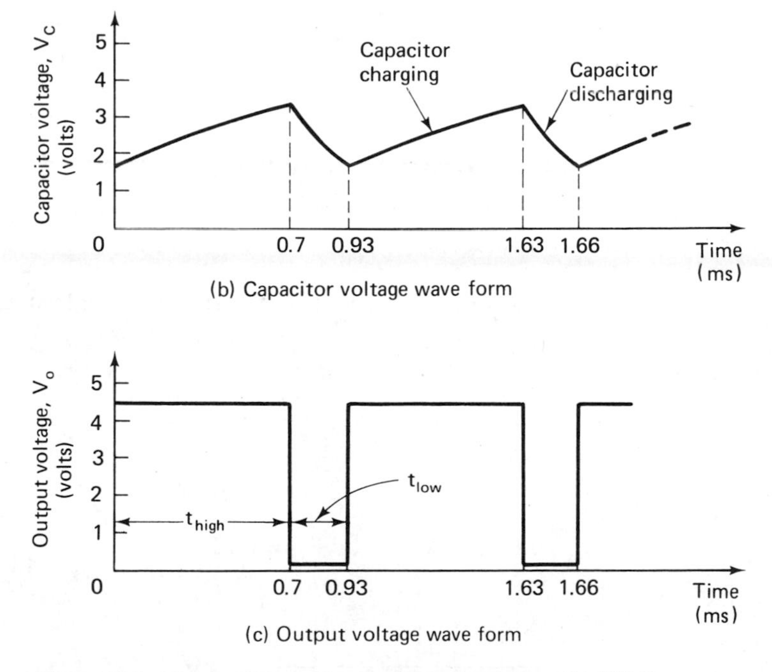

(b) Capacitor voltage wave form

(c) Output voltage wave form

Figure 13-6 Connections and wave forms to operate the 555 timer as a free-running multivibrator.

put is high (at approximately 4.5 V if $V_{CC} = 5$ V). When the capacitor voltage V_C reaches $\frac{2}{3}V_{CC}$, comparator 1 (in Fig. 13-1) causes the output voltage to go low as shown in Fig. 13-6c. When V_C discharges to $\frac{1}{3}V_{CC}$, comparator 2 causes the output voltage to go high and the cycle repeats. The capacitor voltage and output voltage waveforms are shown in Fig. 13-6.

13-3.2 Frequency of Oscillation. The output stays high during the time interval that C charges from $\frac{1}{3}V_{CC}$ to $\frac{2}{3}V_{CC}$ as shown in Fig. 13-6(b) and (c). This time interval is given by

$$t_{\text{high}} = 0.695(R_A + R_B)C \tag{13-1}$$

The output is low during the time interval that C discharges from $\frac{2}{3}V_{CC}$ to $\frac{1}{3}V_{CC}$ and is given by:

$$t_{\text{low}} = 0.695R_B C \tag{13-2}$$

Thus the total period of oscillation, T, is

$$T = t_{\text{high}} + t_{\text{low}} = 0.695(R_A + 2R_B)C \tag{13-3}$$

The free-running frequency of oscillation, f, is

$$f = \frac{1}{T} = \frac{1.44}{(R_A + 2R_B)C} \tag{13-4}$$

Figure 13-7 is a plot of Eq. (13-4) for different values of $(R_A + 2R_B)$ and quickly shows what combinations of resistance and capacitance are needed to design an astable multivibrator.

Example 13-1: Calculate (a) t_{high} (b) t_{low} and (c) the free-running frequency for the timer circuit of Fig. 13-6(a).
Solution: (a) By Eq. (13-1),

$$t_{\text{high}} = 0.695(6.8\text{ k}\Omega + 3.3\text{ k}\Omega)(0.1\ \mu\text{F}) = 0.7\text{ ms}$$

(b) By Eq. (13-2),

$$t_{\text{low}} = 0.695(3.3\text{ k}\Omega)(0.1\ \mu\text{F}) = 0.23\text{ ms}$$

(c) By Eq. (13-4),

$$f = \frac{1.44}{(6.8\text{ k}\Omega) + (2)(3.3\text{ k}\Omega)(0.1\ \mu\text{F})} = 1.07\text{ kHz}$$

The answer of part c agrees with results obtainable from Fig. 13-7.

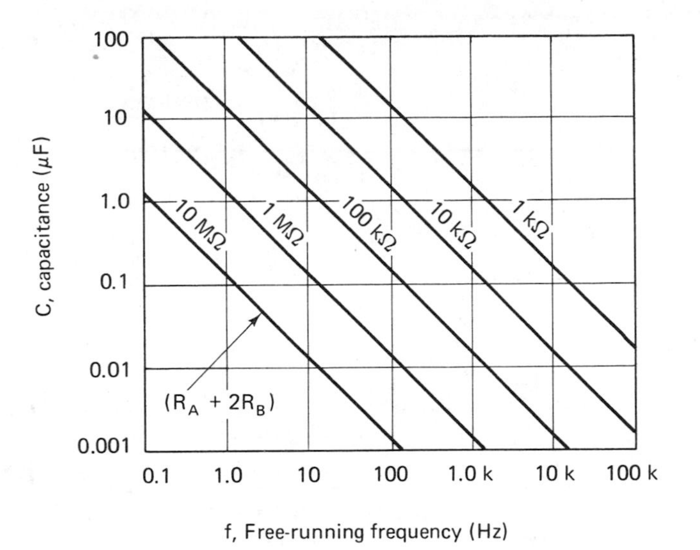

$$f = \frac{1}{T} = \frac{1.44}{(R_A + 2R_B)C}$$

Figure 13-7 Free-running frequency for astable operation.

13-3.3 Duty Cycle. The ratio of time when the output is low, t_{low}, to the total period T is called *duty cycle, D*. In equation form,

$$D = \frac{t_{low}}{T} = \frac{R_B}{R_A + 2R_B} \tag{13-5}$$

Example 13-2: Calculate the duty cycle for the values given in Fig. 13-6(a).
Solution: By Eq. (13-5),

$$D = \frac{3.3\ \text{k}\Omega}{6.8\ \text{k}\Omega + 2(3.3\ \text{k}\Omega)} = 0.25$$

This checks with Fig. 13-6c, which shows that the timer's output is low for approximately 25% of the total period, T. Equation (13-5) shows that it is impossible to obtain a duty cycle of $\frac{1}{2}$ or 50%. As presented, the circuit of Fig. 13-6(a) is not capable of producing a square wave. The only way D in Eq. (13-5) can equal $\frac{1}{2}$ is for R_A to equal 0. Then there would be a short between V_{CC} and pin 7. However, R_A must be large enough so that when the discharge transistor is "on," current through it is limited to 0.2 A. Considering $V_{CE\ sat}$, the minimum value of R_A in ohms is given by

$$R_A \simeq \frac{V_{CC}}{0.2\ \text{A}} \tag{13-6a}$$

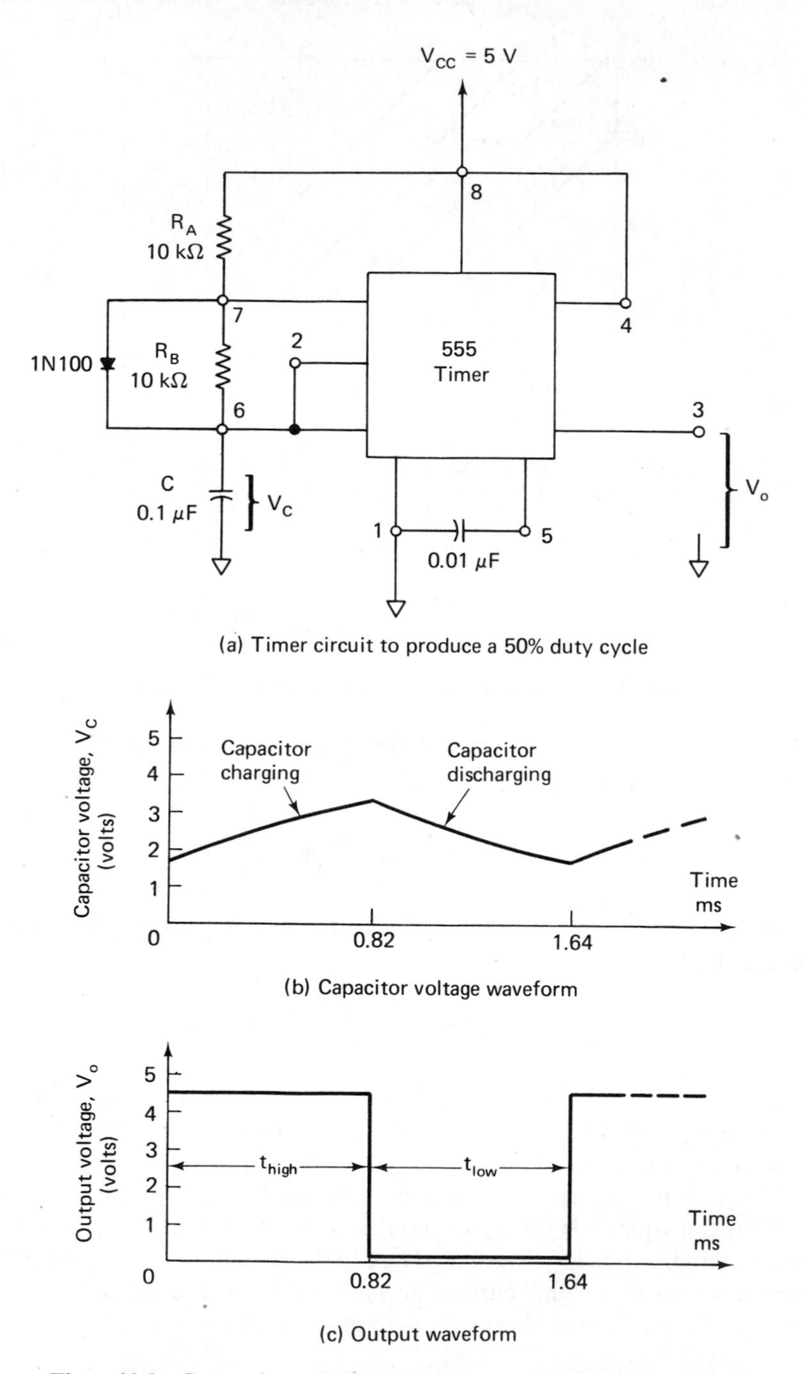

Figure 13-8 Connecting a diode across R_B to produce duty cycles $\geqq 50\%$.

or

$$R_A \cong 5\,V_{CC} \tag{13-6b}$$

The conclusion drawn from Eq. (13-6b) is that R_A cannot be equal to 0. Therefore, to extend the duty cycle an alternative solution must be found.

13-3.4 Extending the Duty Cycle. The duty cycle for the circuit of Fig. 13-6(a) can never be equal to or greater than 50%, as discussed in Section 13-3.3. By connecting a diode in parallel with R_B in Fig. 13-8(a), a duty cycle of 50% or greater can be obtained. Now the capacitor charges through R_A and the diode but discharges through R_B. The times for the output waveform are

$$t_{high} = 0.695 R_A C \tag{13-7a}$$

$$t_{low} = 0.695 R_B C \tag{13-7b}$$

$$T = 0.695(R_A + R_B)C \tag{13-7c}$$

Equations (13-7a) and (13-7b) show that if $R_A = R_B$, then the duty cycle is 50%, as shown in Figs. 13-8(b) and (c).

13-4 Applications of the 555 as an Astable Multivibrator

13-4.1 Tone-Burst Oscillator. With the switch in Fig. 13-9 set to the "continuous" position, the B 555 timer functions as a free-running multivibrator. The frequency can be varied from about 1.3 kHz to 14 kHz by the 10-kΩ potentiometer. If the potentiometer is replaced by a thermistor or photoconductive cell, the oscillating frequency will be proportional to temperature or light intensity respectively.

The A 555 timer oscillates at a slower frequency. The 1-MΩ potentiometer sets the lowest frequency at about 1.5 Hz. Lower frequencies are possible by replacing the 1 μF capacitor with a larger value. When the connecting switch is thrown to the "burst" position, output pin 3 of the *A* timer alternately places a ground or high voltage on reset pin 4 of the B 555 timer. When pin 4 of the *B* timer is grounded, it cannot oscillate, and when ungrounded the timer oscillates. This causes the *B* timer to oscillate in bursts. The output of the tone-burst generator is V_o and is taken from pin 3 of timer *B*. V_o can drive either an audio amplifier or a stepdown transformer directly to a speaker.

The 556 IC timer contains two 555 timers in a single 14-pin dual-in-line package. The tone-burst generator can be made with one 556.

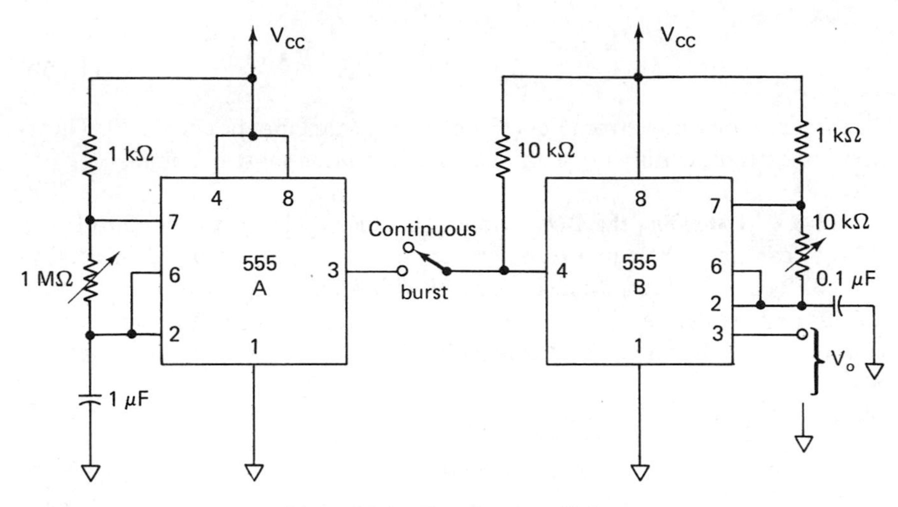

Figure 13-9 Tone-burst oscillator.

13-4.2 Variable Duty Cycle Oscillator. By adding another diode, resistor, and potentiometer to the 50% duty cycle circuit of Fig. 13-8(a), a variable duty-cycle square wave generator can be constructed. The result is shown in Fig. 13-10, where independent charge and discharge paths for capacitor C are established by diodes D_A and D_B. The charge path for C is from V_{CC} through R_A and D_A. The discharge path for C is through D_B, R_B, and pin 7. The charge and discharge time t_{high} and t_{low} are given by Eqs. (13-7a) and

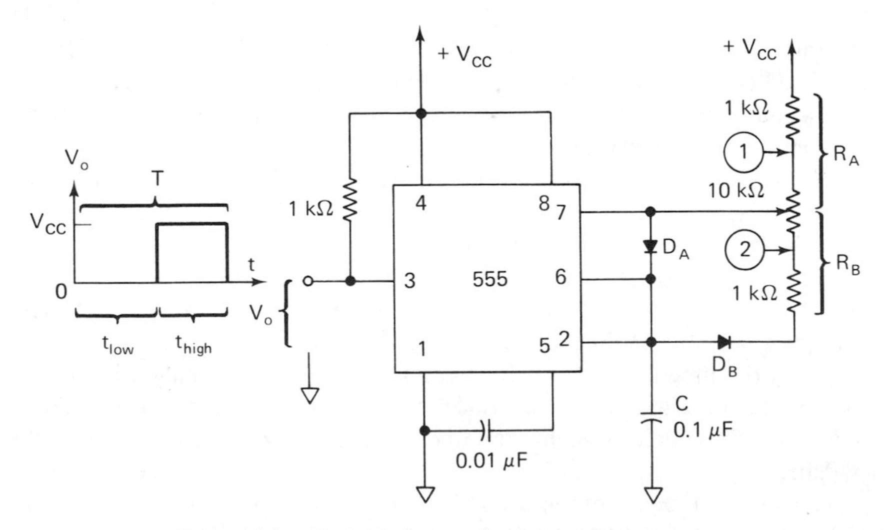

Figure 13-10 Variable duty cycle square wave generator.

(13-7b) respectively. The period, T, and duty cycle are given by:

$$T = 0.7(R_A + R_B)C \tag{13-8a}$$

$$\text{Duty cycle} = \frac{R_B}{R_A + R_B} \tag{13-8b}$$

Duty cycles from 1% to 99% can be realized.

Example 13-3: Calculate the duty cycle for Fig. 13-10 if the wiper of the 10-kΩ potentiometer is set at (a) position 1 (b) position 2.
Solution: (a) $R_A = 1$ kΩ and $R_B = 11$ kΩ, so

$$D = \frac{11\text{ k}\Omega}{1\text{ k}\Omega + 11\text{ k}\Omega} \cong 0.92 \text{ or } 92\%$$

which means t_{low} is 92% of the total period T.
(b) $R_A = 11$ kΩ and $R_B = 1$ kΩ, so

$$D = \frac{1\text{ k}\Omega}{11\text{ k}\Omega + 1\text{ k}\Omega} \cong 0.082 \text{ or } 8.3\%$$

In this position t_{low} is only 8.3% of the total period T.

13-5 One-Shot or Monostable Operation

13-5.1 Introduction. Not all applications require a continuous repetitive wave such as that obtained from a free-running multivibrator. Many applications need to operate only for a specified length of time. These circuits require a one-shot or monostable multivibrator. Figure 13-11(a) is a circuit diagram using the 555 for monostable operation. When a negative-going pulse is applied to pin 2, the output goes high and terminal 7 removes a short circuit from capacitor C. The voltage across C rises at a rate determined by R_A and C. When the capacitor voltage reaches $\frac{2}{3}V_{CC}$, comparator 1 in Fig. 13-1 causes the output to switch from high to low. The input and output voltage waveforms are shown in Fig. 13-11(a). The output is high for a time given by

$$t_{high} = 1.1R_AC \tag{13-9}$$

Figure 13-11(b) is a plot of Eq. (13-9) and quickly shows the wide range of output pulses that are obtainable and the required values of R_A and C.

Example 13-4: If $R_A = 9.1$ kΩ, find C for an output pulse duration of 1 ms.

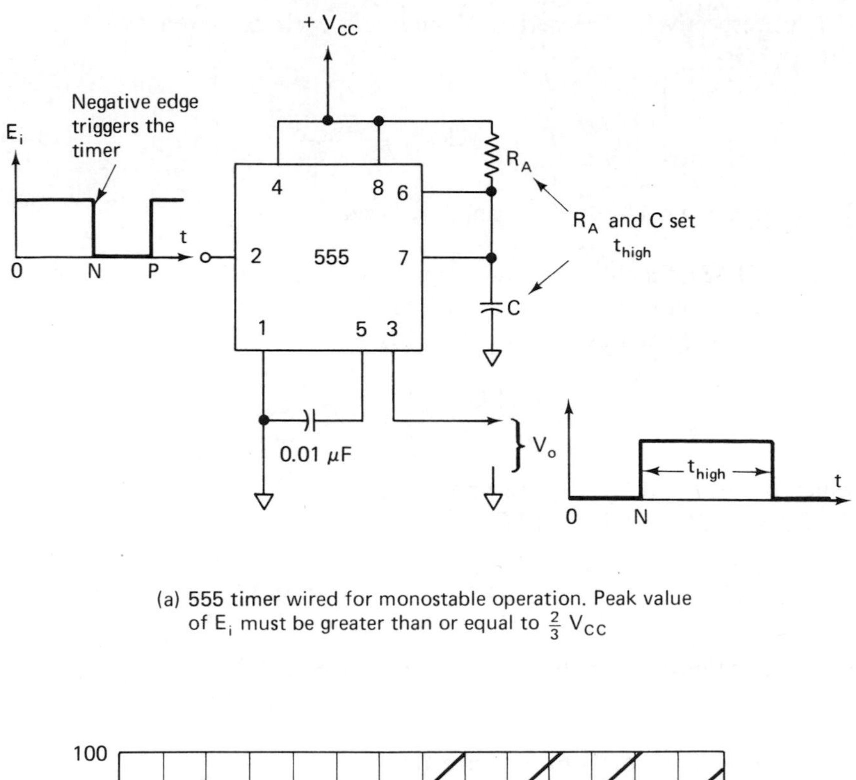

(a) 555 timer wired for monostable operation. Peak value of E_i must be greater than or equal to $\frac{2}{3}$ V_{CC}

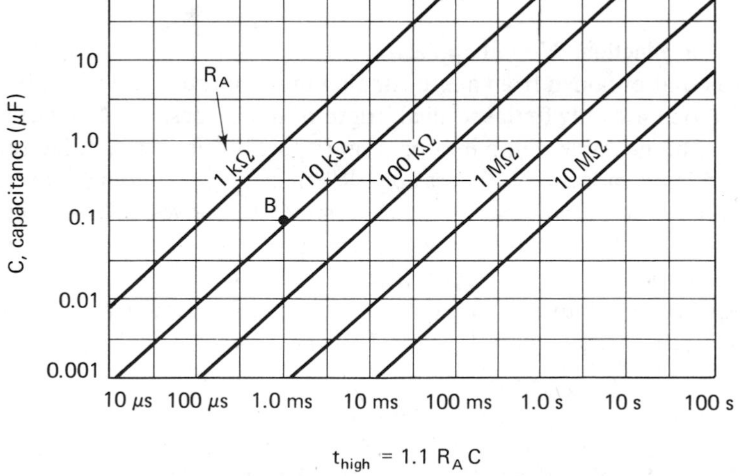

(b) Design aid to determine output pulse duration

Figure 13-11 Monostable operation.

Solution: Rearrange Eq. (13-9):

$$C = \frac{t_{high}}{1.1R_A} = \frac{1 \times 10^{-3}\ \text{sec}}{1.1(9.1 \times 10^3)\Omega} = 0.1\ \mu\text{F}$$

This answer checks with that obtainable at point B in Fig. 13-11(b). For the 555 timer to trigger properly in this type of operation, the width of the trigger pulse must be less than t_{high} and a trigger input pulse network is needed so that the output does not switch on the positive-going edge of the trigger pulse (point P).

13-5.2 Input Pulse Circuit. Figure 13-12 shows the multivibrator wired for monostable operation (as in Fig. 13-11a). R_i, C_i, and diode D are needed to generate a single output pulse for one input pulse.

Resistor R_A and capacitor C determine the time that the output is high as given by Eq. (13-9). Resistor R_i is connected between V_{CC} and pin 2 to ensure (1) that the output is low and (2) that C_i is charged to V_{CC} until the negative trigger pulse occurs. The time contant of R_i and C_i should be small with respect to the output timing interval t_{high}. Diode D prevents the 555 timer from triggering on the positive-going edges of E_i. Wave forms for the input pulse, E_i, the pulse at pin 2, V_2, and the output pulse, V_o, are all shown in Fig. 13-12.

Example 13-5: (a) If $R_A = 10\ \text{k}\Omega$ and $C = 0.2\ \mu\text{F}$ in Fig. 13-12, find t_{high}. (b) What is the time constant of R_i and C_i in Fig. 13-12?

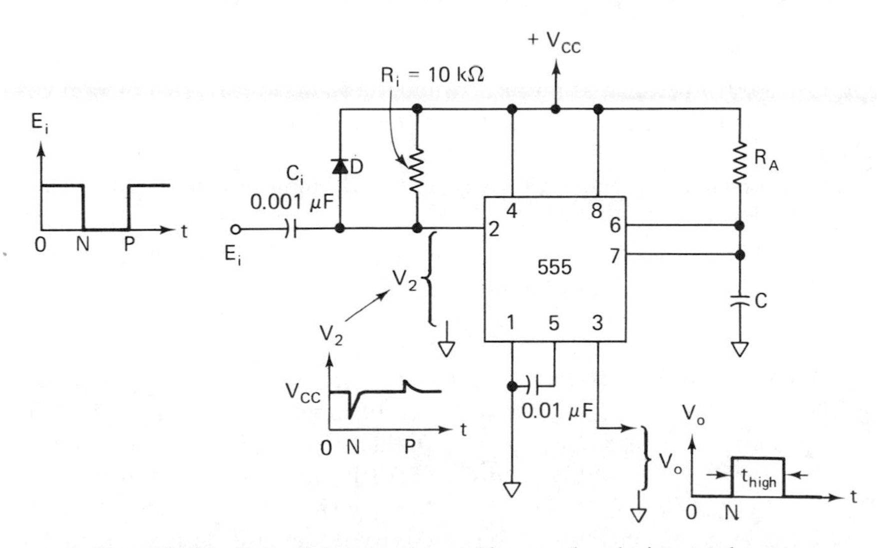

Figure 13-12 For satisfactory monostable operation the input pulse network of R_i, C_i, and D is needed.

Solution: (a) By Eq. (13-9),

$$t_{high} = 1.1(10 \times 10^3)(0.2 \times 10^{-6}) = 2.2 \text{ ms}$$

(b) Time constant $= R_iC_i = (10 \times 10^3)(0.001 \times 10^{-6}) = 0.01$ ms.

As with astable operation, the reset terminal pin 4 is normally tied to the supply voltage, V_{CC}. If pin 4 is grounded at any time, the timing cycle is stopped. When the reset terminal is grounded, both output pin 3 and discharge terminal 7 go to ground potential. Thus the output goes low and any charge accumulated by the timing capacitor C is removed. As long as the reset terminal is grounded, these conditions remain.

13-6 Applications of the 555 as a One-Shot Multivibrator

13-6.1 Water Level Fill Control. In Fig. 13-13(a), the start switch is closed and the output of the 555 is low. When the start switch is closed, the output goes high to actuate the pump. The time interval the output is high is given by Eq. (13-9). Upon completion of the timing interval, the output of the 555 returns to its low state, turning the pump off. The height of the water level is set by the timing interval which is set by R_A and C. In the event of a potential overflow, the overfill switch must place a ground on reset pin 4, which causes the timer's output to go low and stops the pump.

13-6.2 Touch Switch. The 555 is wired as a one-shot multivibrator in Fig. 13-13(b) to perform as a touch switch. A 22-MΩ resistor to pin 2 holds the 555 in its idle state. If you scuff your feet to build up a static charge, the 555 will produce a single shot output pulse when you touch the finger plate. If the electrical noise level is high (due, for example to fluorescent lights) the 555 may oscillate when you touch the finger plate. Reliable and consistent triggering will occur if a thumb is placed on a ground plate and fingers of the same hand tap the finger plate. An isolated power supply or batteries should be used for safety.

13-6.3 Frequency Divider. Fig. 13-11(a) can be used as a frequency divider if the timing interval is adjusted to be longer than the period of the input signal E_i. For example, suppose frequency of E_i is 1 kHz so that its period is 1 ms. If $R_A = 10$ kΩ and $C = 0.1$ μF, the timing interval given by Eq. (13-9) is $t_{high} = 1.1$ ms. Therefore, the one shot will be triggered by the first negative going pulse of E_i, but the output will still be high when the second negative-going pulse occurs. The one-shot will, however, be retriggered on the third negative-going pulse. In this example, the one-shot triggers

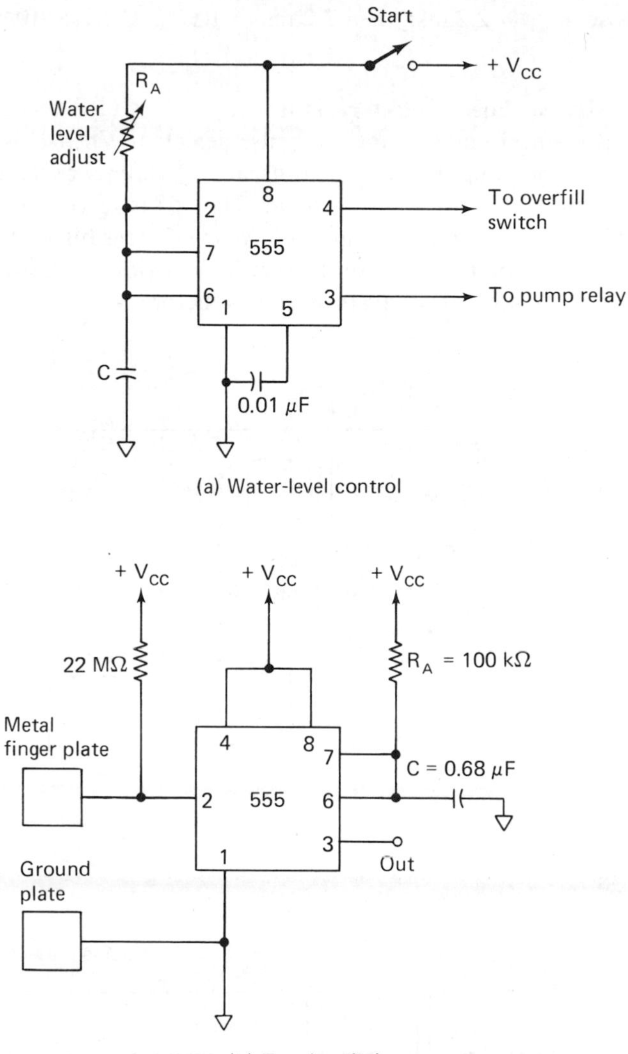

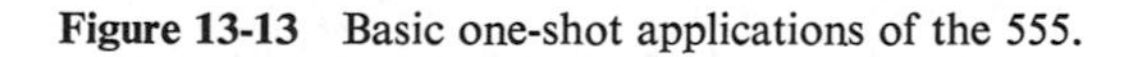
Figure 13-13 Basic one-shot applications of the 555.

on every other pulse of E_i, so there is only one output for every two input pulses; thus E_i is divided by 2.

Example 13-6: (a) Calculate the timing interval in Fig. 13-11(a) if $R_A = 10$ kΩ and $C = 0.1$ μF (b) What value of R_A should be installed to divide a 1-kHz input signal by 3?

Solution: (a) By Eq. (13-9), $t_{high} = 1.1\ (10 \times 10^3)(0.1 \times 10^{-6}) = 1.1$ ms. (b) t_{high} should exceed two periods of E_i, or 2 ms, and be less than 3 periods, or

3 ms. Choose $t_{high} = 2.2$ ms; then $2.2 \text{ ms} = 1.1R_A \times 0.1 \times 10^{-6}$ F; $R_A =$ 20 kΩ.

13-6.4 Missing Pulse Detector. Transistor Q is added to the 555 one-shot in Fig. 13-14(a) to make a missing pulse detector. When E_i is at ground potential (0 V), the emitter diode of transistor Q clamps capacitor voltage V_C to a few tenths of a volt above ground. The 555 is forced into its timing state with a high output voltage V_o at pin 3. When E_i goes high, the transistor cuts off and capacitor C begins to charge. This action is shown by waveshapes in Fig. 13-14(b). If E_i again goes low before the 555 completes its

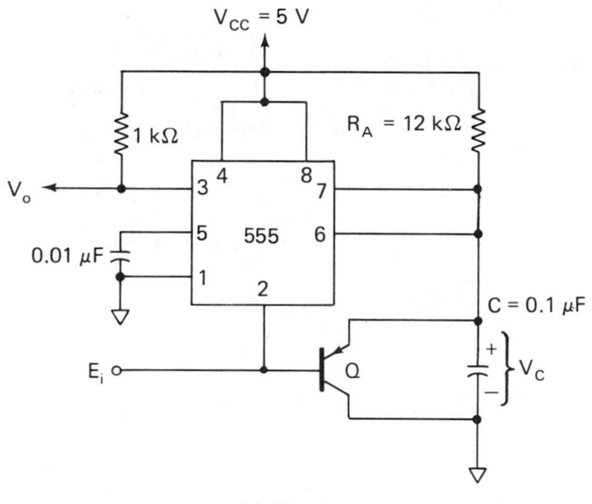

(a) Circuit

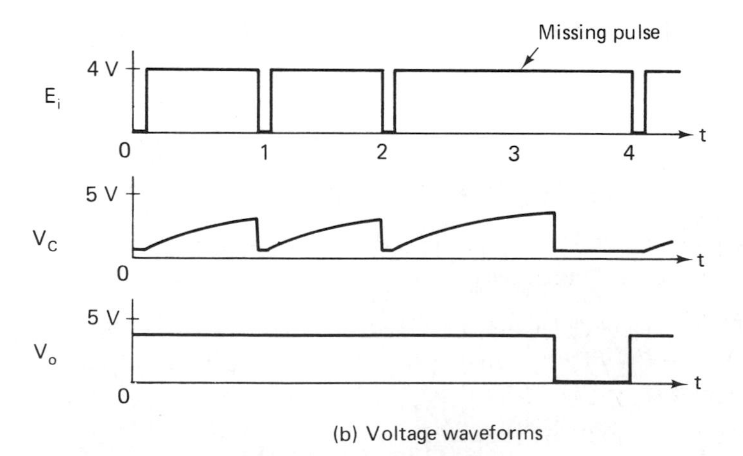

(b) Voltage waveforms

Figure 13-14 Missing pulse detector.

timing cycle, the voltage across C is reset to about 0 V. If, however, E_i does *not* go low before the 555 completes its timing cycle, the 555 enters its normal state and output V_o goes low. This is exactly what happens if the $R_A C$ timing interval is slightly longer than the period of E_i and E_i suddenly misses a pulse. This type of circuit can detect a missing heartbeat. If E_i pulses are generated from a rotating wheel this circuit tells when wheel speed drops below a predetermined value. Thus the missing pulse detector circuit also performs speed control and measurement.

13-7 Introduction to Counter Timers

When a timer circuit is connected as an oscillator and is used to drive a counter, the resultant circuit is *a counter timer*. Typically, the counter has many separate output terminals. One output terminal gives one pulse for each period T of the oscillator. A second output terminal gives one output pulse for every two periods ($2T$) of the oscillator. A third output terminal gives one output pulse for every four oscillator periods ($4T$), and so on depending on the design of the counter. Thus each output terminal is rated in terms of the basic oscillator period T.

Some counters are designed so that their outputs can be connected together. The resultant output pulse is the *sum* of the individual output pulses. For example, if the first, second, and third output terminals are wired together, the result is one output pulse for every $1T + 2T + 4T = 7T$ oscillator periods. A counter with this capability is said to be *programmable*, because the user can program the counter to give one output pulse for any combination of timer outputs. One such programmable timer/counter is Exar's XR 2240. This integrated-circuit device is representative of the timer/counter family and some of its features will be studied next.

13-8 The XR 2240 Programmable Timer/Counter

13-8.1 Circuit Description. As shown in Fig. 13-15 and Appendix 5, the XR 2240 consists of one modified 555 timer, one 8-bit binary counter, and a control circuit. They are all contained in a single 16-pin dual-in-line package.

A positive-going pulse applied to *trigger* input 11 starts the 555 time base oscillator.

A positive-going pulse on *reset* pin 10 stops the 555 time base oscillator. The threshold voltage for both trigger and reset terminals is about +1.4 V.

The time base period T for one cycle of the 555 oscillator is set by an external RC network connected to the *timing* pin 13. T is calculated from

$$T = RC \tag{13-10}$$

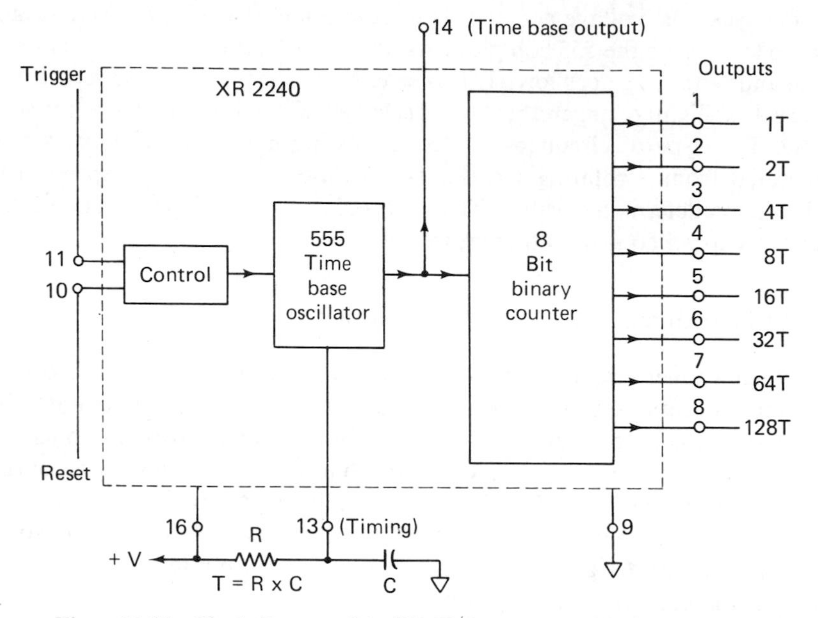

Figure 13-15 Block diagram of the XR 2240 programmable timer/counter.

where R is in ohms, C is in Farads, T is in seconds. R can range from 1 kΩ to 10 MΩ and C from 0.005 μF to 1000 μF. Thus, the period of the 555 can range from microseconds to hours.

Output of the 555 time base oscillator is available for measurement at pin 14 and also drives the 8-bit binary counter. Operation of the counter is discussed in Section 13-8.2.

13-8.2 Counter Operation. A simplified schematic of the 8-bit binary counter is shown in Fig. 13-16. Output of the 555 time base oscillator is shown as a switch. One side of the switch is connected to ground while the other side is wired to a 20-kΩ pull-up resistor. A regulated plus voltage is available at pin 15. Each negative going edge from the 555 steps the 8 bit counter up by one count.

Normally, the 2240 is in its *reset* condition. That is, all 8 output pins (pins 1 to 8) act like open circuits as shown by the output switch models in Fig. 13-16. Pull-up resistors (10 kΩ) should be installed, as shown, to those terminals that are going to be used. Outputs 1 and 4 will then be high in the reset condition.

When the 2240 is triggered (pulse applied to pin 11), all output switches of the counter are closed by the control circuit and outputs 1 to 8 go low. Thus, the counter begins its count with all outputs essentially grounded. At the end of every time base period, the 555 steps the counter once. The counter's T switch on terminal 1 opens after the first time base period (output 1

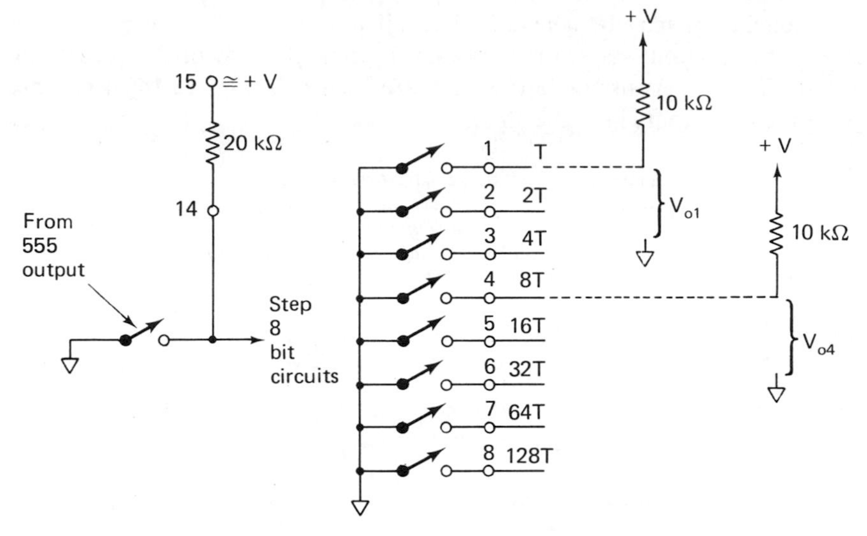

(a) Simplified outputs of the 2240

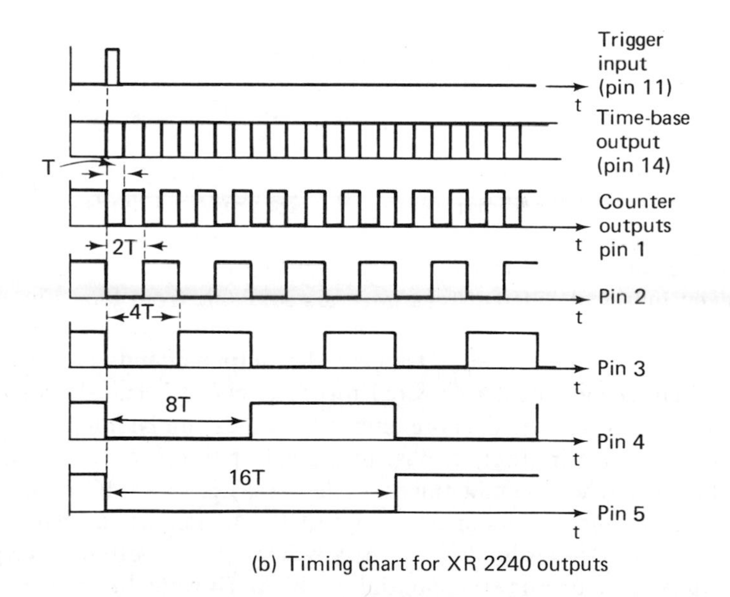

(b) Timing chart for XR 2240 outputs

Figure 13-16 Counter operation.

goes high) and closes after the second time base period. This counting action of the timer is shown in Fig. 13-16(b).

Output pin 2 is labeled $2T$ in Fig. 13-16(a). It is seen from Fig. 13-16(b) that the output on pin 2 has stayed low for two time base periods ($2T$). Thus, the second output stays low for twice the time interval of the first output.

This conclusion may be generalized to all outputs of the binary counter; that is, each output stays low for twice the time interval of the preceeding output. Time intervals for pins 1 to 5 are shown in Fig. 13-16(b) and are given for all outputs in Table 13-1.

Table 13-1 OUTPUT TERMINAL TIME CHART

Terminal Number	*Time output stays low after trigger pulse*
1	T
2	$2T$
3	$4T$
4	$8T$
5	$16T$
6	$32T$
7	$64T$
8	$128T$

Example 13-7: After triggering, how long will the following output terminals stay low? (a) Pin 3; (b) pin 4; (c) pin 7; (d) pin 8. $R = 100$ kΩ, and $C =$ 0.01 μF.

Solution: By Eq. (13-10), the time base period is

$$T = (100 \times 10^3)(0.01 \times 10^{-6}) = 1 \text{ ms}$$

From Table 13-1, (a) $t_{low} = 4(1 \text{ ms}) = 4$ ms; (b) $t_{low} = 8(1 \text{ ms}) = 8$ ms; (c) $t_{low} = 64(1 \text{ ms}) = 64$ ms; (d) $t_{low} = 128(1 \text{ ms}) = 128$ ms.

The conclusion to be drawn from Example 13-7 is that after triggering, there are eight pulses of different time intervals available from the counter timer.

13-8.3 Programming the Outputs. The output circuits are designed to be used either individually or wired together, which is called *wired-or*. The term wire-or means two or more output terminals can be jumpered together with a common wire (output bus) to a single pull-up resistor as shown in Fig. 13-17(a). The resultant timing cycle for V_o is found by redrawing the individual timing of pins 4 and 5 in Fig. 13-17(b). Here we see that as long as either pin 4 *or* pin 5 is low, V_o will be low. Only when both outputs go high (output switches open) will the output go high. Thus the timing cycle for the output bus is found simply by calculating the sum, T_{sum}, of the individual outputs.

Example 13-8: Calculate the timing cycle for (a) Fig. 13-17a (b) a circuit where pins 3, 6 and 7 are jumpered to a common bus. Let $T = 1$ second

Solution: (a) $T_{sum} = 8T + 16T = 24T = 24 \times 1 \text{ s} = 24$ s; (b) $T_{sum} = 4T + 32T + 64T = 100T = 100 \times 1 \text{ s} = 100$ s.

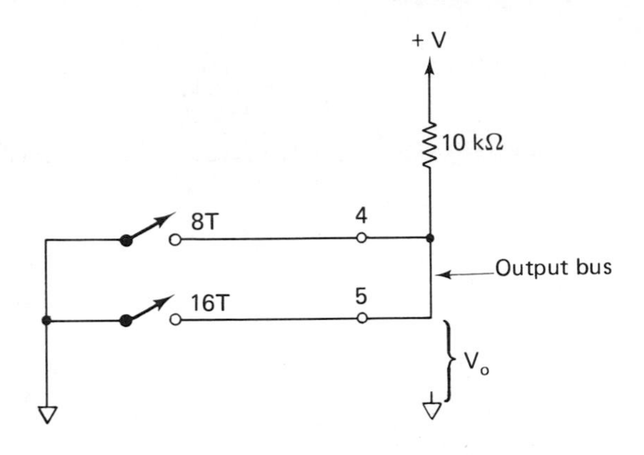

(a) Pins 4 and 5 are wired together to program 24T

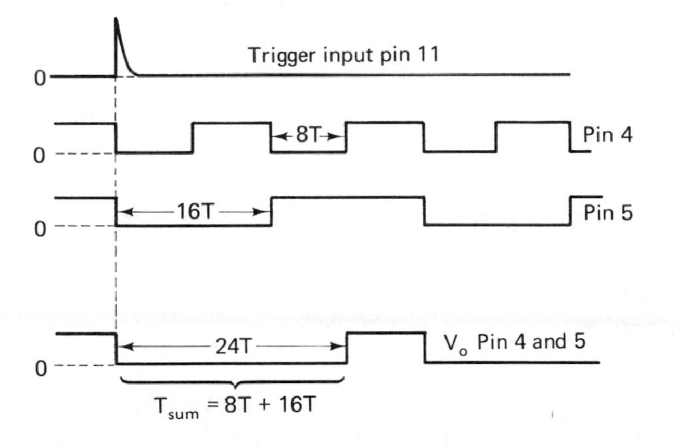

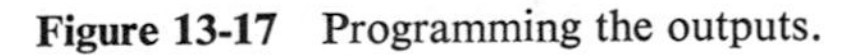
(b) Common bus V_o stays low as long as either pin 4 or pin 5 stays low

Figure 13-17 Programming the outputs.

By using switches instead of jumper wires, T_{sum} can be easily changed or *programmed* for any desired timing cycle from T to $255T$.

13-9 Timer/Counter Applications

13-9.1 Timing Applications. The 2240 is wired for monostable operation in the programmable timer application of Fig. 13-18. When the trigger input goes high, the output bus goes low for a timing cycle period equal to T_{sum}.

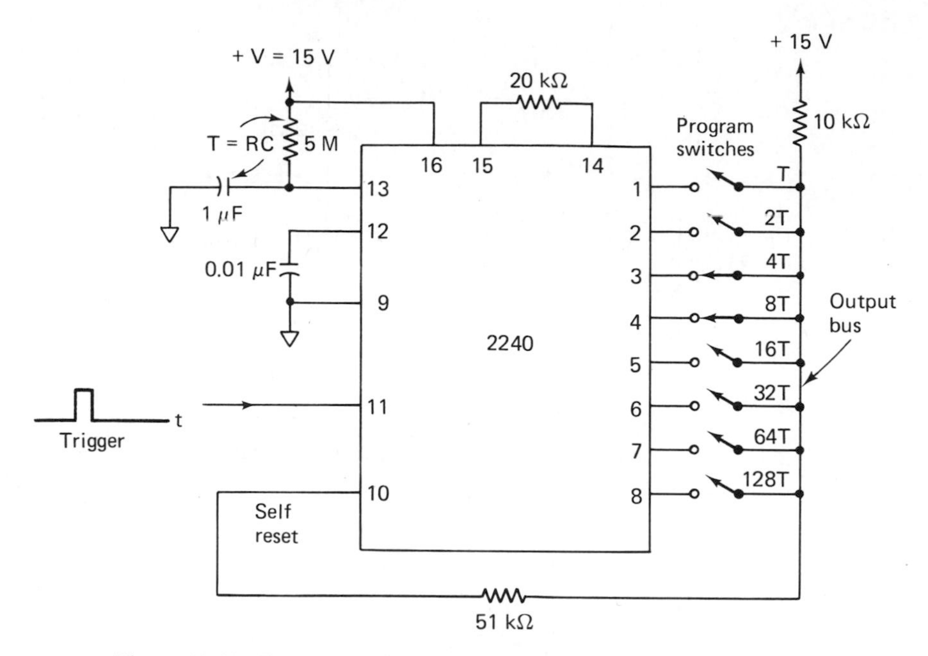

Figure 13-18 Programmable timer 5 seconds to 21 min + 15 sec in 5 second intervals.

(See Section 13-8.3.) At the end of the timing cycle, the output bus goes high. The connection from output bus via a 51 kΩ resistor to reset pin 10, forces the timer to reset itself when the output bus goes high. Thus after each trigger pulse, the 2240 generates a timing interval selected by the program switches.

Example 13-9: In Fig. 13-18, $C = 1.0\ \mu$F and $R = 5$ MΩ to establish a time base period given by Eq. (13-10) to be 5 s. What is (a) the timing cycle for switch positions shown in Fig. 13-18, (b) the minimum programmable timing cycle, and (c) the maximum programmable timing cycle?

Solution: (a) $T_{\text{sum}} = 4T + 8T = 12T = 12 \times 5\text{ s} = 60\text{ s} = 1\text{ min}$; (b) minimum timing cycle is $1T = 5$ s; (c) with all program switches closed,

$$T_{\text{sum}} = T + 2T + 4T + 8T + 16T + 32T + 64T + 128T = 225T$$

$$225T = 225 \times 5\text{ s} = 1275\text{ s} = 21\text{ min and }15\text{ s}$$

13-9.2 Free-Running Oscillator, Synchronized Outputs. The 2240 operates as a free-running oscillator in the circuit of Fig. 13-19. The reset terminal is grounded so that the 2240 will stay in its timing cycle once it is started. When power is applied, R_R and C_R couple a positive-going pulse into trigger input 11 to start the internal time base oscillator running.

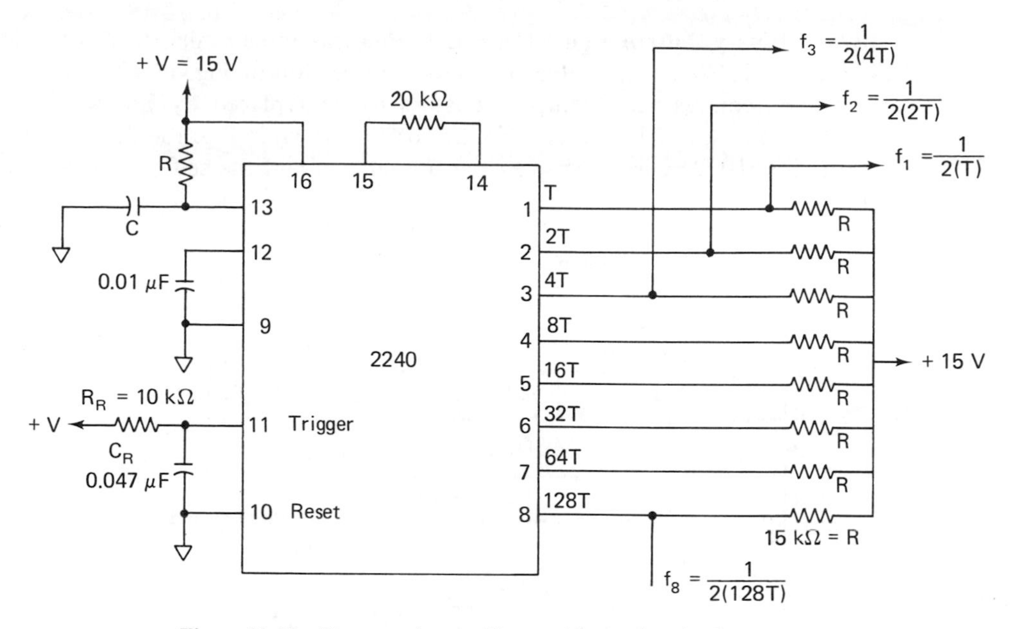

Figure 13-19 Free-running oscillator with synchronized outputs.

Each output is wired through an external control switch to an individual pull-up resistor. A square wave output voltage is available at each counter output. Their frequencies have a binary relationship. That is, the frequency available at each pin is one-half the frequency present at the preceeding pin. The wave shapes are identical to those in Fig. 13-16(b). Observe that the *period* of the f_1 frequency at pin 1 is twice the time base period rating T or $2(T)$. Thus, $f_1 = 1/2T$. At pin 4, the period is $2(8T)$ and $f_4 = 1/16T$.

Example 13-10: In Fig. 13-19, $T = 2.5$ ms; what frequencies are present at (a) output 1, (b) output 2, (c) output 3, and (d) output 4?
Solution: Tabulating calculations,

Pin No.	*Time Base Rating*	*Period*	*Frequency*
1	T	$2T = 5$ s	200 Hz
2	$2T$	$4T = 10$ ms	100 Hz
3	$4T$	$8T = 20$ ms	50 Hz
4	$8T$	$16T = 40$ ms	25 Hz

The connections to pins 10 and 11 may be removed to allow the oscillator to be started with a positive-going trigger pulse at pin 11. To stop oscillation, apply a positive-going pulse to reset pin 10.

13-9.3 Binary Pattern Signal Generator. Pulse patterns similar to those shown in Fig. 13-20 are generated by a modified version of Fig. 13-19. The modification requires the 8 output resistors to be replaced by program switches and a single 10-kΩ resistor similar to that shown in Fig. 13-18. Also eliminate the 51-kΩ resistor between the output bus and the self reset terminal.

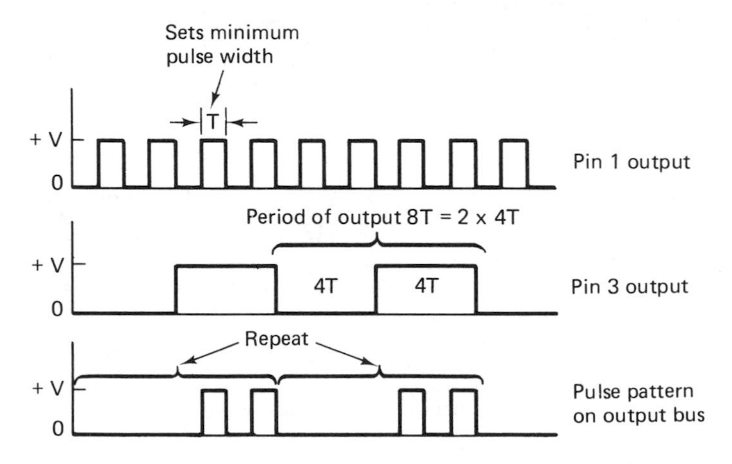

Figure 13-20 Binary pattern signal generator with outputs T and $4T$ connected to output bus.

The output is a train of pulses (as shown in Appendix 5, Fig. 22) that depends on which program switches are closed. The period of the pulse pattern is set by the highest program switch that is closed, and the pulse width is set by the lowest program switch that is closed. For example, if the $4T$ (pin 3) and $1T$ (pin 1) switches are closed, the pulse pattern is repeated every $2 \times 4T = 8T$ seconds (see Fig. 13-20). The minimum pulse width is $1T$. To determine the actual pulse pattern, refer to the timing chart in Fig. 13-16(b). If switches $1T$ and $4T$ are closed, there is an output pulse only when there are high output pulses from each line. The repeating pulse patterns are shown in Fig. 13-20.

13-9.4 Frequency Synthesizer. The output bus in Fig. 13-21(a) is capable of generating any one of 255 related frequencies. Each frequency is selected by closing the desired program switches to program a particular frequency at output V_o.

To understand circuit operation, assume the output bus goes high. This will drive reset pin 10 high and couple a positive-going pulse into trigger pin 11. The reset terminal going positive resets the 2240 (all outputs low). The positive pulse on pin 10 retriggers the 2240 time base oscillator, to begin generation of a time period that depends on which program switches are

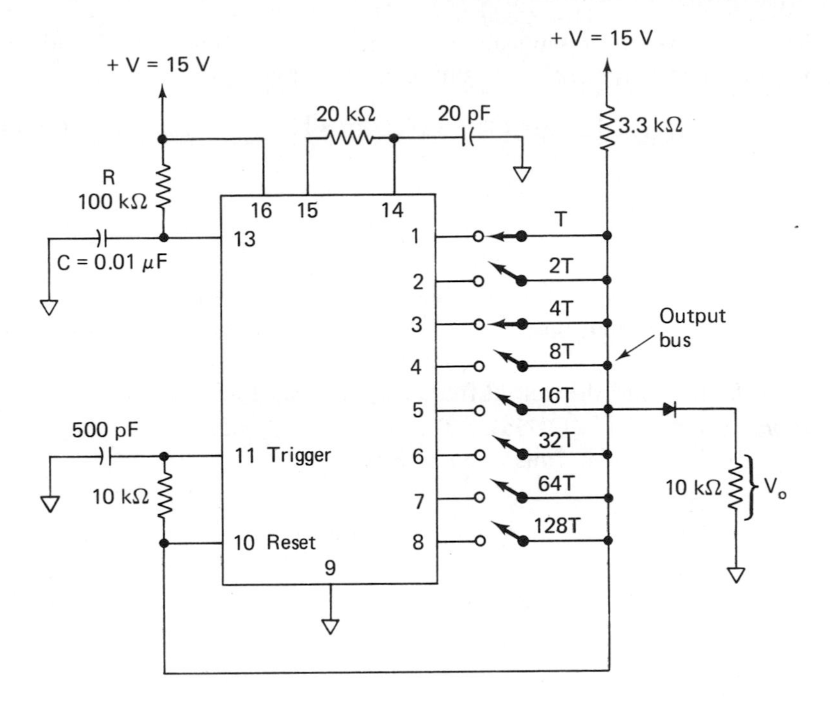

(a) Frequency synthesizer connections

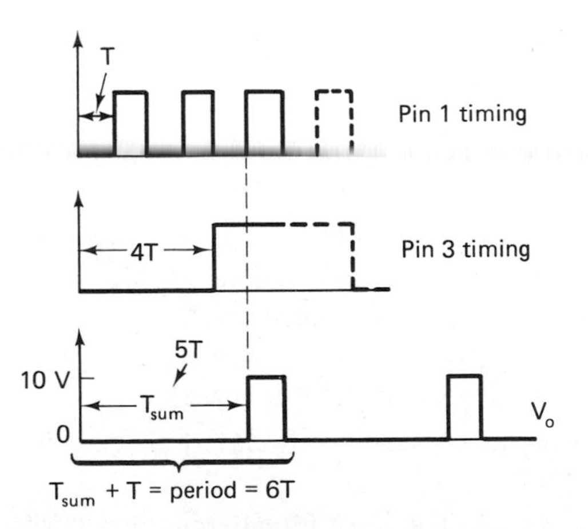

(b) Output voltage for program switches 1 and 4 closed

Figure 13-21 Frequency synthesizer, $T = 1$ ms, $f = 166$ Hz.

closed. For example, assume that switches T and $4T$ are closed in Fig. 13-21(a). The timing for these switches is shown in Fig. 13-21(b). The output bus stays low for $4T$ from pin 3 plus $1T$ from pin 1 before going high (to initiate a reset-retrigger sequence noted above). The time period and frequency of the output signal V_o is thus expressed by

$$\text{period} = (T_{sum} + T) \tag{13-11a}$$

and

$$f = \frac{1}{\text{period}} \tag{13-11b}$$

where T_{sum} is found by adding the time base rating for each output terminal connected to the output bus.

Example 13-11: Find the output frequency for Fig. 13-21(a).
Solution: From Eq. (13-11a), $T_{sum} = 1T + 4T = 5T$, and period $=$ $(T_{sum} + T) = 6T = 6 \times 1$ ms $= 6$ ms. By Eq. (13-11b),

$$f = \frac{1}{6 \times 10^{-3}\ \text{s}} = 166\ \text{Hz}$$

Other representative applications of the XR 2240 are shown in Appendix 5.

Problems

13-1 What are the operating modes of the 555 timer?

13-2 In Fig. 13-6(a), $R_A = R_B = 10$ kΩ, $C = 0.1$ μF. Find (a) t_{high}, (b) t_{low} and (c) frequency of oscillation.

13-3 Using the graph of Fig. 13-7, estimate the free-running frequency of oscillation, f, if $(R_A + 2R_B) = 1$ MΩ and $C = 0.02$ μF.

13-4 What is the duty cycle in Problem 13-2?

13-5 In Example 13-1, R_A and R_B are increased by a factor of 10 to 68 kΩ and 33 kΩ. Find the new frequency of oscillation.

13-6 In Fig. 13-8, R_A and R_B are each reduced to 5 kΩ. What is the effect on (a) the duty cycle and (b) the period T of the output?

13-7 In Fig. 13-9, at what value should the 10-kΩ resistor be set for a 2-kHz output from the B 555?

13-8 What is the duty cycle for Fig. 13-10 if the 10-kΩ potentiometer is set in the middle (5 kΩ on either side of the wiper)?

13-9 In Fig. 13-9(a), $R_A = 100$ kΩ and $C = 0.1$ μF. Find t_{high}.

13-10 R_A is changed to 20 kΩ in Example 13-5. Find t_{high}.

13-11 In Example 13-6(b), what value of R_A is required to divide a 1-kHz signal by 2?

13-12 Refer to Example 13-7, how long will the following output terminals stay low (a) pin 1, (b) pin 2, (c) pin 5, and (d) pin 6?

13-13 In Fig. 13-17(a), *T* is set for 1 ms and pins 2, 4, 6, and 8 are connected to the output bus. Find the timing interval.

13-14 In Problem 13-13, the odd-numbered pins 1, 3, 5, and 7 are connected to the output bus. Find the timing interval.

13-15 In Example 13-9, *C* is changed to 0.1 μF and *R* to 500 kΩ. Find (a) the time base period, (b) the timing cycle for switch positions shown in Fig. 13-18, and (c) the maximum timing cycle.

13-16 In Example 13-10, what frequencies are present at pins (a) 5, (b) 6, (c) 7, and (d) 8?

13-17 In Fig. 13-21, only switches to pins 1, 2, 3, and 4 are closed. Find the output frequency.

Bibliography

AN 273A, More Value Out of Integrated Operational Amplifier Data Sheets, Motorola Semiconductor Products Inc., Phoenix, Ariz. (1970).

Applications Manual for Computing Amplifiers for Modeling, Measuring, Manipulating and Much Else, Philbrick Researchers, Inc., Nimrod Press, Inc., Boston (1966).

BECKETT F., *A Practical Approach to Differential Amplifiers and Measurements*, Tekscope, Tektronix Inc., Beaverton, Ore. (1972, 1973).

BIRMAN P., *Operational Power Supply Technology*, Kepco Inc., Flushing, N.Y. (1968).

CLAYTON G. B., *Operational Amplifiers*, Butterworth & Co. (Publishers) Ltd., London (1971).

COUGHLIN, ROBERT F., *Principles and Applications of Semiconductors and Circuits*, Prentice-Hall, Inc., Englewood Cliffs, N.J. (1971).

COUGHLIN, ROBERT F. and DRISCOLL, FREDERICK F., *Semiconductor Fundamentals*, Prentice-Hall, Inc., Englewood Cliffs, N.J. (1976).

CROMWELL L., et al, *Biomedical Instrumentation and Measurements*, Prentice-Hall, Inc., Englewood Cliffs, N.J. (1973).

Data Sheet, Timer, 555, Signetics, 811 East Arques Ave., Sunnyvale, Ca. (1972).

DEMROW, ROBERT J., "Narrowing the Margin of Error," *Electronics*, April 15, 1968.

DEMROW, ROBERT J., "Protecting Data from the Ground Up," *Electronics*, April 29, 1968.

DRISCOLL, FREDERICK F., *Analysis of Electric Circuits*, Prentice-Hall, Inc., Englewood Cliffs, N.J., (1973).

DRISCOLL, FREDERICK F. and COUGHLIN, ROBERT F., *Solid State Devices and Applications*, Prentice-Hall, Inc., Englewood Cliffs, N.J., 1975.

EHRSAM, W., *Audio Power Generation Using IC Op Amps*, Motorola Semiconductor Products Inc., Phoenix, Ariz. (1968).

GODDEN, A. K., "Amplify Biological Signals with IC's," *Electronic Design 17*, August 16, 1969.

GRAEME, J., "Have a 50 V Swing on a 30 V Supply," *EDN/EEE*, January 1, 1972.

Handbook of Operational Amplifier Applications, Burr-Brown Research Corporation, Tucson, Ariz. (1963).

HEATER, J. C., "Monolithic Timer Makes Convenient Touch Switch," *EDN*, Boston, Mass., April, 1973.

Hewlett-Packard Company, Palo Alto, Calif., *Floating Measurements and Guarding* (1970).

HOOVER, M. V., *A Medley of Linear ICs . . . and Some of Their Applications*, ST-4777A, RCA Solid State Division, Somerville, N.J. (1972).

The Linear Integrated Circuits Data Catalog, Fairchild Semiconductor, Mountain View, Ca. (1973).

Model 4450 Multiplier/Divider, Data Sheet, Teledyne Philbrick Nexus, Dedham, Mass. (1970).

MORRISON, R., *Grounding and Shielding Techniques in Instrumentation*, John Wiley & Sons, Inc., New York (1967).

National Semiconductor Corporation, *Linear Applications* (1972).

PAYNTER, H. M., *A Palimpsest on the Electronic Analog Art*, Geo. A. Philbrick Researches Inc. (1965).

POULIOT, F., *Simplify Amplifier Selection, Electronic Design* **16**, August 2, 1973.

RENSCHLER E., *Analysis and Basic Operation of the MC 1595*, AN 489, Motorola Semiconductor Products Inc., Phoenix, Ariz. (1970).

RENSCHLER E., *The MC 1539 Op Amp and Its Applications*, AN 439, Motorola Semiconductor Products Inc., Phoenix, Ariz. (1968).

Signetics, Digital, Linear, MOS Applications, Signetics Corporation, Sunnyvale, Ca. (1971).

Signetics Linear Volume 1 Data Book, Signetics Corporation, Sunnyvale, Ca. (1972).

SMITH, J. I., *Modern Operational Circuit Design*, John Wiley & Sons, Inc., New York (1971).

Total Linears, Raytheon Semiconductor, Mountain View, Ca. (no date).

VANDER KOOI, M. K., "Simple IC Meter Amplifier Circuit Measures 100 Nanoamps, Full-Scale, *EDN/EEE*, April 15, 1972.

VANDER KOOI, M. K., L144 *Programmable Micro-power Triple Op Amp*, Silconix Inc., Santa Clara, Ca. (1974).

VILLANUCCI, R. et al., *Electronic Techniques: Shop Practices and Construction*, Prentice-Hall, Inc., Englewood Cliffs, N.J. (1974).

Voltage Regulator Applications Handbook, Fairchild Semiconductor Corp., Mountain View, Ca. (1974).

WISEMAN, L. L., *A High Voltage Monolithic Operational Amplifier*, Motorola Semiconductor Products, Inc., Phoenix, Ariz. (1967).

Appendix

1

μA 741 Frequency-Compensated Operational Amplifier*

*Courtesy of Fairchild Semiconductor, a Division of Fairchild Camera and Instrument Corporation.

GENERAL DESCRIPTION — The μA741 is a high performance monolithic Operational Amplifier constructed using the Fairchild Planar* epitaxial process. It is intended for a wide range of analog applications. High common mode voltage range and absence of "latch-up" tendencies make the μA741 ideal for use as a voltage follower. The high gain and wide range of operating voltage provides superior performance in integrator, summing amplifier, and general feedback applications.

- **NO FREQUENCY COMPENSATION REQUIRED**
- **SHORT CIRCUIT PROTECTION**
- **OFFSET VOLTAGE NULL CAPABILITY**
- **LARGE COMMON-MODE AND DIFFERENTIAL VOLTAGE RANGES**
- **LOW POWER CONSUMPTION**
- **NO LATCH UP**

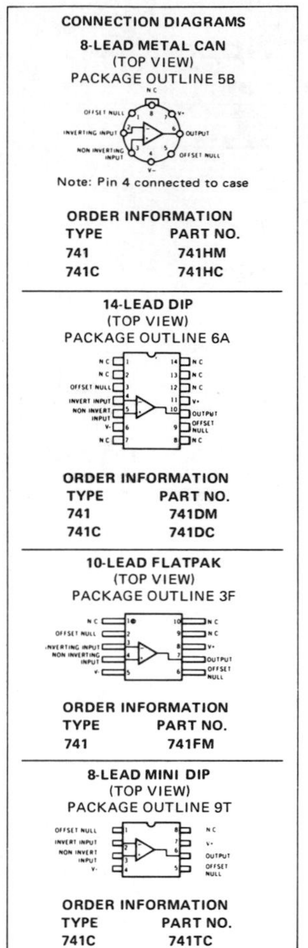

ABSOLUTE MAXIMUM RATINGS

Supply Voltage	
Military (741)	±22 V
Commercial (741C)	±18 V
Internal Power Dissipation (Note 1)	
Metal Can	500 mW
DIP	670 mW
Mini DIP	310 mW
Flatpak	570 mW
Differential Input Voltage	±30 V
Input Voltage (Note 2)	±15 V
Storage Temperature Range	
Metal Can, DIP, and Flatpak	−65°C to +150°C
Mini DIP	−55°C to +125°C
Operating Temperature Range	
Military (741)	−55°C to +125°C
Commercial (741C)	0°C to +70°C
Lead Temperature (Soldering)	
Metal Can, DIP, and Flatpak (60 seconds)	300°C
Mini DIP (10 seconds)	260°C
Output Short Circuit Duration (Note 3)	Indefinite

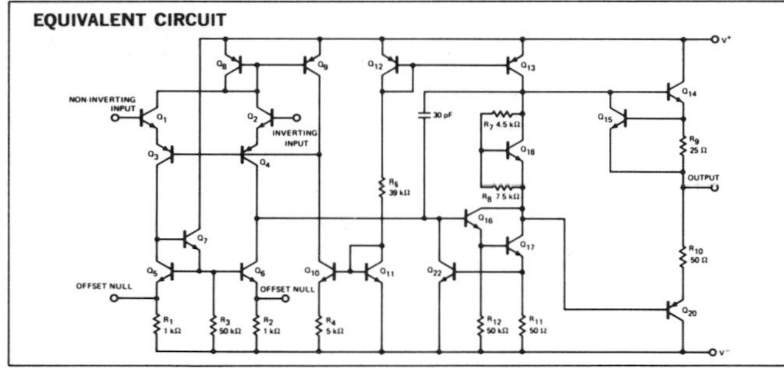

Notes on following pages.

*Planar is a patented Fairchild process.

ELECTRICAL CHARACTERISTICS (V_S = ±15 V, T_A = 25°C unless otherwise specified)

PARAMETERS (see definitions)		CONDITIONS	MIN.	TYP.	MAX.	UNITS
Input Offset Voltage		$R_S \leq 10$ kΩ		1.0	5.0	mV
Input Offset Current				20	200	nA
Input Bias Current				80	500	nA
Input Resistance			0.3	2.0		MΩ
Input Capacitance				1.4		pF
Offset Voltage Adjustment Range				±15		mV
Large Signal Voltage Gain		$R_L \geq 2$ kΩ, V_{OUT} = ±10 V	50,000	200,000		
Output Resistance				75		Ω
Output Short Circuit Current				25		mA
Supply Current				1.7	2.8	mA
Power Consumption				50	85	mW
Transient Response (Unity Gain)	Risetime	V_{IN} = 20 mV, R_L = 2 kΩ, $C_L \leq 100$ pF		0.3		μs
	Overshoot			5.0		%
Slew Rate		$R_L \geq 2$ kΩ		0.5		V/μs

The following specifications apply for $-55°C \leq T_A \leq +125°C$:

PARAMETERS	CONDITIONS	MIN.	TYP.	MAX.	UNITS
Input Offset Voltage	$R_S \leq 10$ kΩ		1.0	6.0	mV
Input Offset Current	T_A = +125°C		7.0	200	nA
	T_A = −55°C		85	500	nA
Input Bias Current	T_A = +125°C		0.03	0.5	μA
	T_A = −55°C		0.3	1.5	μA
Input Voltage Range		±12	±13		V
Common Mode Rejection Ratio	$R_S \leq 10$ kΩ	70	90		dB
Supply Voltage Rejection Ratio	$R_S \leq 10$ kΩ		30	150	μV/V
Large Signal Voltage Gain	$R_L \geq 2$ kΩ, V_{OUT} = ±10 V	25,000			
Output Voltage Swing	$R_L \geq 10$ kΩ	±12	±14		V
	$R_L \geq 2$ kΩ	±10	±13		V
Supply Current	T_A = +125°C		1.5	2.5	mA
	T_A = −55°C		2.0	3.3	mA
Power Consumption	T_A = +125°C		45	75	mW
	T_A = −55°C		60	100	mW

TYPICAL PERFORMANCE CURVES FOR 741

OPEN LOOP VOLTAGE GAIN AS A FUNCTION OF SUPPLY VOLTAGE

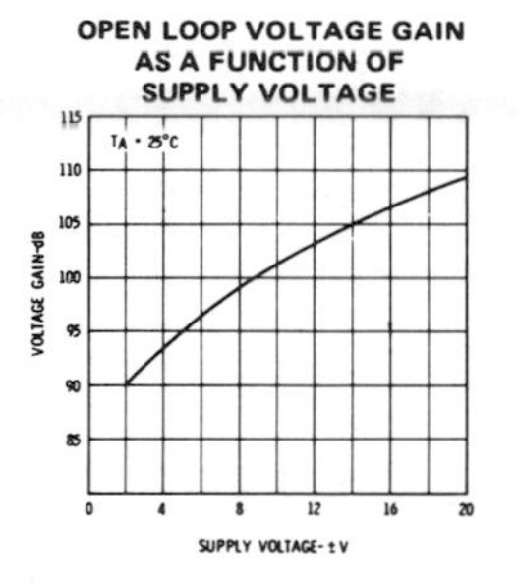

OUTPUT VOLTAGE SWING AS A FUNCTION OF SUPPLY VOLTAGE

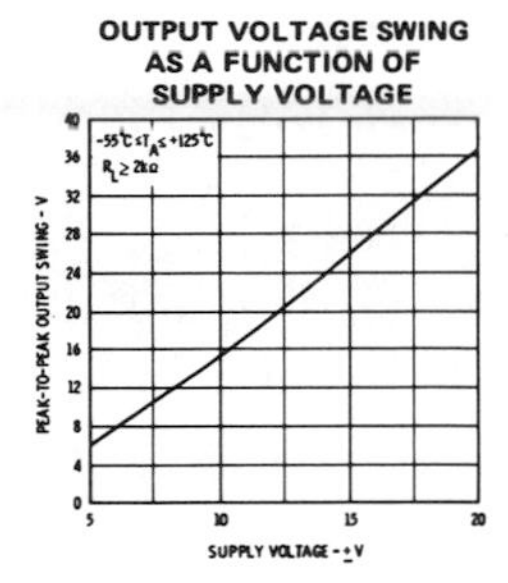

INPUT COMMON MODE VOLTAGE RANGE AS A FUNCTION OF SUPPLY VOLTAGE

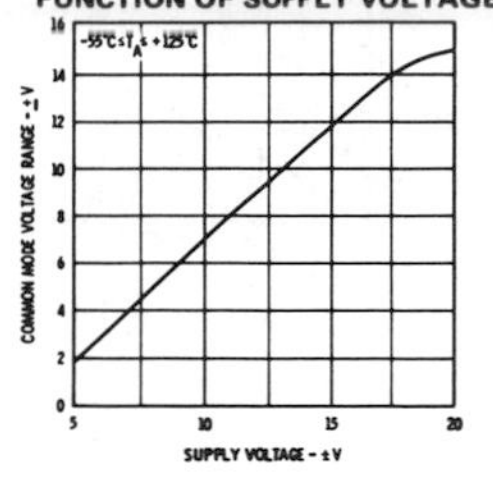

741C

ELECTRICAL CHARACTERISTICS (V_S = ±15 V, T_A = 25°C unless otherwise specified)

PARAMETERS (see definitions)	CONDITIONS	MIN.	TYP.	MAX.	UNITS
Input Offset Voltage	$R_S \leq 10\ k\Omega$		2.0	6.0	mV
Input Offset Current			20	200	nA
Input Bias Current			80	500	nA
Input Resistance		0.3	2.0		MΩ
Input Capacitance			1.4		pF
Offset Voltage Adjustment Range			±15		mV
Input Voltage Range		±12	±13		V
Common Mode Rejection Ratio	$R_S \leq 10\ k\Omega$	70	90		dB
Supply Voltage Rejection Ratio	$R_S \leq 10\ k\Omega$		30	150	μV/V
Large Signal Voltage Gain	$R_L \geq 2\ k\Omega$, V_{OUT} = ±10 V	20,000	200,000		
Output Voltage Swing	$R_L \geq 10\ k\Omega$	±12	±14		V
	$R_L \geq 2\ k\Omega$	±10	±13		V
Output Resistance			75		Ω
Output Short Circuit Current			25		mA
Supply Current			1.7	2.8	mA
Power Consumption			50	85	mW
Transient Response (Unity Gain) — Risetime	V_{IN} = 20 mV, R_L = 2 kΩ, $C_L \leq 100$ pF		0.3		μs
Transient Response (Unity Gain) — Overshoot			5.0		%
Slew Rate	$R_L \geq 2\ k\Omega$		0.5		V/μs

The following specifications apply for 0°C ≤ T_A ≤ +70°C:

PARAMETERS	CONDITIONS	MIN.	TYP.	MAX.	UNITS
Input Offset Voltage				7.5	mV
Input Offset Current				300	nA
Input Bias Current				800	nA
Large Signal Voltage Gain	$R_L \geq 2\ k\Omega$, V_{OUT} = ±10 V	15,000			
Output Voltage Swing	$R_L \geq 2\ k\Omega$	±10	±13		V

TYPICAL PERFORMANCE CURVES FOR 741C

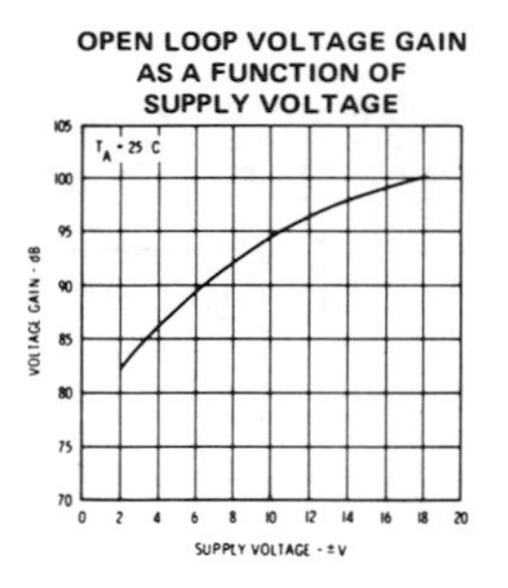

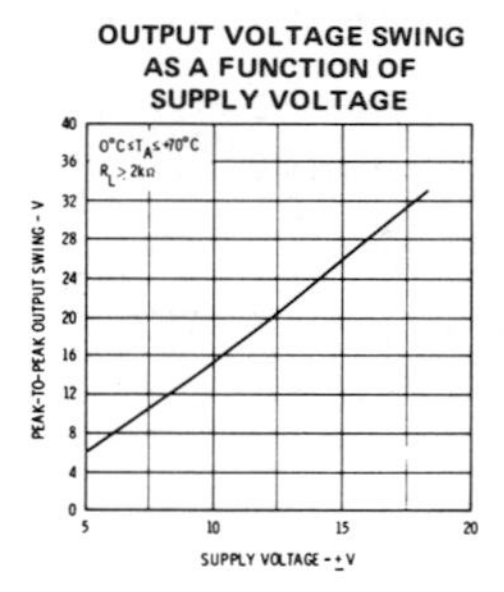

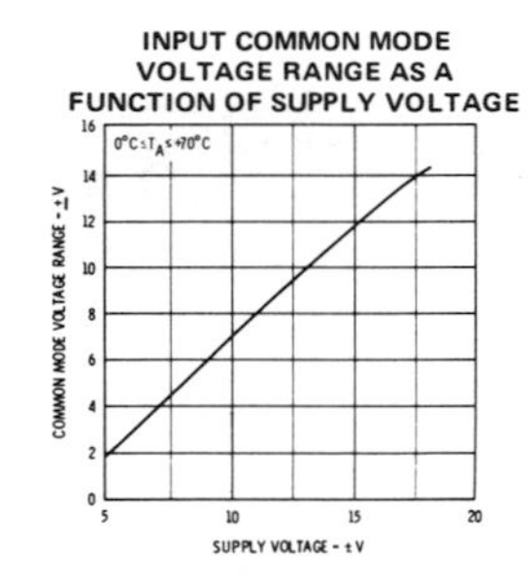

NOTES:

1. Rating applies to ambient temperatures up to 70°C. Above 70°C ambient derate linearly at 6.3 mW/°C for the Metal Can, 8.3 mW/°C for the DIP, 5.6 mW/°C for the Mini DIP and 7.1 mW/°C for the Flatpak.
2. For supply voltages less than ±15 V, the absolute maximum input voltage is equal to the supply voltage.
3. Short circuit may be to ground or either supply. Rating applies to +125°C case temperature or 75°C ambient temperature.

TYPICAL PERFORMANCE CURVES FOR 741

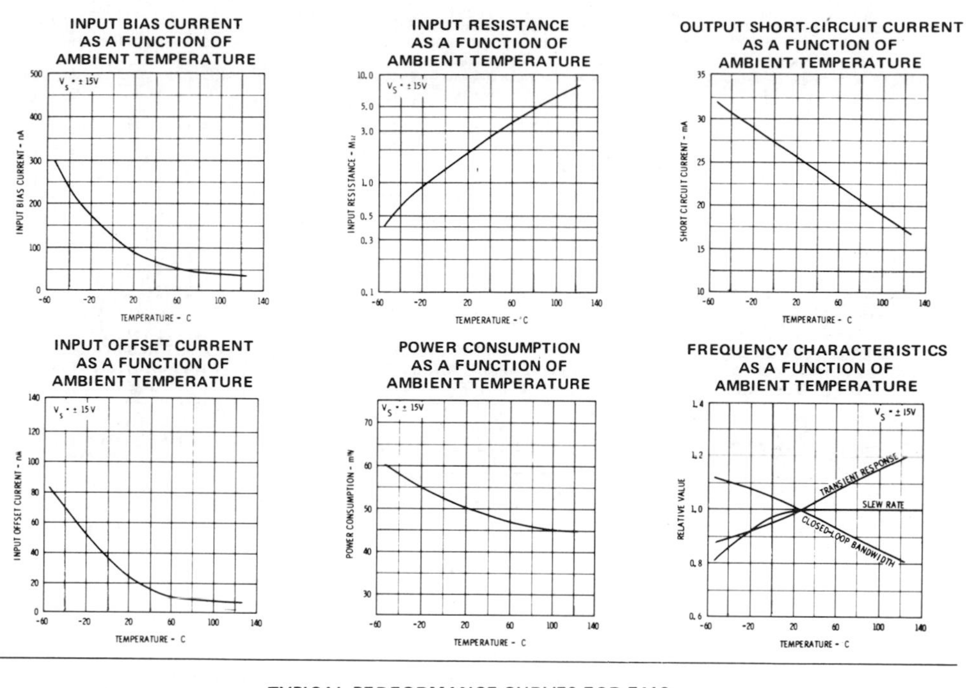

TYPICAL PERFORMANCE CURVES FOR 741C

INPUT BIAS CURRENT AS A FUNCTION OF AMBIENT TEMPERATURE

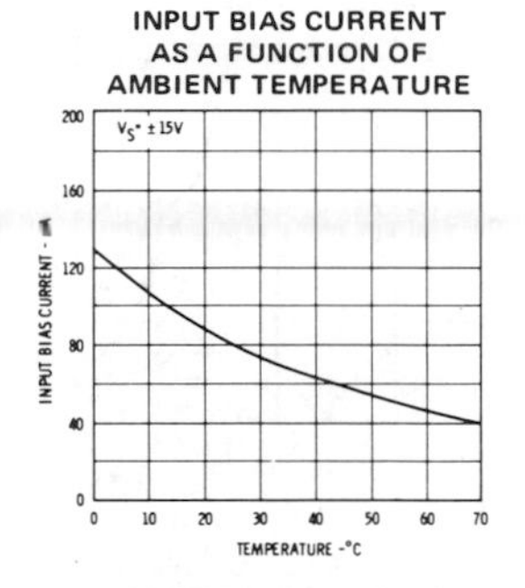

INPUT RESISTANCE AS A FUNCTION OF AMBIENT TEMPERATURE

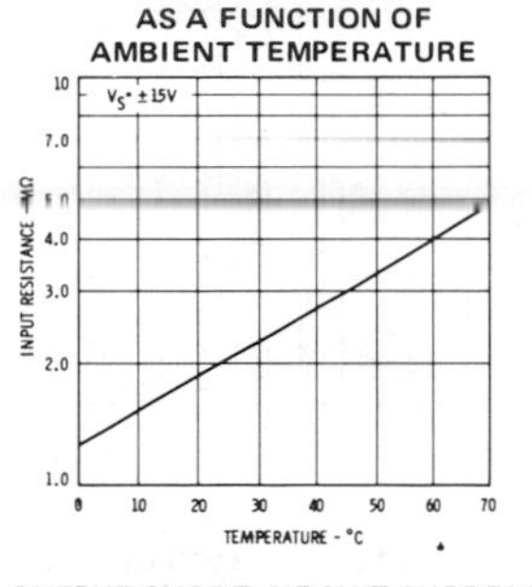

INPUT OFFSET CURRENT AS A FUNCTION OF AMBIENT TEMPERATURE

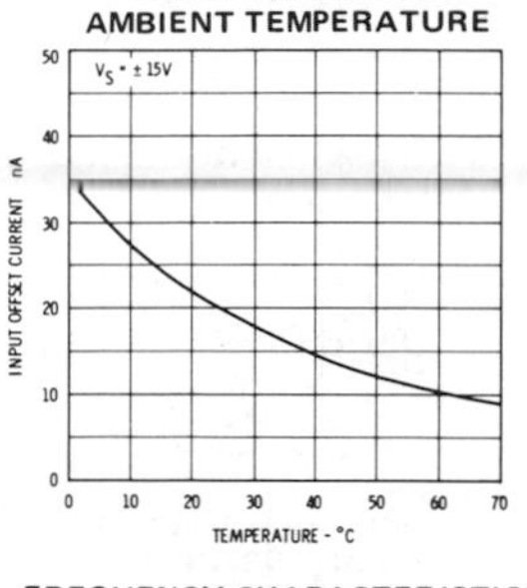

POWER CONSUMPTION AS A FUNCTION OF AMBIENT TEMPERATURE

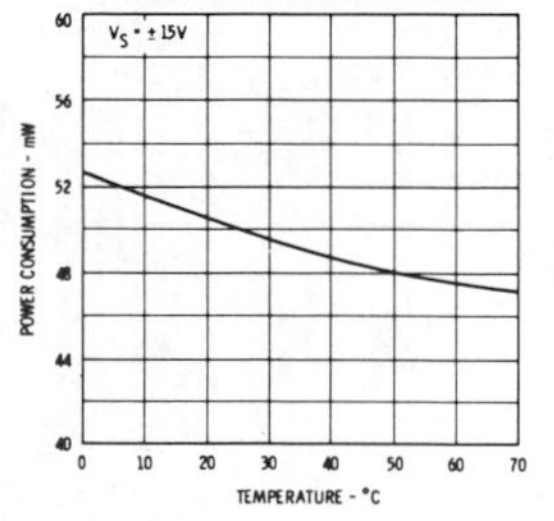

OUTPUT SHORT-CIRCUIT CURRENT AS A FUNCTION OF AMBIENT TEMPERATURE

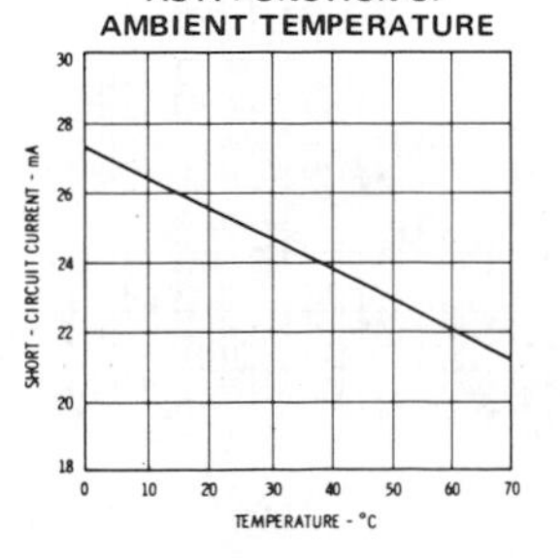

FREQUENCY CHARACTERISTICS AS A FUNCTION OF AMBIENT TEMPERATURE

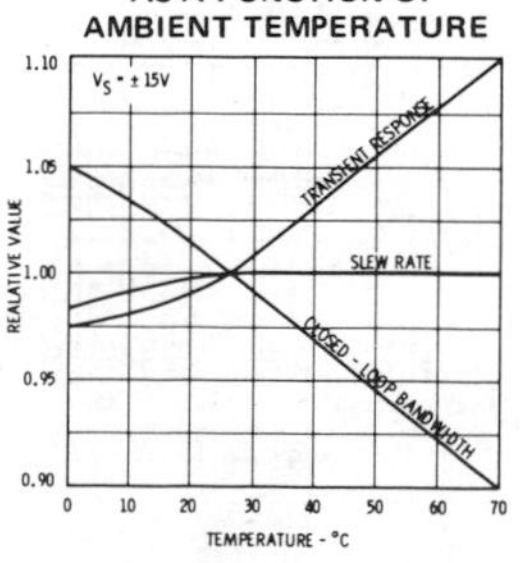

TYPICAL PERFORMANCE CURVES FOR 741 AND 741C (Cont'd)

POWER CONSUMPTION AS A FUNCTION OF SUPPLY VOLTAGE

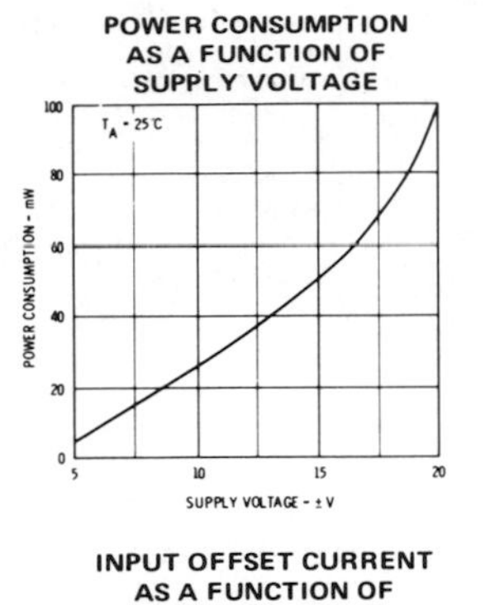

OPEN LOOP VOLTAGE GAIN AS A FUNCTION OF FREQUENCY

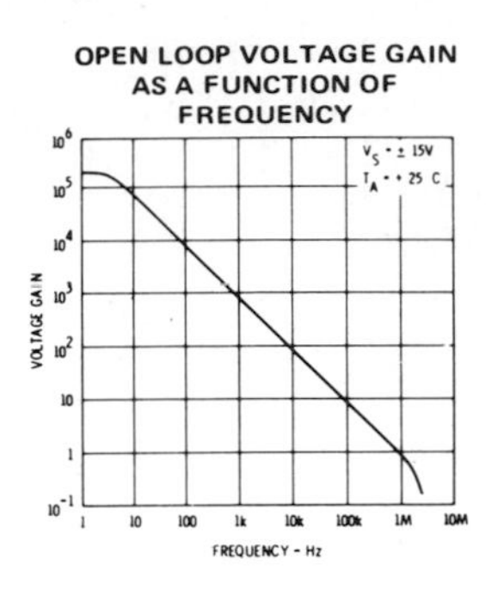

OPEN LOOP PHASE RESPONSE AS A FUNCTION OF FREQUENCY

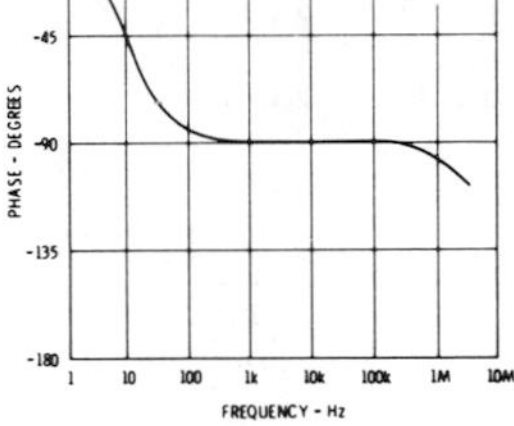

INPUT OFFSET CURRENT AS A FUNCTION OF SUPPLY VOLTAGE

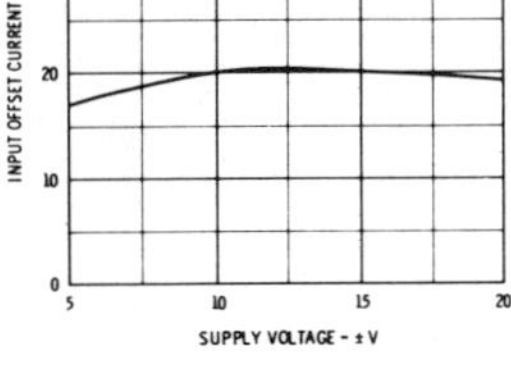

INPUT RESISTANCE AND INPUT CAPACITANCE AS A FUNCTION OF FREQUENCY

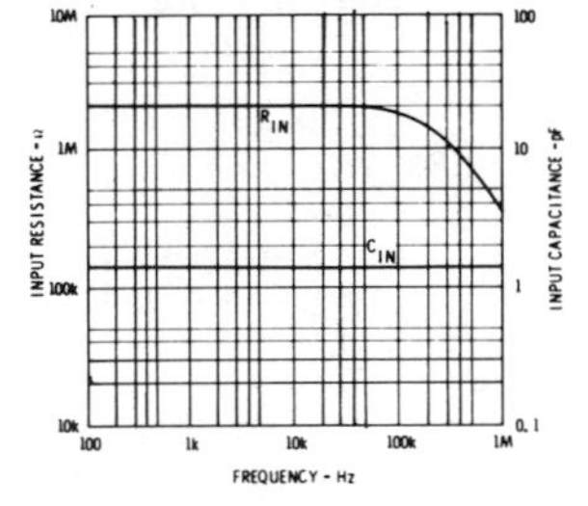

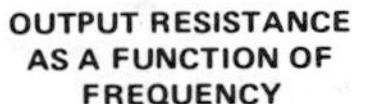

OUTPUT RESISTANCE AS A FUNCTION OF FREQUENCY

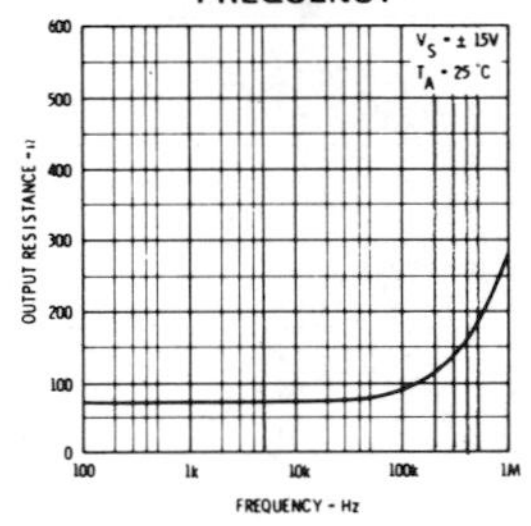

OUTPUT VOLTAGE SWING AS A FUNCTION OF LOAD RESISTANCE

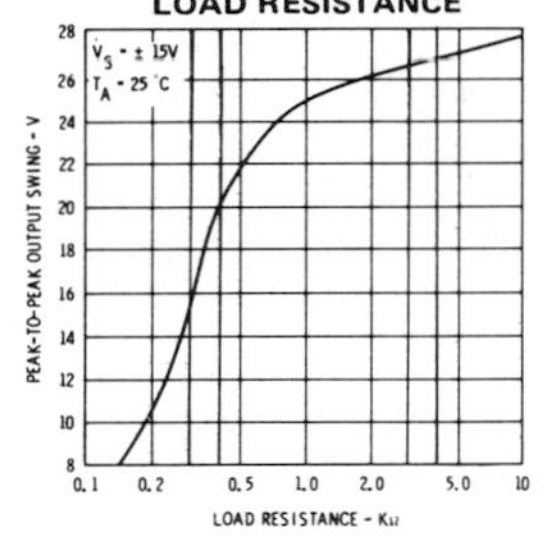

OUTPUT VOLTAGE SWING AS A FUNCTION OF FREQUENCY

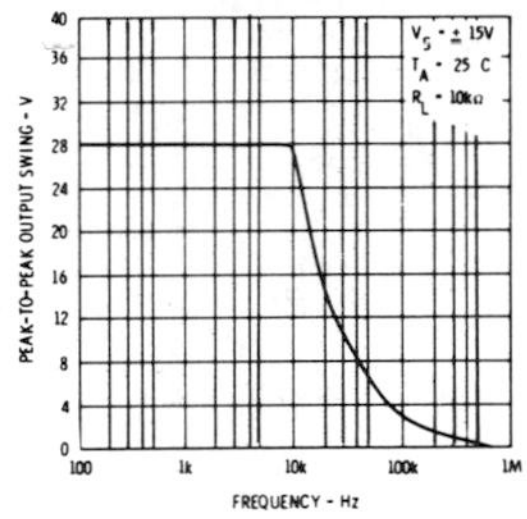

ABSOLUTE MAXIMUM POWER DISSIPATION AS A FUNCTION OF AMBIENT TEMPERATURE

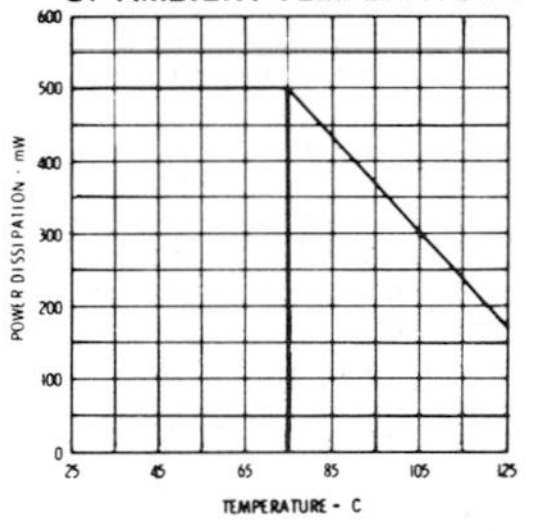

INPUT NOISE VOLTAGE AS A FUNCTION OF FREQUENCY

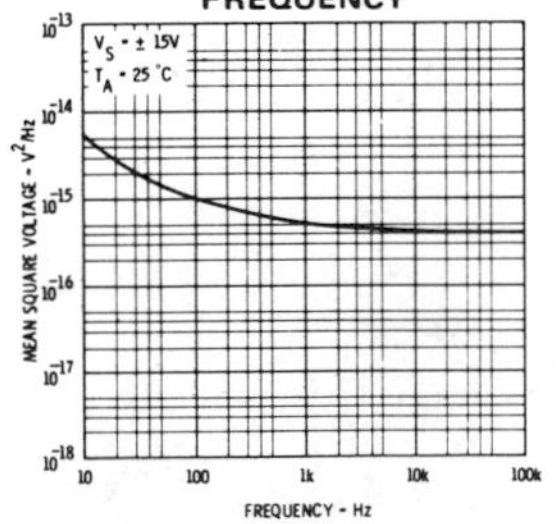

INPUT NOISE CURRENT AS A FUNCTION OF FREQUENCY

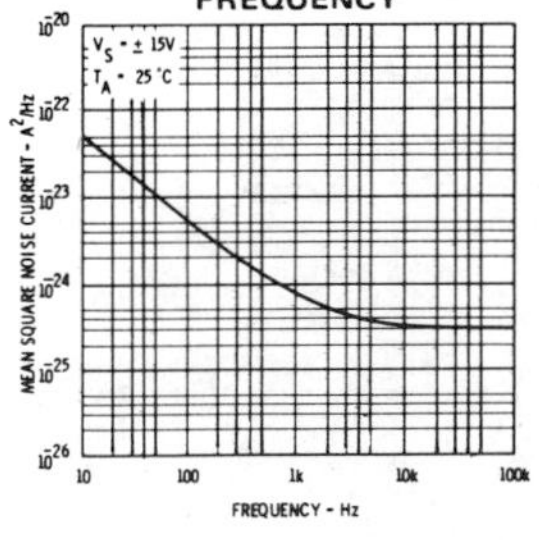

BROADBAND NOISE FOR VARIOUS BANDWIDTHS

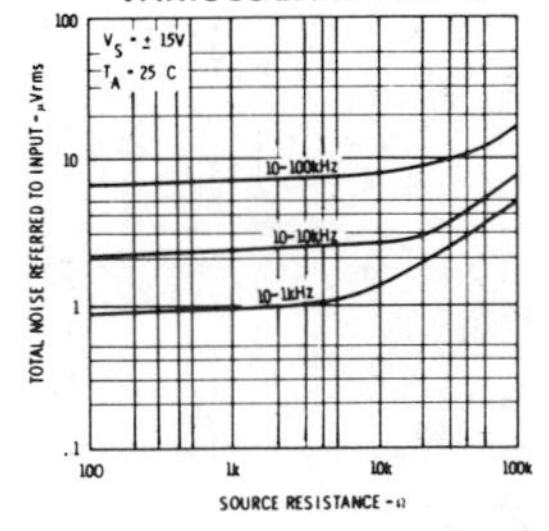

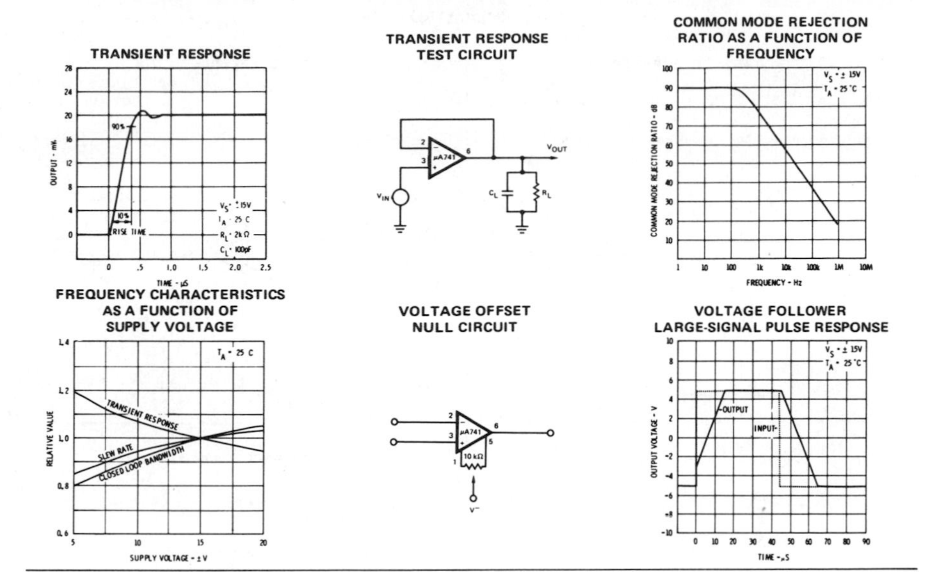

TYPICAL APPLICATIONS

UNITY-GAIN VOLTAGE FOLLOWER

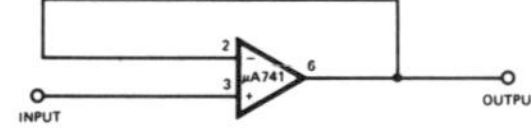

R_{IN} = 400 MΩ

C_{IN} = 1 pF

R_{OUT} << 1 Ω

B.W. = 1 MHz

NON-INVERTING AMPLIFIER

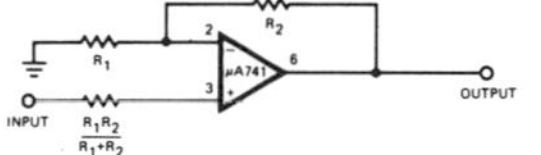

GAIN	R_1	R_2	B.W.	R_{IN}
10	1 kΩ	9 kΩ	100 kHz	400 MΩ
100	100 Ω	9.9 kΩ	10 kHz	280 MΩ
1000	100 Ω	99.9 kΩ	1 kHz	80 MΩ

INVERTING AMPLIFIER

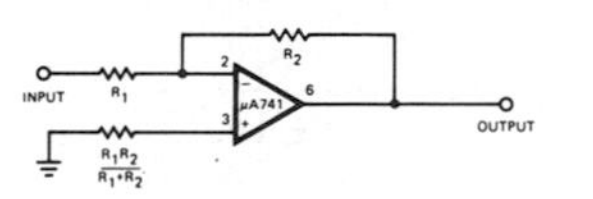

GAIN	R_1	R_2	B.W.	R_{IN}
1	10 kΩ	10 kΩ	1 MHz	10 kΩ
10	1 kΩ	10 kΩ	100 kHz	1 kΩ
100	1 kΩ	100 kΩ	10 kHz	1 kΩ
1000	100 Ω	100 kΩ	1 kHz	100 Ω

CLIPPING AMPLIFIER

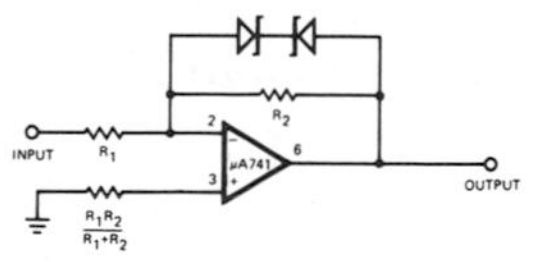

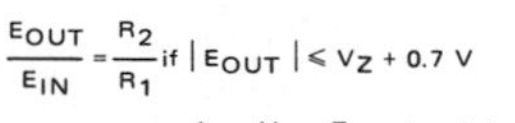

$$\frac{E_{OUT}}{E_{IN}} = \frac{R_2}{R_1} \text{ if } |E_{OUT}| \leq V_Z + 0.7 \text{ V}$$

where V_Z = Zener breakdown voltage

TYPICAL APPLICATIONS (Cont'd)

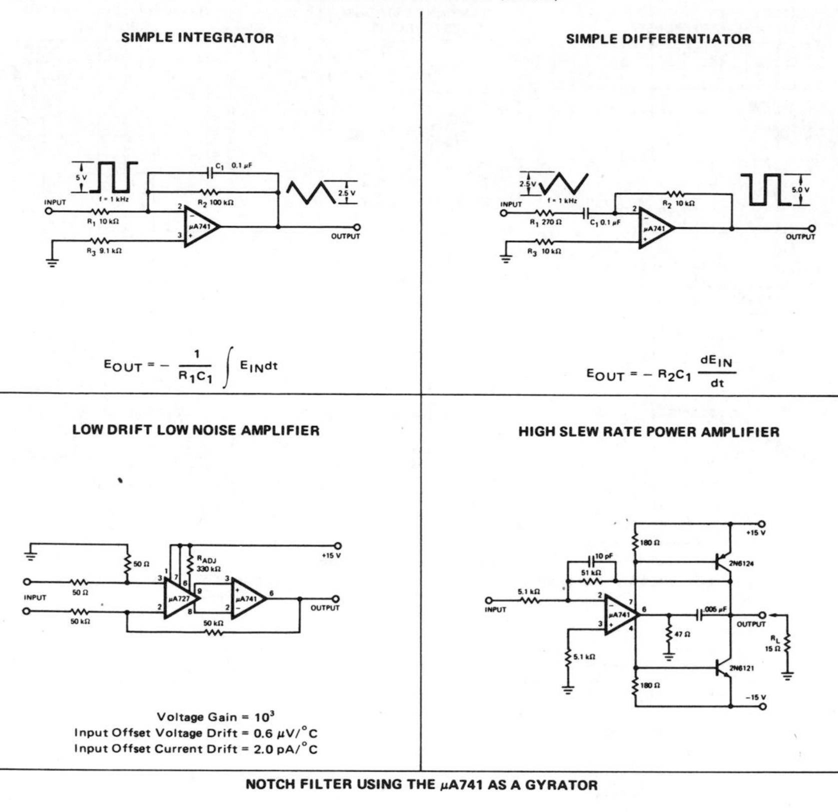

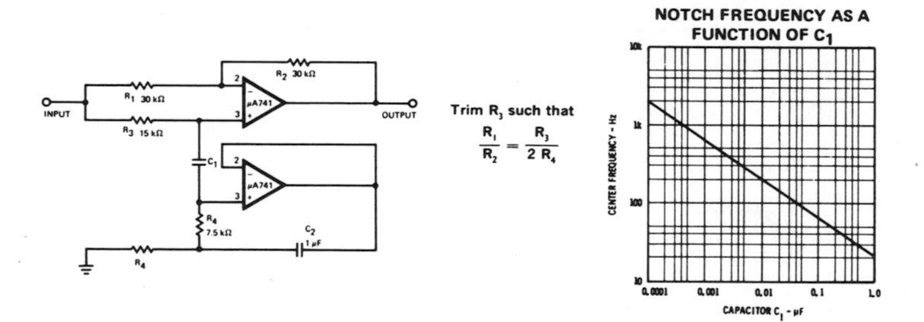

PHYSICAL DIMENSIONS

(H) 5B 8-LEAD METAL CAN
in accordance with JEDEC (TO-99) outline

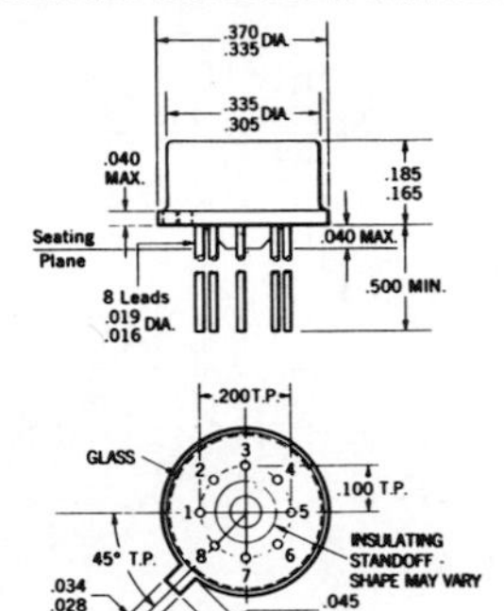

NOTES:
All dimensions in inches
Leads are gold-plated kovar
Package weight is 1.22 gram

(D) 6A 14-LEAD HERMETIC DIP
in accordance with JEDEC (TO-116) outline

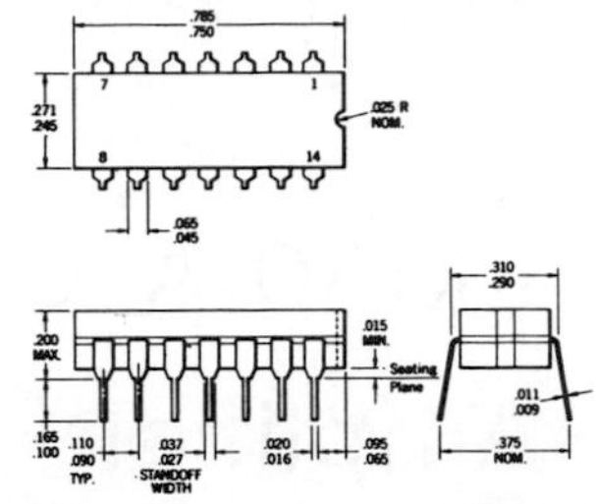

NOTES:
All dimensions in inches
Leads are intended for insertion in hole rows on .300" centers
They are purposely shipped with "positive" misalignment to facilitate insertion
Board-drilling dimensions should equal your practice for .020 inch diameter lead
Leads are tin-plated kovar
Package weight is 2.0 grams

(T) 9T 8-LEAD MINI DIP

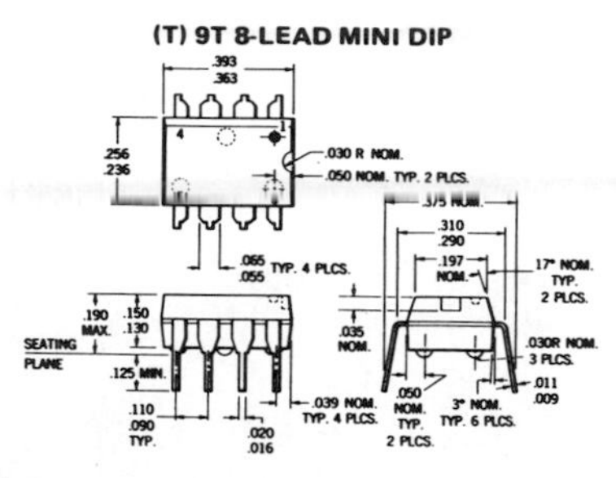

NOTES:
All dimensions in inches
Leads are intended for insertion in hole rows on .300" centers
They are purposely shipped with "positive" misalignment to facilitate insertion
Board-drilling dimensions should equal your practice for .020 inch diameter lead
Leads are tin or gold-plated kovar
Package weight is 0.6 gram

(F) 3F 10-LEAD FLATPAK
in accordance with JEDEC (TO-91) outline

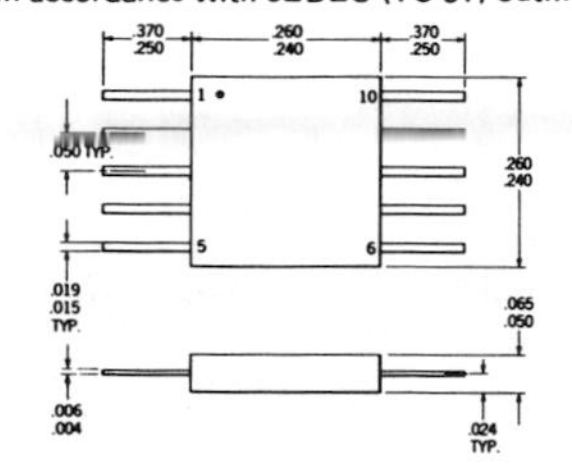

NOTES:
All dimensions in inches
Leads are gold-plated kovar
Package weight is 0.26 gram

Appendix

2

LM301A Operational Amplifier*

*Courtesy of National Semiconductor Corporation.

general description

The LM301A is a general-purpose operational amplifier which features improved performance over the 709C and other popular amplifiers. Advanced processing techniques make possible an order of magnitude reduction in input currents, and a redesign of the biasing circuitry reduces the temperature drift of input current.

This amplifier offers many features which make its application nearly foolproof: overload protection on the input and output, no latch-up when the common mode range is exceeded, freedom from oscillations and compensation with a single 30 pF capacitor. It has advantages over internally compensated amplifiers in that the compensation can be tailored to the particular application. For example, as a summing amplifier, slew rates of 10 V/µs and bandwidths of 10 MHz can be realized. In addition, the circuit can be used as a comparator with differential inputs up to ±30V; and the output can be clamped at any desired level to make it compatible with logic circuits.

The LM301A provides better accuracy and lower noise than its predecessors in high impedance circuitry. The low input currents also make it particularly well suited for long interval integrators or timers, sample and hold circuits and low frequency waveform generators. Further, replacing circuits where matched transistor pairs buffer the inputs of conventional IC op amps, it can give lower offset voltage and drift at reduced cost.

typical applications

Fast Summing Amplifier

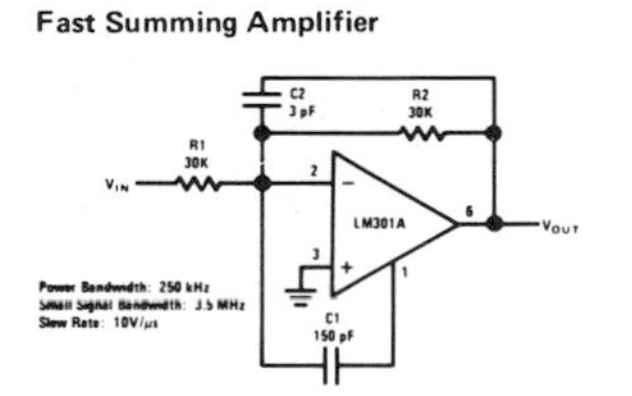

Fast Voltage Follower

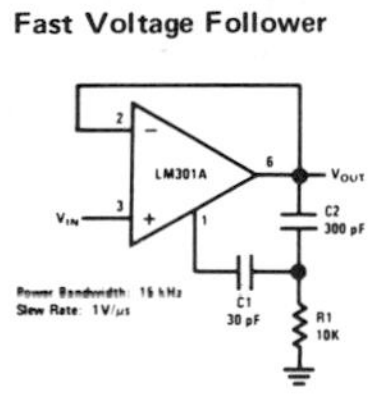

Standard Compensation and Offset Balancing Circuit

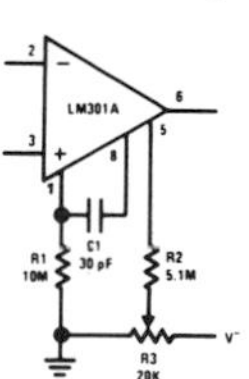

Integrator with Bias Current Compensation

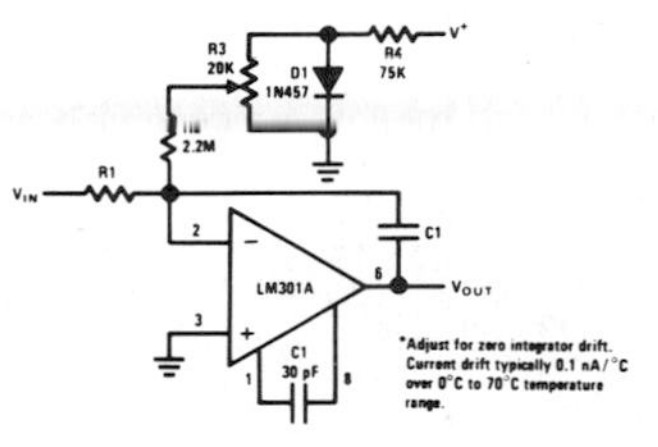

Low Frequency Square Wave Generator

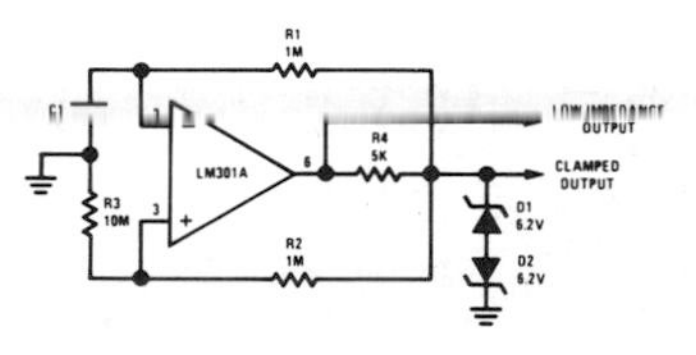

Bilateral Current Source

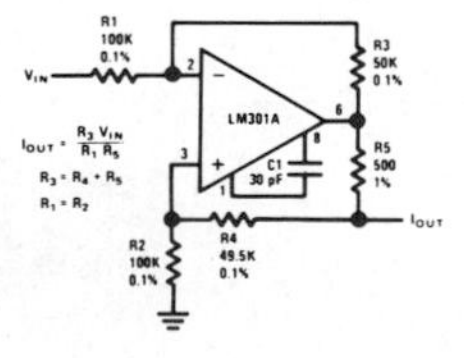

Voltage Comparator for Driving DTL or TTL Integrated Circuits

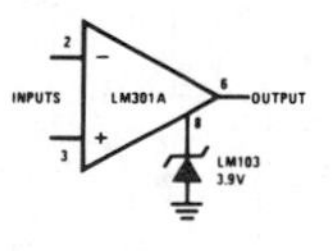

Double-Ended Limit Detector

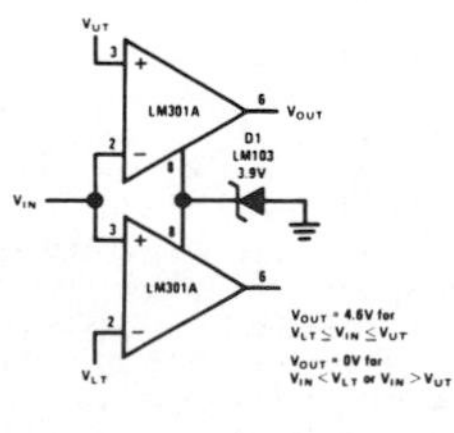

absolute maximum ratings

Supply Voltage	±18V
Power Dissipation (Note 1)	500 mW
Differential Input Voltage	±30V
Input Voltage (Note 2)	±15V
Output Short-Circuit Duration (Note 3)	Indefinite
Operating Temperature Range	0°C to 70°C
Storage Temperature Range	−65°C to 150°C
Lead Temperature (Soldering, 60 sec)	300°C

electrical characteristics (Note 4)

PARAMETER	CONDITIONS	MIN	TYP	MAX	UNITS
Input Offset Voltage	T_A = 25°C, $R_S \leq$ 50 kΩ		2.0	7.5	mV
Input Offset Current	T_A = 25°C		3	50	nA
Input Bias Current	T_A = 25°C		70	250	nA
Input Resistance	T_A = 25°C	0.5	2		MΩ
Supply Current	T_A = 25°C, V_S = ±15V		1.8	3.0	mA
Large Signal Voltage Gain	T_A = 25°C, V_S = ±15V V_{OUT} = ±10V, $R_L \geq$ 2 kΩ	25	160		V/mV
Input Offset Voltage	$R_S \leq$ 50 kΩ			10	mV
Average Temperature Coefficient of Input Offset Voltage			6.0	30	μV/°C
Input Offset Current				70	nA
Average Temperature Coefficient of Input Offset Current	25°C $\leq T_A \leq$ 70°C 0°C $\leq T_A \leq$ 25°C		0.01 0.02	0.3 0.6	nA/°C nA/°C
Input Bias Current				300	nA
Large Signal Voltage Gain	V_S = ±15V, V_{OUT} = ±10V $R_L \geq$ 2 kΩ	15			V/mV
Output Voltage Swing	V_S = ±15V, R_L = 10 kΩ R_L = 2 kΩ	±12 ±10	±14 ±13		V V
Input Voltage Range	V_S = ±15V	±12			V
Common Mode Rejection Ratio	$R_S \leq$ 50 kΩ	70	90		dB
Supply Voltage Rejection Ratio	$R_S \leq$ 50 kΩ	70	96		dB

Note 1: For operating at elevated temperatures, the device must be derated based on a 100°C maximum junction temperature and a thermal resistance of 150°C/W junction to ambient or 45°C/W junction to case.

Note 2: For supply voltages less than ±15V, the absolute maximum input voltage is equal to the supply voltage.

Note 3: Continuous short circuit is allowed for case temperatures to 70°C and ambient temperatures to 55°C.

Note 4: These specifications apply for 0°C $\leq T_A <$ 70°C, ±5V, $\leq V_S \leq$ ±15V and C1 = 30 pF unless otherwise specified.

guaranteed performance

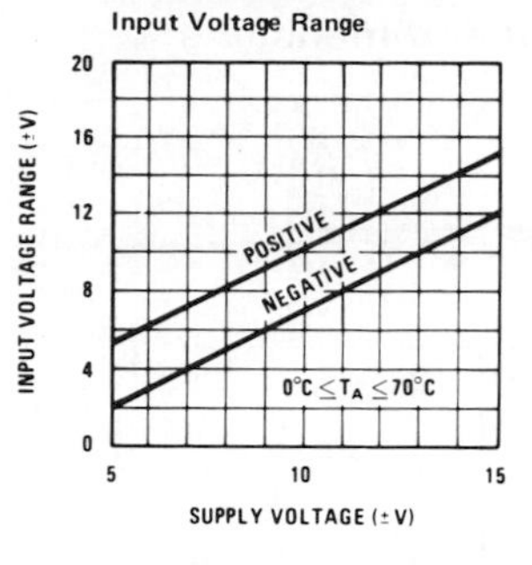

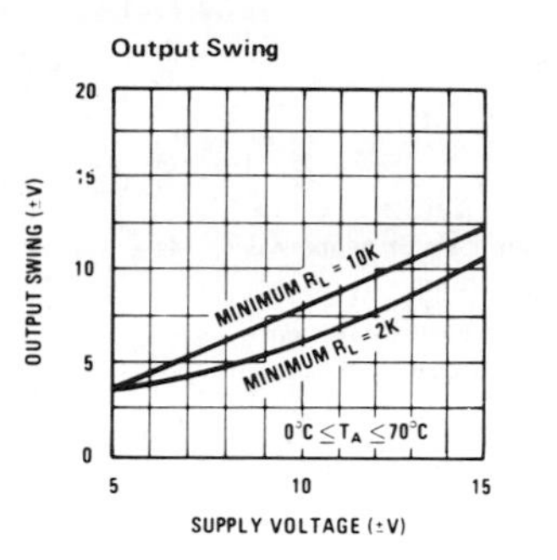

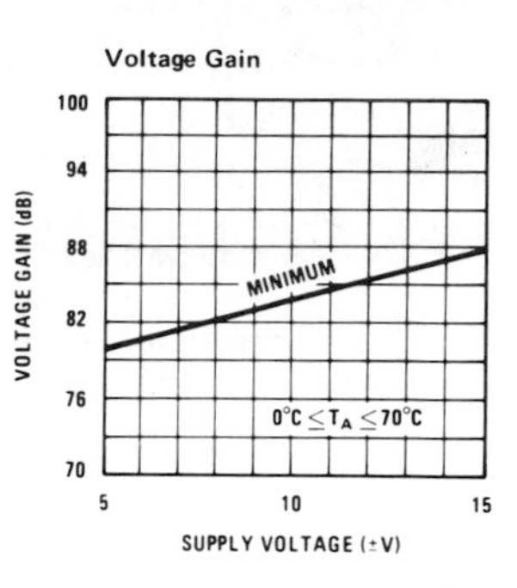

typical performance

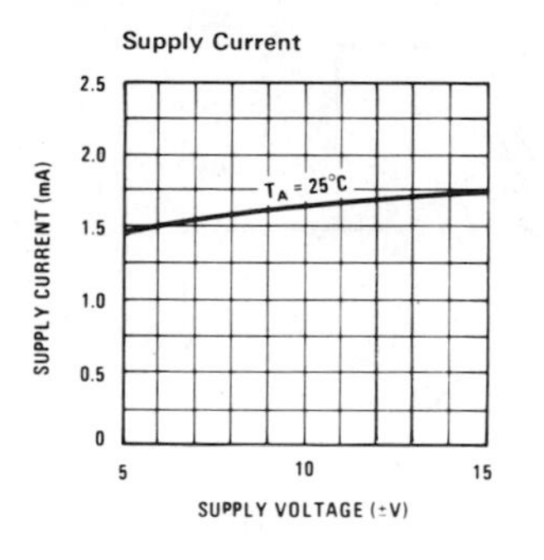

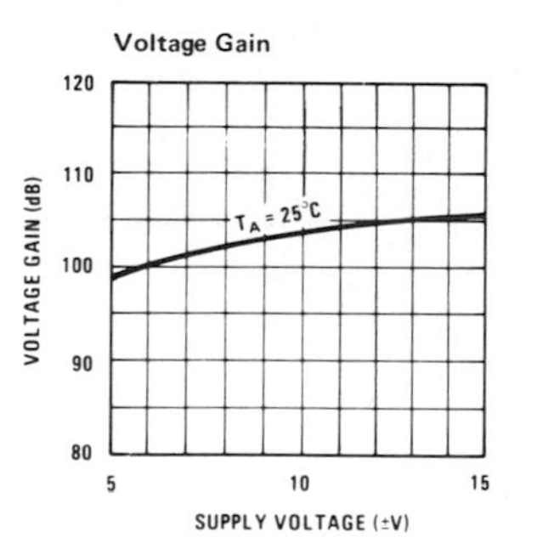

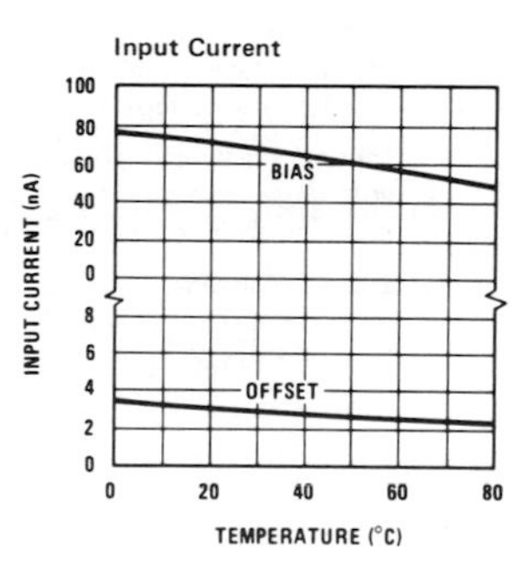

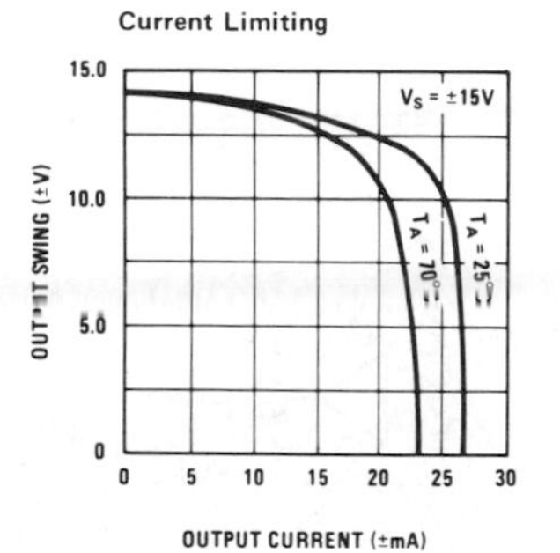

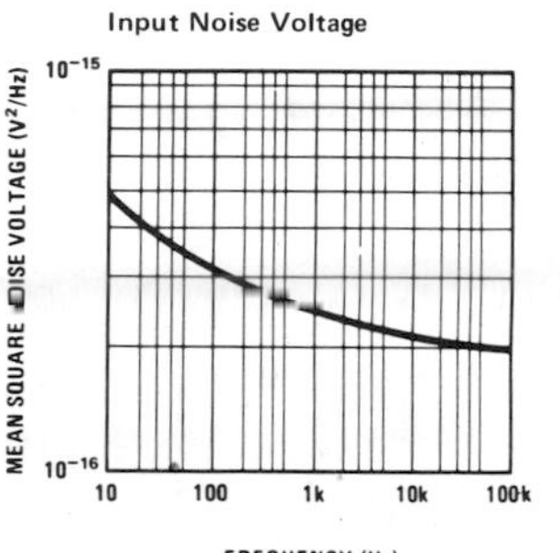
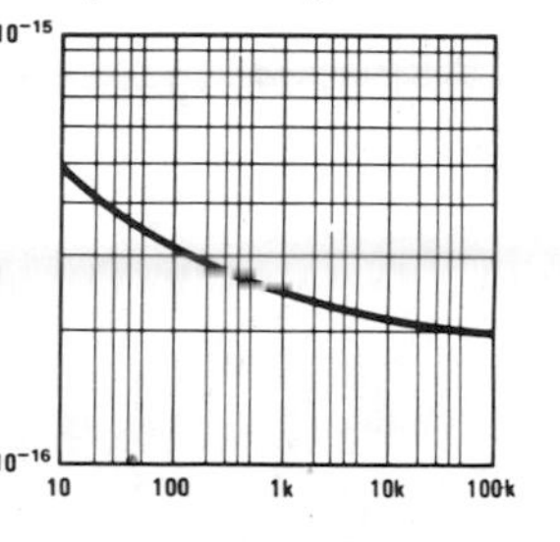

Input Noise Current

MEAN SQUARE NOISE CURRENT (A²/Hz)

FREQUENCY (Hz)

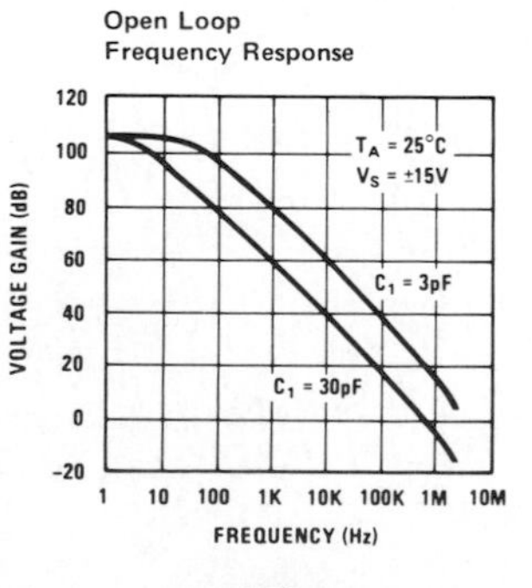

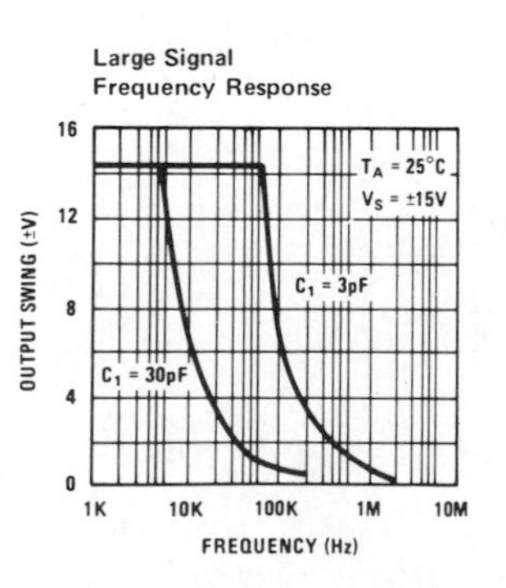

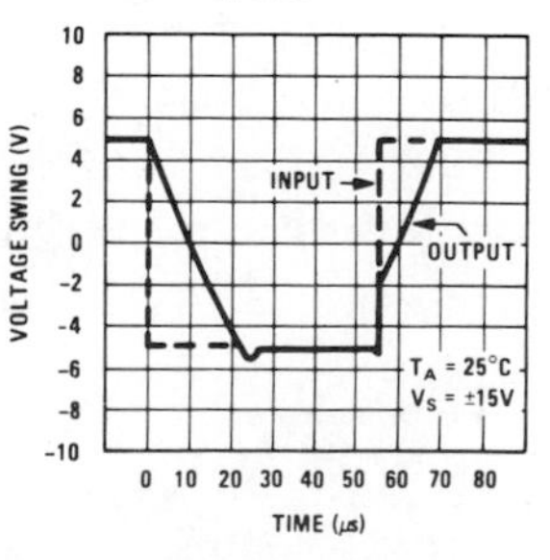

definition of terms

Input Offset Voltage: That voltage which must be applied between the input terminals through two equal resistances to obtain zero output voltage.

Input Offset Current: The difference in the currents into the two input terminals when the output is at zero.

Input Voltage Range: The range of voltages on the input terminals for which the offset specifications apply.

Input Bias Current: The average of the two input currents.

Common Mode Rejection Ratio: The ratio of the input voltage range to the peak-to-peak change in input offset voltage over this range.

Input Resistance: The ratio of the change in input voltage to the change in input current on either input with the other grounded.

Supply Current: The current required from the power supply to operate the amplifier with no load and the output at zero.

Output Voltage Swing: The peak output voltage swing, referred to zero, that can be obtained without clipping.

Large-Signal Voltage Gain: The ratio of the output voltage swing to the change in input voltage required to drive the output from zero to this voltage.

Power Supply Rejection: The ratio of the change in input offset voltage to the change in power supply voltages producing it.

physical dimensions

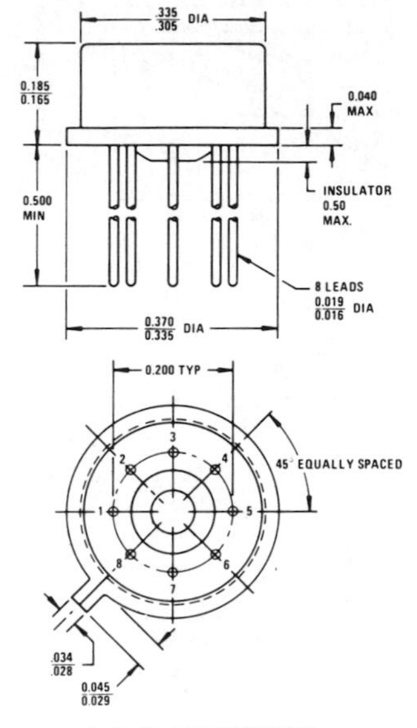

Order Number LM301AH

connection diagram

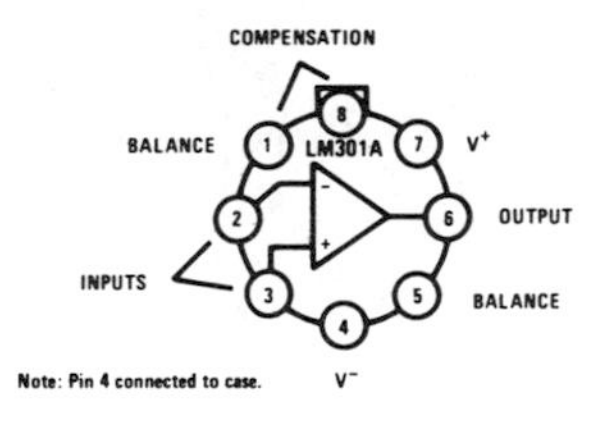

Appendix 3

LM311 Voltage Comparator*

*Courtesy of National Semiconductor Corporation.

general description

The LM311 is a voltage comparator that has input currents more than a hundred times lower than devices like the LM306 or LM710C. It is also designed to operate over a wider range of supply voltages: from standard ±15V op amp supplies down to the single 5V supply used for IC logic. Its output is compatible with RTL, DTL and TTL as well as MOS circuits. Further, it can drive lamps or relays, switching voltages up to 40V at currents as high as 50 mA. Outstanding characteristics include:

- Maximum input current: 250 nA
- Maximum offset current: 50 nA
- Differential input voltage range: ±30V
- Power consumption: 135 mW at ±15V

Both the input and the output of the LM311 can be isolated from system ground, and the output can drive loads referred to ground, the positive supply or the negative supply. Offset balancing and strobe capability are provided and outputs can be wire OR'ed. Although slower than the LM306 and LM710C (200 ns response time vs 40 ns) the device is also much less prone to spurious oscillations. The LM311 has the same pin configuration as the LM306 and LM710C.

schematic diagram and auxiliary circuits

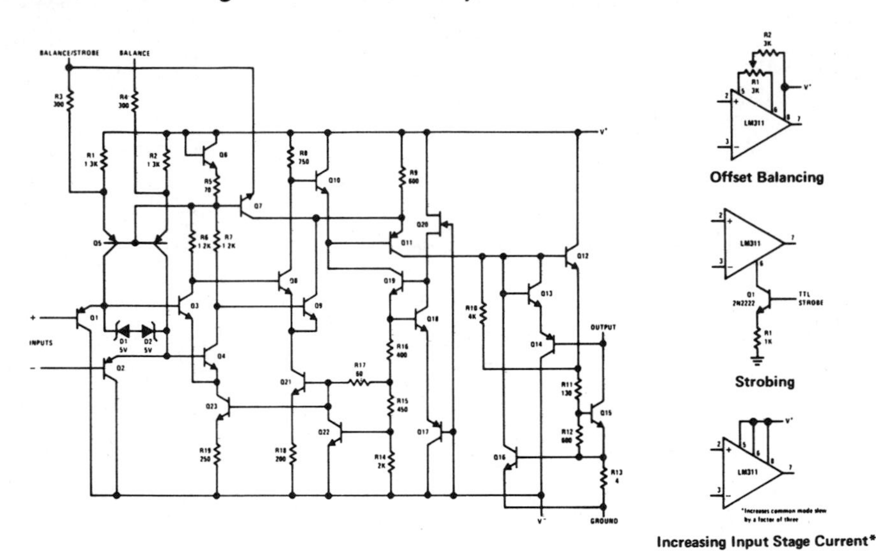

Offset Balancing

Strobing

Increasing Input Stage Current*

typical applications

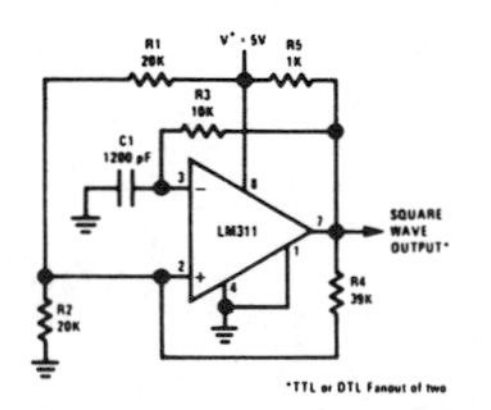

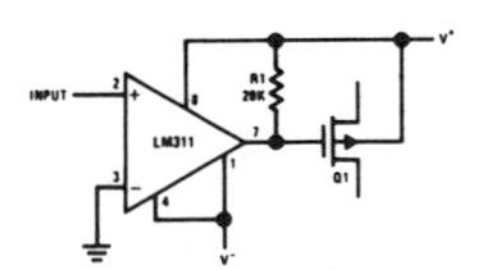

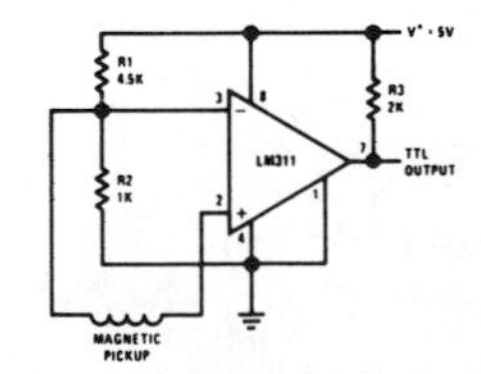

100 kHz Free Running Multivibrator

Zero Crossing Detector Driving MOS Switch

Detector for Magnetic Transducer

absolute maximum ratings

Total Supply Voltage (V_{84})	36V
Output to Negative Supply Voltage (V_{74})	40V
Ground to Negative Supply Voltage (V_{14})	30V
Differential Input Voltage	±30V
Input Voltage (Note 1)	±15V
Power Dissipation (Note 2)	500 mW
Output Short Circuit Duration	10 sec
Operating Temperature Range	0°C to 70°C
Storage Temperature Range	−65°C to 150°C
Lead Temperature (soldering, 10 sec)	300°C

electrical characteristics (Note 3)

PARAMETER	CONDITIONS	MIN	TYP	MAX	UNITS
Input Offset Voltage (Note 4)	$T_A = 25°C$		2.0	7.5	mV
Input Offset Current (Note 4)	$T_A = 25°C$		6.0	50	nA
Input Bias Current	$T_A = 25°C$		100	250	nA
Voltage Gain	$T_A = 25°C$		200		V/mV
Response Time (Note 5)	$T_A = 25°C$		200		ns
Saturation Voltage	$V_{IN} \leq -10$ mV, $I_{OUT} = 50$ mA $T_A = 25°C$		0.75	1.5	V
Output Leakage Current	$V_{IN} \geq 10$ mV, $V_{OUT} = 35V$ $T_A = 25°C$		0.2	50	nA
Input Offset Voltage (Note 4)				10	mV
Input Offset Current (Note 4)				70	nA
Input Bias Current				300	nA
Input Voltage Range			±14		V
Saturation Voltage	$V_{IN} \leq -10$ mV, $I_{SINK} \leq 8$ mA		0.23	0.4	V
Positive Supply Current	$T_A = 25°C$		5.1	7.5	mA
Negative Supply Current	$T_A = 25°C$		4.1	5.0	mA

Note 1. This rating applies for ±15V supplies. The positive input voltage limit is 30V above the negative supply. The negative input voltage limit is equal to the negative supply voltage or 30V below the positive supply, whichever is less.

Note 2. The maximum junction temperature of the LM311 is 85°C. For operating at elevated temperatures, devices in the TO-5 package must be derated based on a thermal resistance of 150°C/W, junction to ambient, or 45°C/W, junction to case. For the flat package, the derating is based on a thermal resistance of 185°C/W when mounted on a 1/16-inch-thick epoxy glass board with ten, 0.03-inch-wide, 2-ounce copper conductors. The thermal resistance of the dual-in-line package is 100°C/W, junction to ambient.

Note 3. These specifications apply for $V_S = \pm 15V$ and $0°C \leq T_A \leq 70°C$, unless otherwise specified.

Note 4. The offset voltages and offset currents given are the maximum values required to drive the output down to 1V or up to 14V with a 7.5 kΩ load. Thus, these parameters define an error band and take into account the worst case effects of voltage gain and input impedance.

Note 5. The response time specified (see definitions) is for a 100 mV input step with 5 mV overdrive.

typical performance

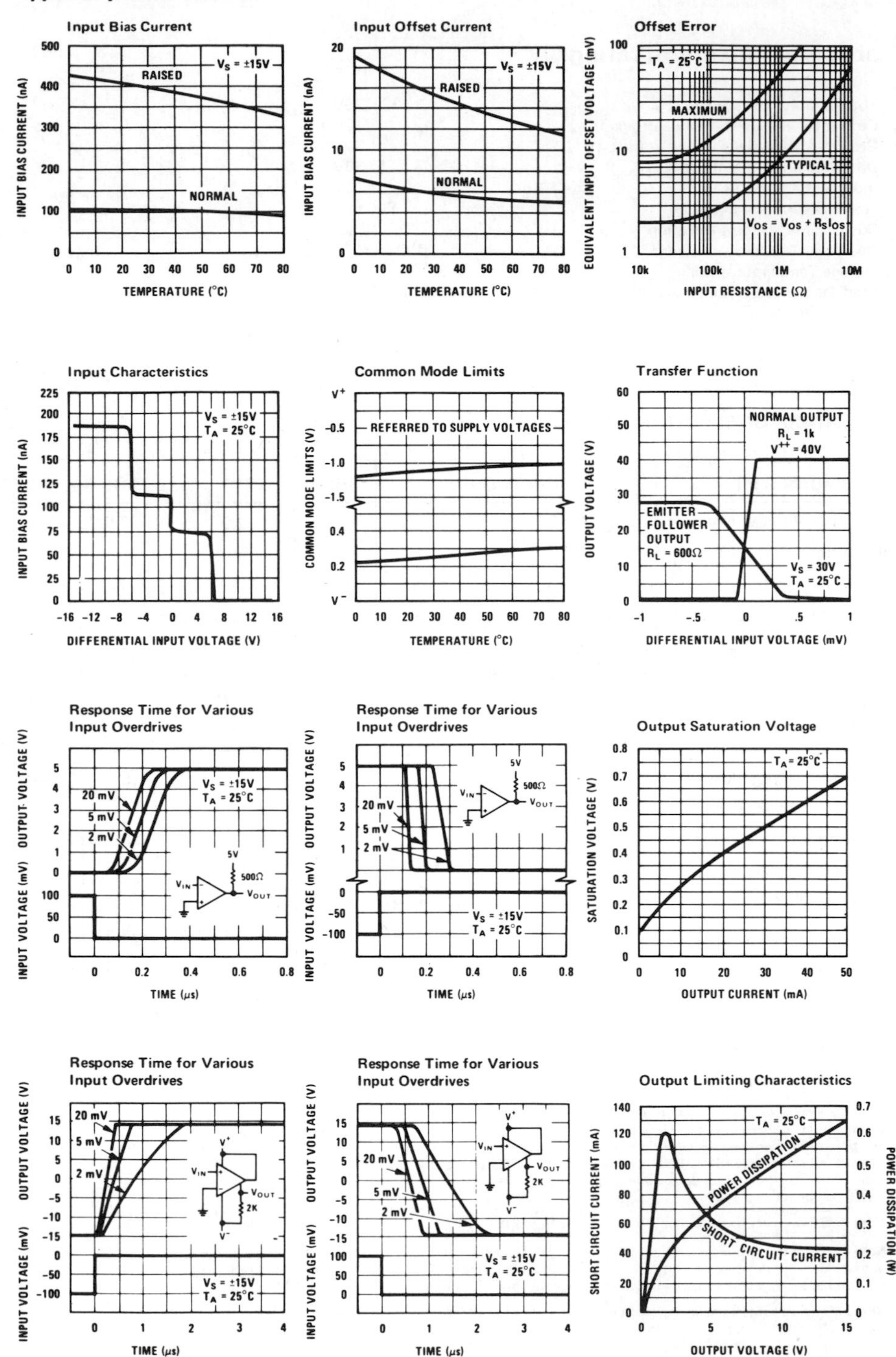

typical performance

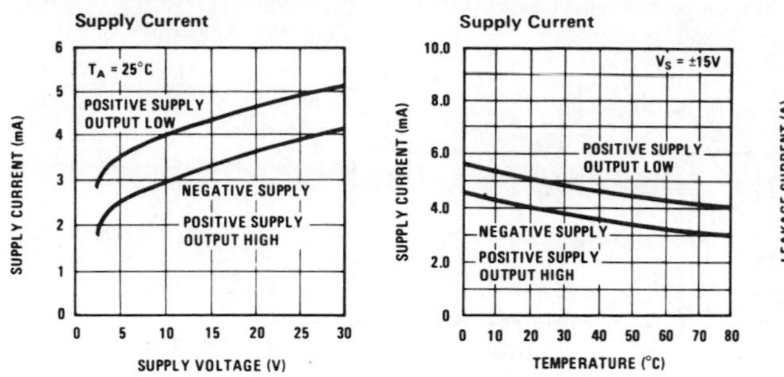

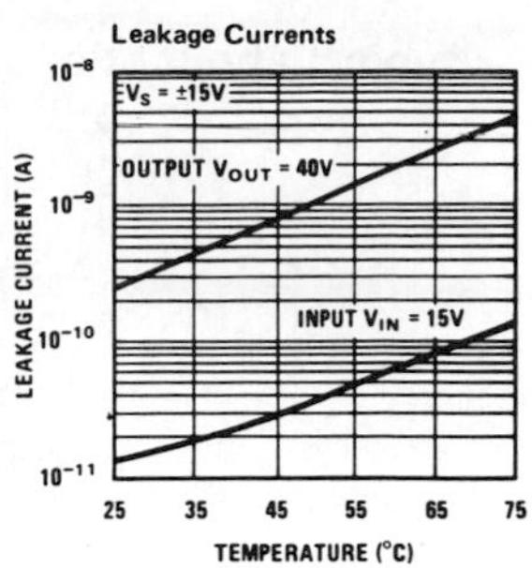

typical applications

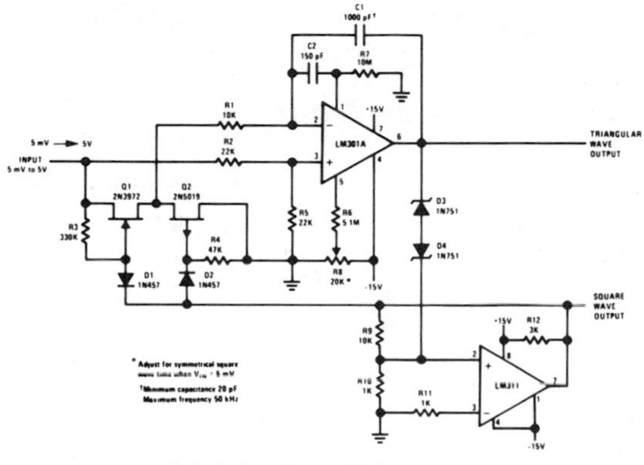

10 Hz to 10 kHz Voltage Controlled Oscillator

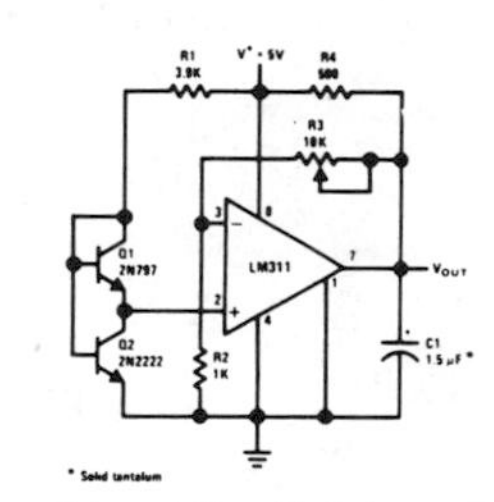

Low Voltage Adjustable Reference Supply

Zero Crossing Detector driving MOS logic

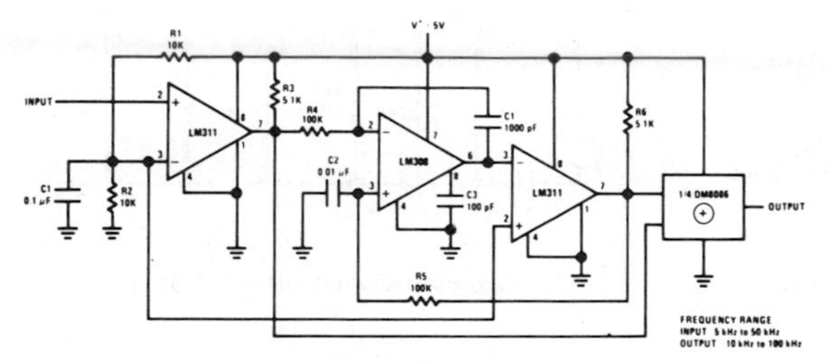

Frequency Doubler

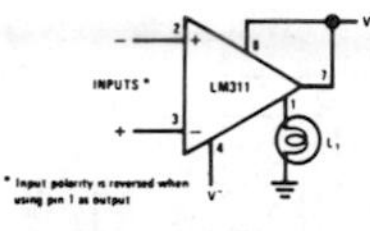

Driving Ground-Referred Load

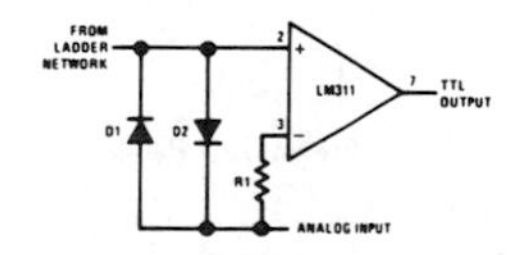

Using Clamp Diodes to Improve Response

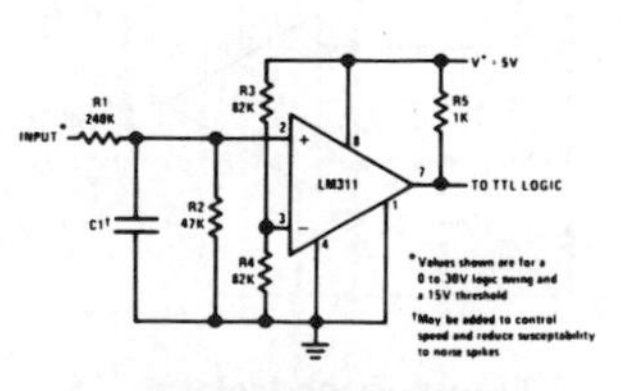

TTL Interface with High Level Logic

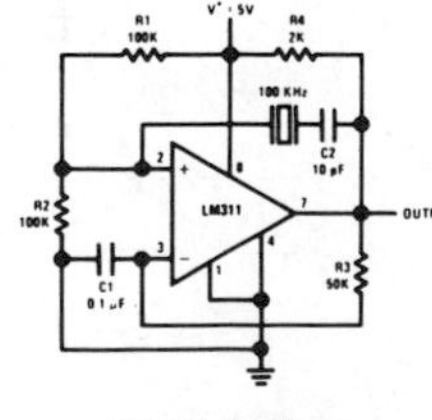

Crystal Oscillator

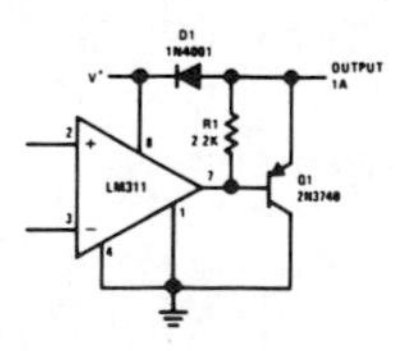

Comparator and Solenoid Driver

typical applications

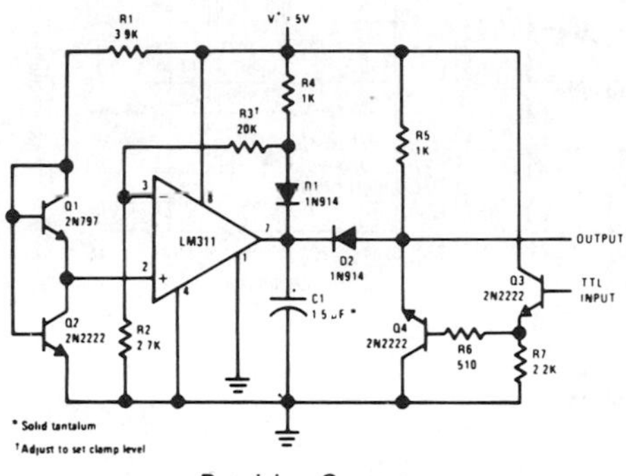

Precision Squarer

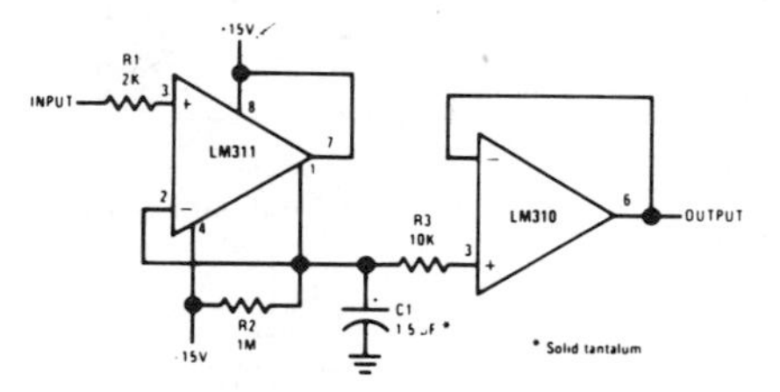

Positive Peak Detector

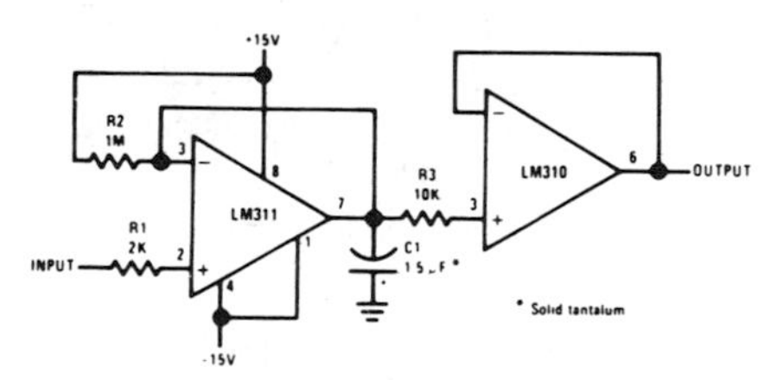

Negative Peak Dectector

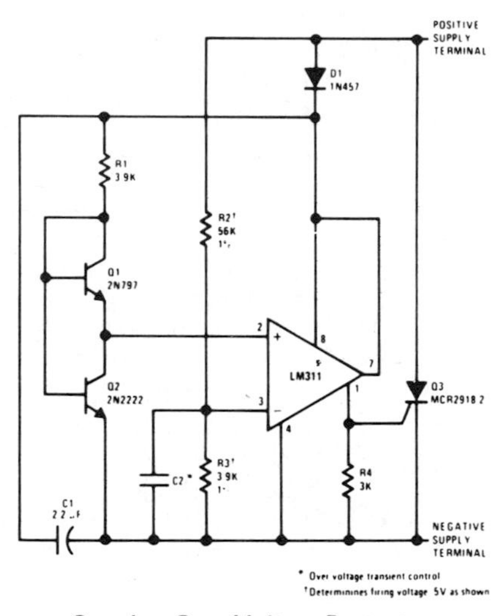

Crowbar Over-Voltage Protector

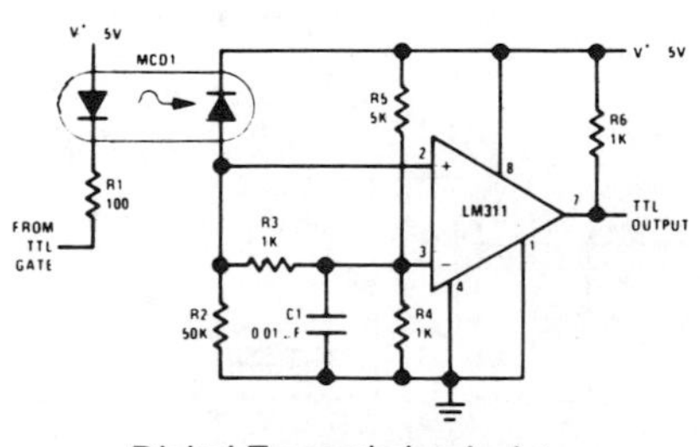

Digital Transmission Isolator

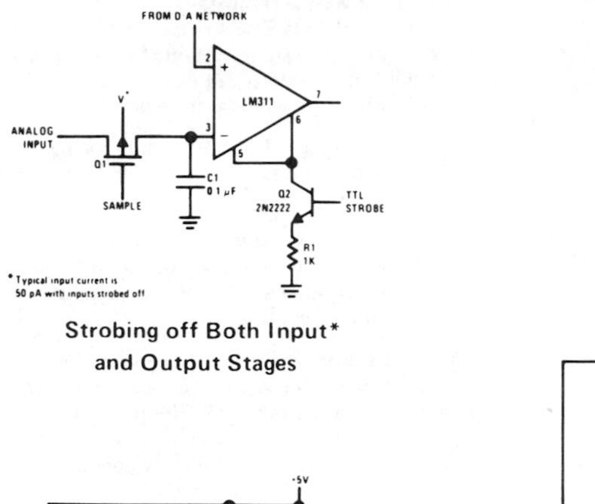

Strobing off Both Input* and Output Stages

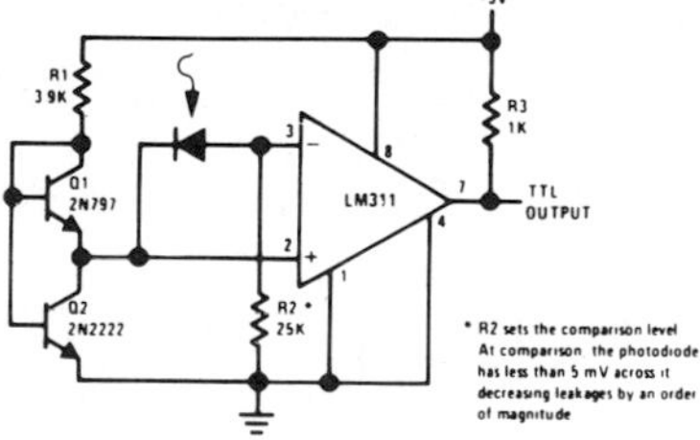

Precision Photodiode Comparator

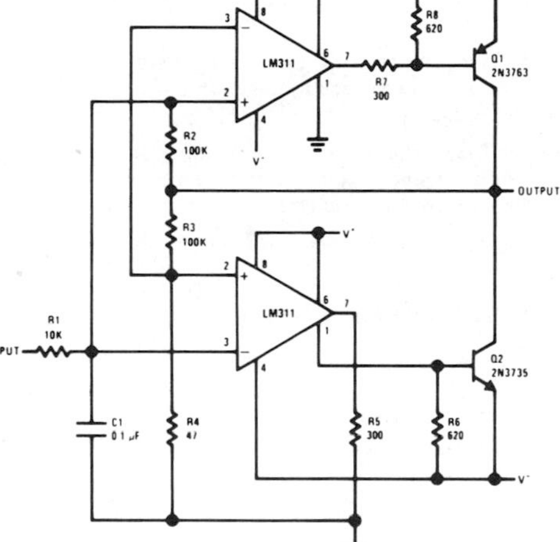

Switching Power Amplifier

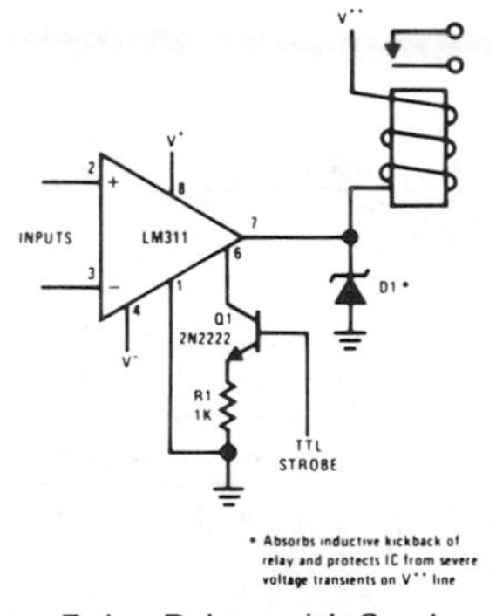

Relay Driver with Strobe

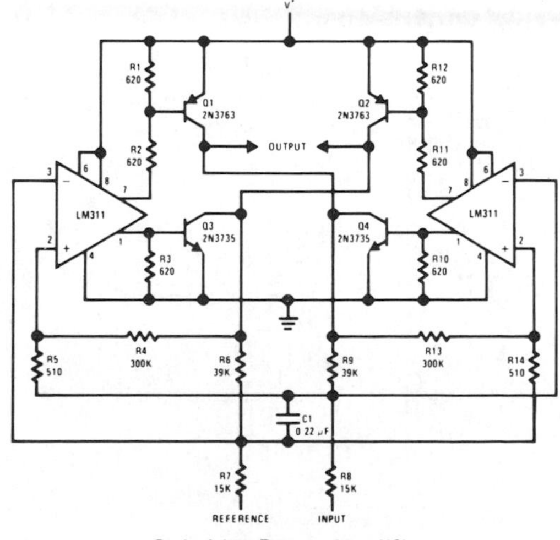

Switching Power Amplifier

definition of terms

Input Offset Voltage: The voltage between the input terminals required to make the output voltage greater than or less than specified voltages.

Input Offset Current: The difference between the two input currents for which the output will be driven higher than or lower than specified voltages.

Input Bias Current: The average of the two input currents.

Input Voltage Range: The range of voltage on the input terminals (common mode) over which all specifications apply.

Voltage Gain: The ratio of the change in output voltage to the change in voltage between the input terminals producing it.

Response Time: The interval between the application of an input step function and the time when the output crosses the logic threshold voltage. The input step drives the comparator from some initial, saturated input voltage to an input level just barely in excess of that required to bring the output from saturation to the logic threshold voltage. This excess is referred to as the voltage overdrive.

Saturation Voltage: The low output voltage level with the input drive equal to or greater than a specified value.

Output Leakage Current: The current into the output terminal with a specified output voltage relative to ground and the input drive equal to or greater than a given value.

Supply Current: The current required from the positive or negative supply to operate the comparator with no output load. The power will vary with input voltage, but is specified as a maximum for the entire range of input voltage conditions.

connection diagrams

Metal Can

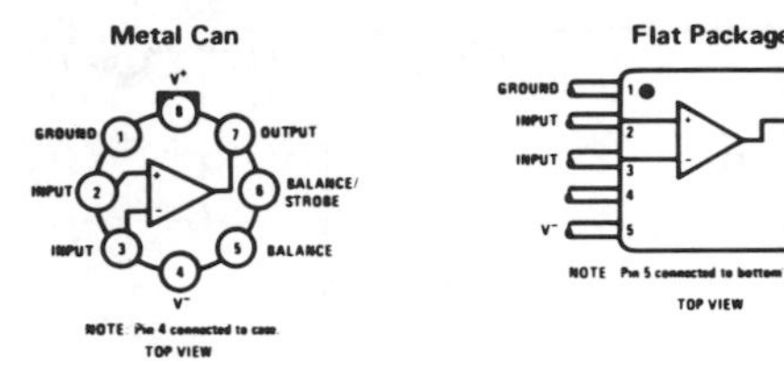

Flat Package

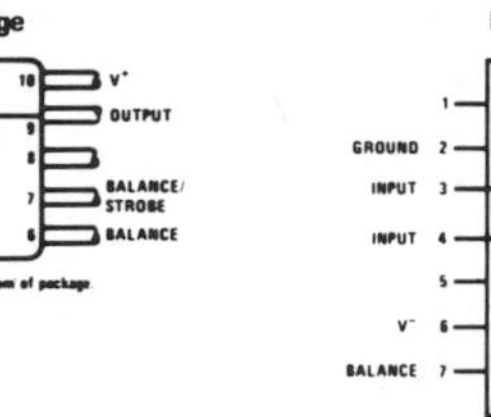

Dual-In-Line

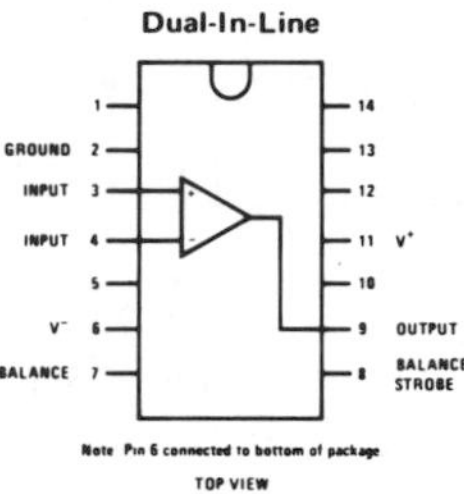

physical dimensions

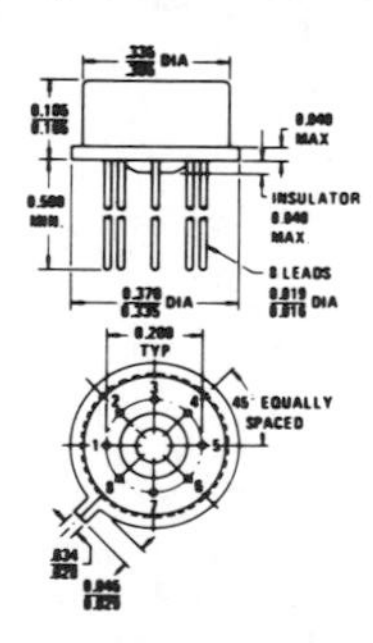

Order Number LM311H

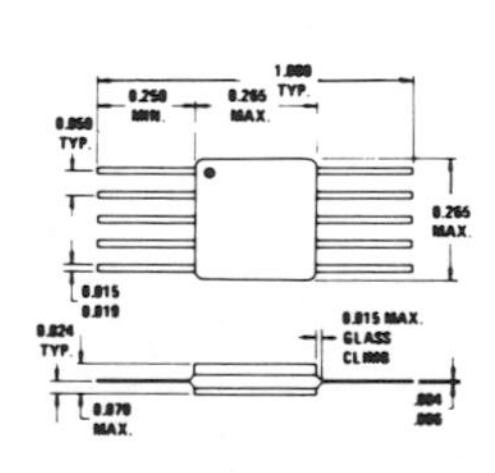

Order Number LM311F

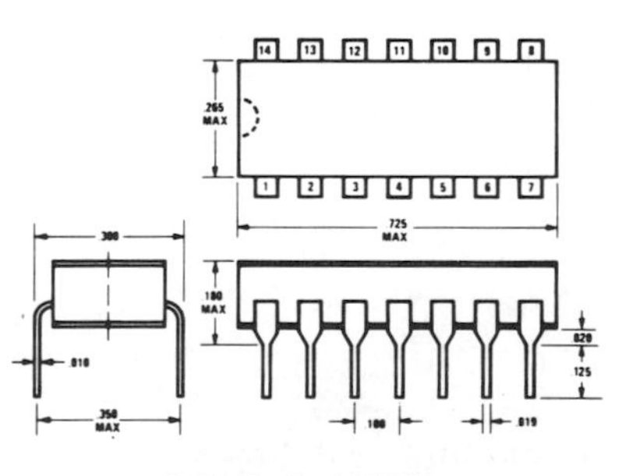

Order Number LM311D

Appendix
4

Timer 555*

*Courtesy of Signetics Corporation, 811 East Arques Avenue, Sunnyvale, California, 94086, copyright 1974.

DESCRIPTION

The NE/SE 555 monolithic timing circuit is a highly stable controller capable of producing accurate time delays, or oscillation. Additional terminals are provided for triggering or resetting if desired. In the time delay mode of operation, the time is precisely controlled by one external resistor and capacitor. For a stable operation as an oscillator, the free running frequency and the duty cycle are both accurately controlled with two external resistors and one capacitor. The circuit may be triggered and reset on falling waveforms, and the output structure can source or sink up to 200mA or drive TTL circuits.

FEATURES

- TIMING FROM MICROSECONDS THROUGH HOURS
- OPERATES IN BOTH ASTABLE AND MONOSTABLE MODES
- ADJUSTABLE DUTY CYCLE
- HIGH CURRENT OUTPUT CAN SOURCE OR SINK 200mA
- OUTPUT CAN DRIVE TTL
- TEMPERATURE STABILITY OF 0.005% PER °C
- NORMALLY ON AND NORMALLY OFF OUTPUT

APPLICATIONS

PRECISION TIMING
PULSE GENERATION
SEQUENTIAL TIMING
TIME DELAY GENERATION
PULSE WIDTH MODULATION
PULSE POSITION MODULATION
MISSING PULSE DETECTOR

PIN CONFIGURATIONS (Top View)

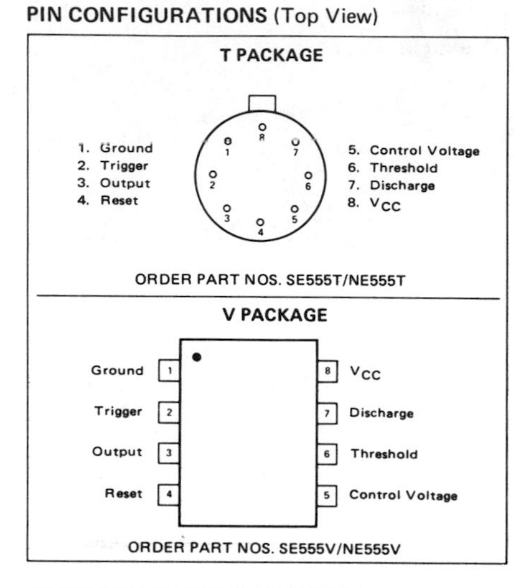

ABSOLUTE MAXIMUM RATINGS

Supply Voltage	+18V
Power Dissipation	600 mW
Operating Temperature Range	
NE555	0°C to +70°C
SE555	–55°C to +125°C
Storage Temperature Range	–65°C to +150°C
Lead Temperature (Soldering, 60 seconds)	+300°C

BLOCK DIAGRAM

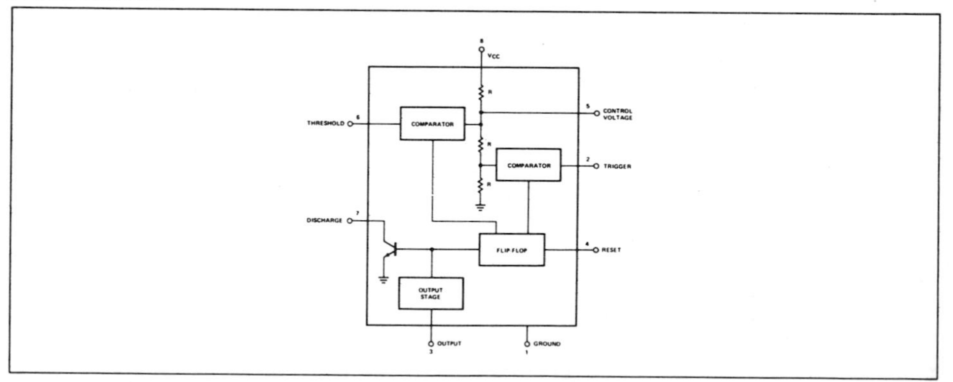

ELECTRICAL CHARACTERISTICS T_A = 25°C, V_{CC} = +5V to +15 unless otherwise specified

PARAMETER	TEST CONDITIONS	SE 555			NE 555			UNITS
		MIN	TYP	MAX	MIN	TYP	MAX	
Supply Voltage		4.5		18	4.5		16	V
Supply Current	V_{CC} = 5V R_L = ∞		3	5		3	6	mA
	V_{CC} = 15V R_L = ∞		10	12		10	15	mA
	Low State, Note 1							
Timing Error(Monostable)	R_A, R_B = 1KΩ to 100KΩ							
Initial Accuracy	C = 0.1 μF Note 2		0.5	2		1		%
Drift with Temperature			30	100		50		ppm/°C
Drift with Supply Voltage			0.05	0.2		0.1		%/Volt
Threshold Voltage			2/3			2/3		X V_{CC}
Trigger Voltage	V_{CC} = 15V	4.8	5	5.2		5		V
Timing Error(Astable)	V_{CC} = 5V	1.45	1.67	1.9		1.67		V
Trigger Current			0.5			0.5		μA
Reset Voltage		0.4	0.7	1.0	0.4	0.7	1.0	V
Reset Current			0.1			0.1		mA
Threshold Current	Note 3		0.1	.25		0.1	.25	μA
Control Voltage Level	V_{CC} = 15V	9.6	10	10.4	9.0	10	11	V
	V_{CC} = 5V	2.9	3.33	3.8	2.6	3.33	4	V
Output Voltage (low)	V_{CC} = 15V							
	I_{SINK} = 10mA		0.1	0.15		0.1	.25	V
	I_{SINK} = 50mA		0.4	0.5		0.4	.75	V
	I_{SINK} = 100mA		2.0	2.2		2.0	2.5	V
	I_{SINK} = 200mA		2.5			2.5		
	V_{CC} = 5V							
	I_{SINK} = 8mA		0.1	0.25				V
	I_{SINK} = 5mA					.25	.35	
Output Voltage Drop (low)								
	I_{SOURCE} = 200mA		12.5			12.5		
	V_{CC} = 15V							
	I_{SOURCE} = 100mA							
	V_{CC} = 15V	13.0	13.3		12.75	13.3		V
	V_{CC} = 5V	3.0	3.3		2.75	3.3		V
Rise Time of Output			100			100		nsec
Fall Time of Output			100			100		nsec

NOTES

1. Supply Current when output high typically 1mA less.
2. Tested at V_{CC} = 5V and V_{CC} = 15V
3. This will determine the maximum value of R_A+ R_B. For 15V operation, the max total R = 20 megohm.

EQUIVALENT CIRCUIT (Shown for One Side Only)

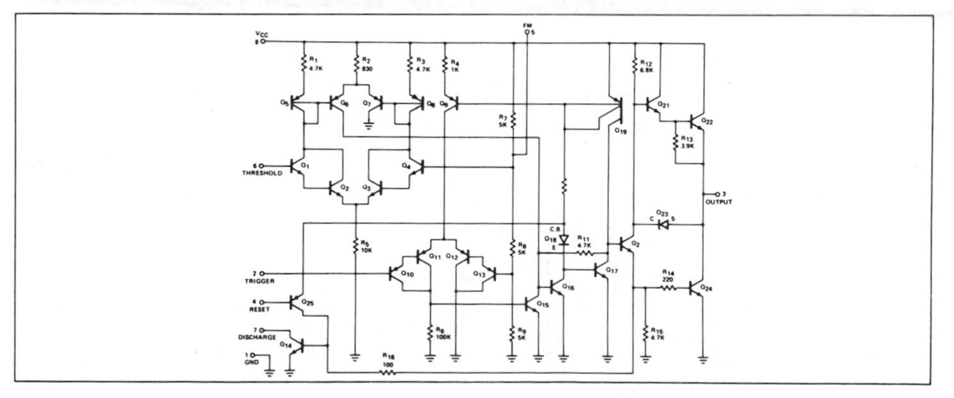

TYPICAL CHARACTERISTICS

MINIMUM PULSE WIDTH REQUIRED FOR TRIGGERING

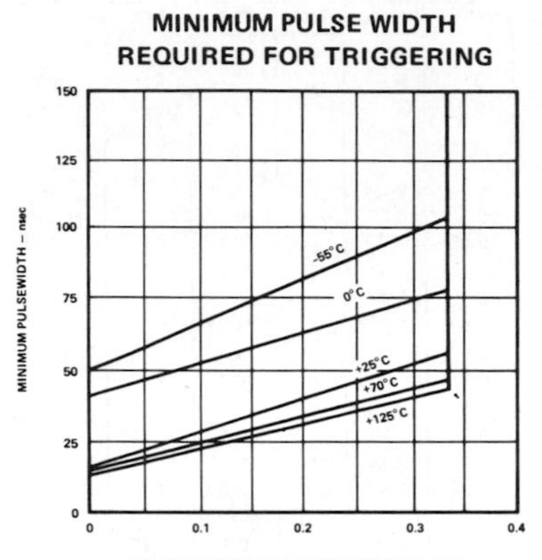

SUPPLY CURRENT vs SUPPLY VOLTAGE

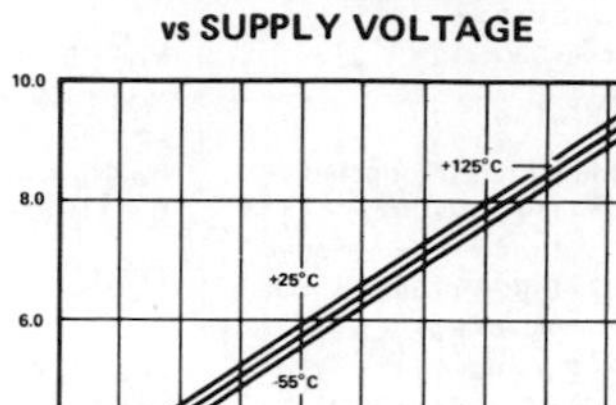
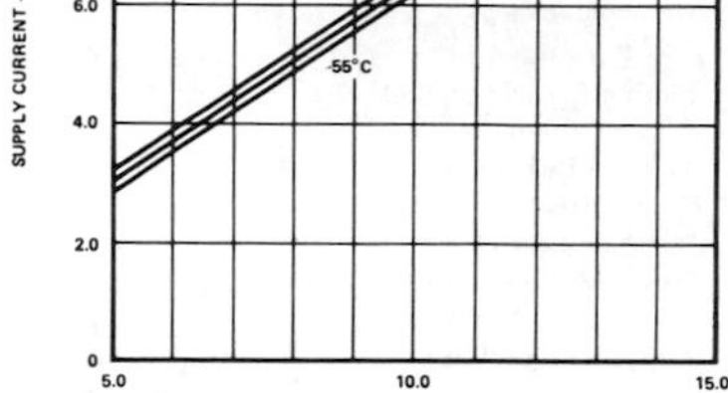

LOW OUTPUT VOLTAGE vs OUTPUT SINK CURRENT

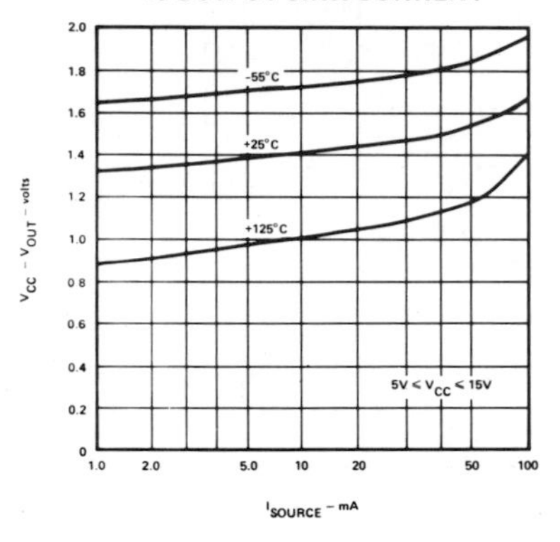

HIGH OUTPUT VOLTAGE vs OUTPUT SOURCE CURRENT

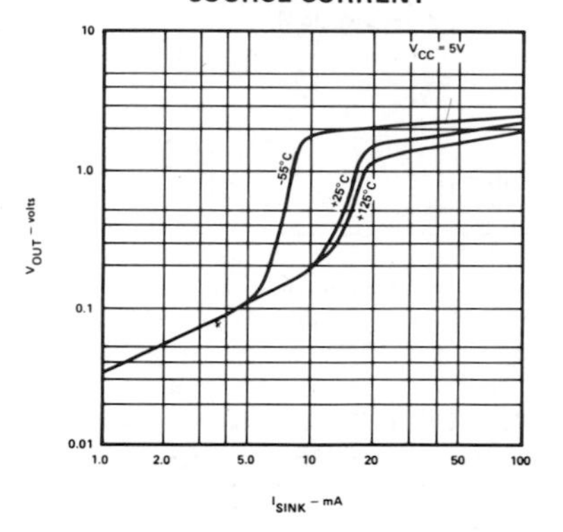

LOW OUTPUT VOLTAGE vs OUTPUT SINK CURRENT

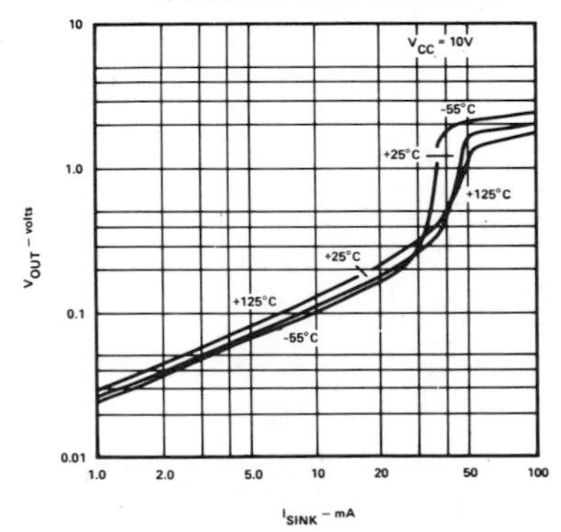

LOW OUTPUT VOLTAGE vs OUTPUT SINK CURRENT

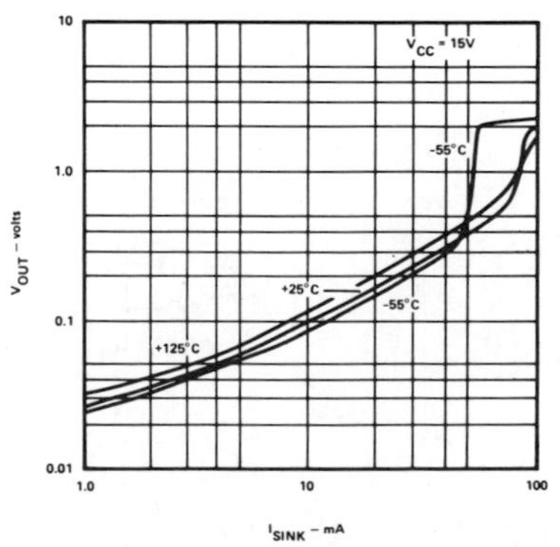

TYPICAL CHARACTERISTICS (Cont'd)

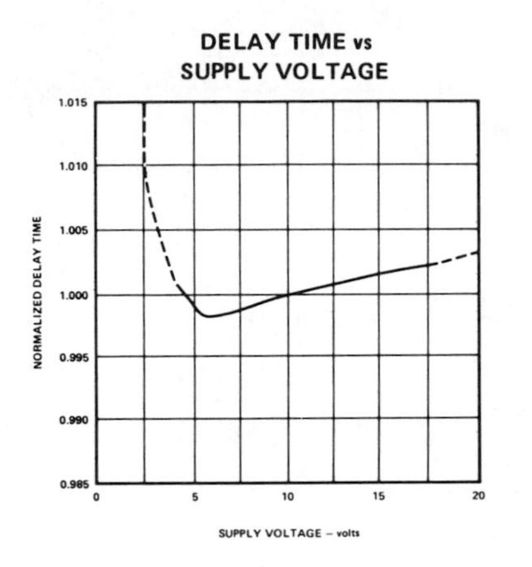

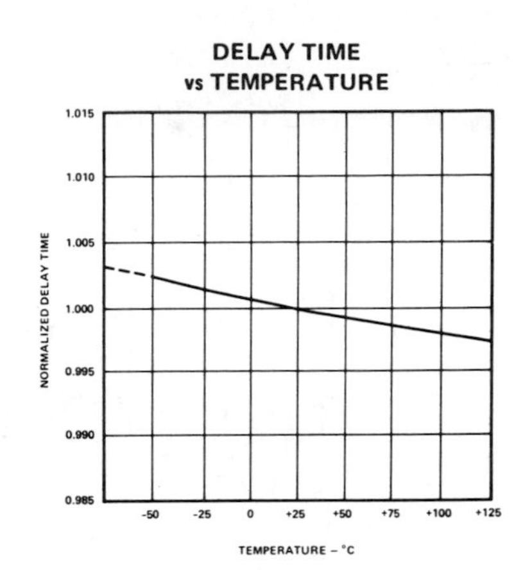

PROPAGATION DELAY vs VOLTAGE LEVEL OF TRIGGER PULSE

PROPAGATION DELAY – nsec

300
250
200
150
100
50
0

-55°C
0°C
+25°C
+70°C
+125°C

0
0.1
0.2
0.3
0.4

LOWEST VOLTAGE LEVEL OF TRIGGER PULSE – X V_{CC}

Appendix 5

Programmable Timer/Counter*

*Courtesy of Exar Integrated Systems, Inc.

The XR-2240 Programmable Timer/Counter is a monolithic controller capable of producing ultra-long time delays without sacrificing accuracy. In most applications, it provides a direct replacement for mechanical or electromechanical timing devices and generates programmable time delays from micro-seconds up to five days. Two timing circuits can be cascaded to generate time delays up to three years.

As shown in Figure 1, the circuit is comprised of an internal time-base oscillator, a programmable 8-bit counter and a control flip-flop. The time delay is set by an external R-C network and can be programmed to any value from 1 RC to 255 RC.

In astable operation, the circuit can generate 256 separate frequencies or pulse-patterns from a single RC setting and can be synchronized with external clock signals. Both the control inputs and the outputs are compatible with TTL and DTL logic levels.

FEATURES

Timing from micro-seconds to days
Programmable delays: 1 RC to 255 RC
Wide supply range: 4V to 15V
TTL and DTL compatible outputs
High accuracy: 0.5%
External Sync and Modulation Capability
Excellent Supply Rejection: 0.2%/V

ABSOLUTE MAXIMUM RATINGS

Supply Voltage	18V
Power Dissipation	
Ceramic Package	750 mW
Derate above +25°C	6 mW/°C
Plastic Package	625 mW
Derate above +25°C	5.0 mW/°C
Operating Temperature	
XR-2240M	−55°C to +125°C
XR-2240C	0°C to +75°C
Storage Temperature	−65°C to +150°C

APPLICATIONS

Precision Timing
Long Delay Generation
Sequential Timing
Binary Pattern Generation
Frequency Synthesis
Pulse Counting/Summing
A/D Conversion
Digital Sample and Hold

PACKAGE INFORMATION

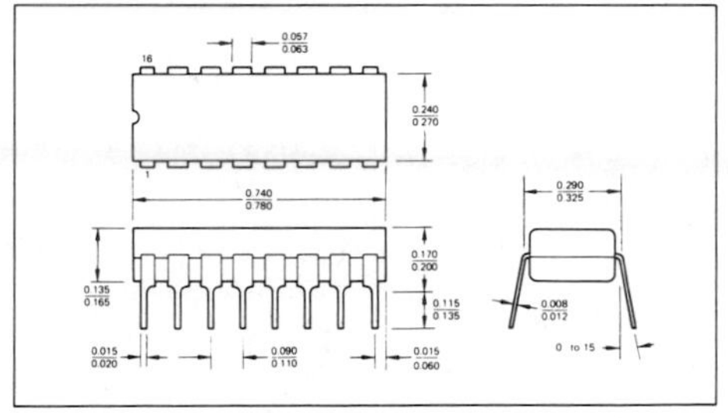

FUNCTIONAL BLOCK DIAGRAM

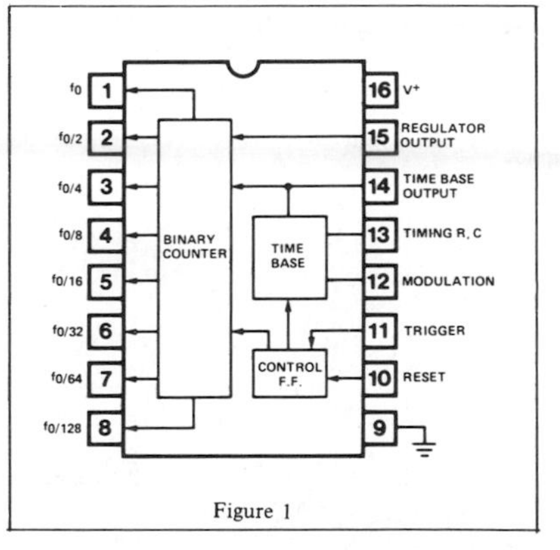

Figure 1

ELECTRICAL CHARACTERISTICS

Test Conditions: See Figure 2, $V^+ = 5V$, $T_A = 25°C$, $R = 10\ k\Omega$, $C = 0.1\ \mu F$, unless otherwise noted.

PARAMETERS	XR-2240			XR-2240C			UNIT	CONDITIONS
	MIN.	TYP.	MAX.	MIN.	TYP.	MAX.		
GENERAL CHARACTERISTICS								
Supply Voltage	4		15	4		15	V	For $V^+ < 4.5V$, Short Pin 15 to Pin 16
Supply Current								
Total Circuit		3.5	6		4	7	mA	$V^+ = 5V$, $V_{TR} = 0$, $V_{RS} = 5V$
		12	16		13	18	mA	$V^+ = 15V$, $V_{TR} = 0$, $V_{RS} = 5V$
Counter Only		1			1.5		mA	See Figure 3
Regulator Output, V_R	4.1	4.4		3.9	4.4		V	Measured at Pin 15, $V^+ = 5V$
	6.0	6.3	6.6	5.8	6.3	6.8	V	$V^+ = 15V$, See Figure 4
TIME BASE SECTION								See Figure 2
Timing Accuracy *		0.5	2.0		0.5	5	%	$V_{RS} = 0$, $V_{TR} = 5V$
Temperature Drift		150	300		200		ppm/°C	$V^+ = 5V$ $0°C \leq T \leq 75°C$
		80			80		ppm/°C	$V^+ = 15V$
Supply Drift		0.05	0.2		0.08	0.3	%/V	$V^+ \geq 8$ Volts, See Figure 11
Max. Frequency	100	130			130		kHz	$R = 1\ k\Omega$, $C = 0.007\ \mu F$
Modulation Voltage Level								Measured at Pin 12
	3.00	3.50	4.0	2.80	3.50	4.20	V	$V^+ = 5V$
		10.5			10.5		V	$V^+ = 15V$
Recommended Range of Timing Components								See Figure 8
Timing Resistor, R	0.001		10	0.001		10	MΩ	
Timing Capacitor, C	0.007		1000	0.01		1000	μF	
TRIGGER/RESET CONTROLS								
Trigger								Measures at Pin 11, $V_{RS} = 0$
Trigger Threshold		1.4	2.0		1.4	2.0	V	
Trigger Current		8			10		μA	$V_{RS} = 0$, $V_{TR} = 2V$
Impedance		25			25		kΩ	
Response Time **		1			1		μsec.	
Reset								
Reset Threshold		1.4	2.0		1.4	2.0	V	
Reset Current		8			10		μA	$V_{TR} = 0$, $V_{RS} = 2V$
Impedance		25			25		kΩ	
Response Time **		0.8			0.8		μsec.	
COUNTER SECTION								See Figure 4, $V^+ = 5V$
Max. Toggle Rate	0.8	1.5			1.5		MHz	$V_{RS} = 0$, $V_{TR} = 5V$ Measured at Pin 14
Input:								
Impedance		20			20		kΩ	
Threshold	1.0	1.4		1.0	1.4		V	
Output:								Measured at Pins 1 thru 8
Rise Time		180			180		nsec.	$R_L = 3k$, $C_L = 10\ pF$
Fall Time		180			180		nsec.	
Sink Current	3	5		2	4		mA	$V_{OL} \leq 0.4V$
Leakage Current		0.01	8		0.01	15	μA	$V_{OH} = 15V$

*Timing error solely introduced by XR-2240, measured as % of ideal time-base period of T = 1.00 RC.
**Propagation delay from application of trigger (or reset) input to corresponding state change in counter output at pin 1.

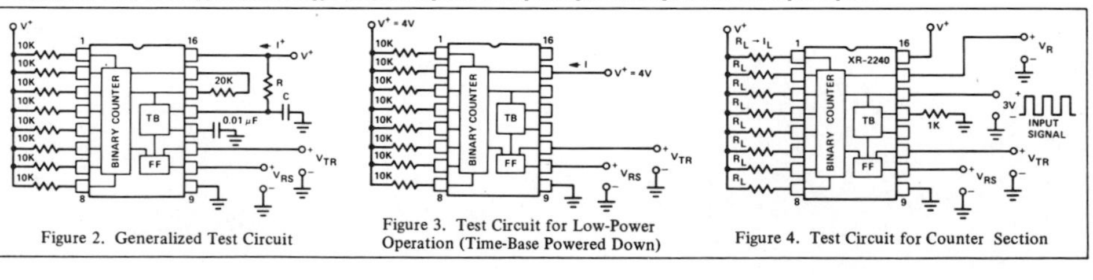

Figure 2. Generalized Test Circuit

Figure 3. Test Circuit for Low-Power Operation (Time-Base Powered Down)

Figure 4. Test Circuit for Counter Section

PRINCIPLE OF OPERATION

The timing cycle for the XR-2240 is initiated by applying a positive-going trigger pulse to pin 11. The trigger input actuates the time-base oscillator, enables the counter section, and sets all the counter outputs to "low" state. The time-base oscillator generates timing pulses with its period, T, equal to 1 RC. These clock pulses are counted by the binary counter section. The timing cycle is completed when a positive-going reset pulse is applied to pin 10.

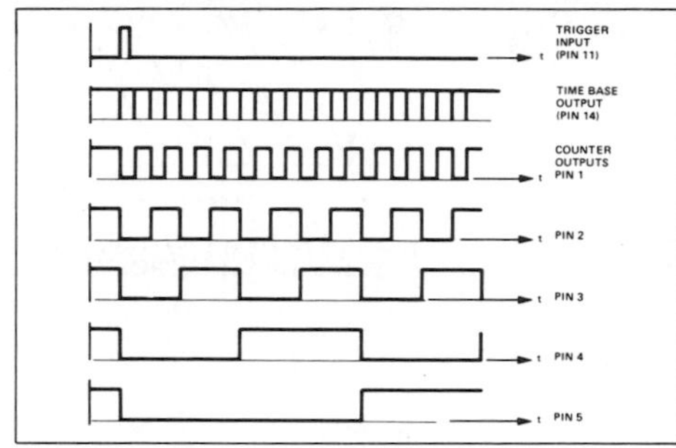

Figure 5. Timing Diagram of Output Waveforms

Figure 5 gives the timing sequence of output waveforms at various circuit terminals, subsequent to a trigger input. When the circuit is at reset state, both the time-base and the counter sections are disabled and all the counter outputs are at "high" state.

In most timing applications, one or more of the counter outputs are connected back to the reset terminal, as shown in Figure 6, with S_1 closed. In this manner, the circuit will start timing when a trigger is applied and will automatically reset itself to complete the timing cycle when a programmed count is completed. If none of the counter outputs are connected back to the reset terminal (switch S_1 open), the circuit would operate in its astable or free-running mode, subsequent to a trigger input.

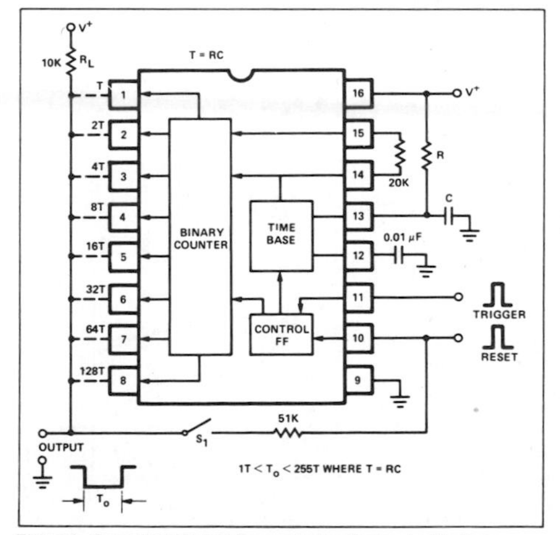

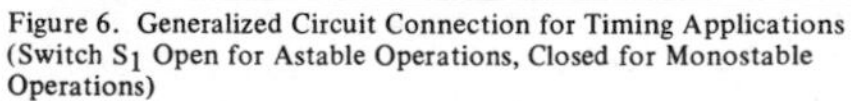

Figure 6. Generalized Circuit Connection for Timing Applications (Switch S_1 Open for Astable Operations, Closed for Monostable Operations)

PROGRAMMING CAPABILITY

The binary counter outputs (pins 1 through 8) are open-collector type stages and can be shorted together to a common pull-up resistor to form a "wired-or" connection. The combined output will be "low" as long as any one of the outputs is low. In this manner, the time delays associated with each counter output can be *summed* by simply shorting them together to a common output bus as shown in Figure 6. For example, if only pin 6 is connected to the output and the rest left open, the total duration of the timing cycle, T_o, would be 32T. Similarly, if pins 1, 5, and 6 were shorted to the output bus, the total time delay would be $T_o = (1+16+32)\ T = 49T$. In this manner, by proper choice of counter terminals connected to the output bus, one can program the timing cycle to be: $1T \leq T_o \leq 255T$, where T = RC.

TRIGGER AND RESET CONDITIONS

When power is applied to the XR-2240 with no trigger or reset inputs, the circuit reverts to "reset" state. Once triggered, the circuit is immune to additional trigger inputs, until the timing cycle is completed or a reset input is applied. If both the reset and the trigger controls are activated simultaneously, trigger overrides reset.

DESCRIPTION OF CIRCUIT CONTROLS

COUNTER OUTPUTS (PINS 1 THROUGH 8)

The binary counter outputs are buffered "open-collector" type stages, as shown in Figure 15. Each output is capable of sinking ≈5 mA of load current. At reset condition, all the counter outputs are at high or non-conducting state. Subsequent to a trigger input, the outputs change state in accordance with the timing diagram of Figure 5.

The counter outputs can be used individually, or can be connected together in a "wired-or" configuration, as described in the Programming section.

RESET AND TRIGGER INPUTS (PINS 10 AND 11)

The circuit is reset or triggered with positive-going control pulses applied to pins 10 and 11. The threshold level for these controls is approximately two diode drops (≈1.4V) above ground.

Minimum pulse widths for reset and trigger inputs are shown in Figure 10. Once triggered, the circuit is immune to additional trigger inputs until the end of the timing cycle.

MODULATION AND SYNC INPUT (PIN 12)

The period T of the time-base oscillator can be modulated by applying a dc voltage to this terminal (see Figure 13). The time-base oscillator can be synchronized to an external clock by applying a sync pulse to pin 12, as shown in Figure 16. Recommended sync pulse widths and amplitudes are also given in the figure.

HARMONIC SYNCHRONIZATION

Time-base can be synchronized with *integer multiples or harmonics* of input sync frequency, by setting the time-base period, T, to be an integer multiple of the sync pulse period, T_s. This can be done by choosing the timing components R and C at pin 13 such that:

$$T = RC = (T_s/m) \text{ where}$$

$$m \text{ is an integer, } 1 \leq m \leq 10.$$

Figure 17 gives the typical pull-in range for harmonic synchronization, for various values of harmonic modulus, m. For $m < 10$, typical pull-in range is greater than ±4% of time-base frequency.

TYPICAL CHARACTERISTICS

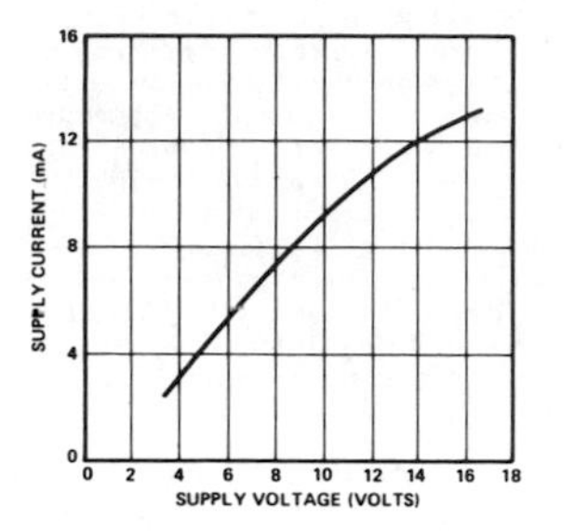

Figure 7. Supply Current vs. Supply Voltage in Reset Condition (Supply Current Under Trigger Condition is ≈0.7 mA less)

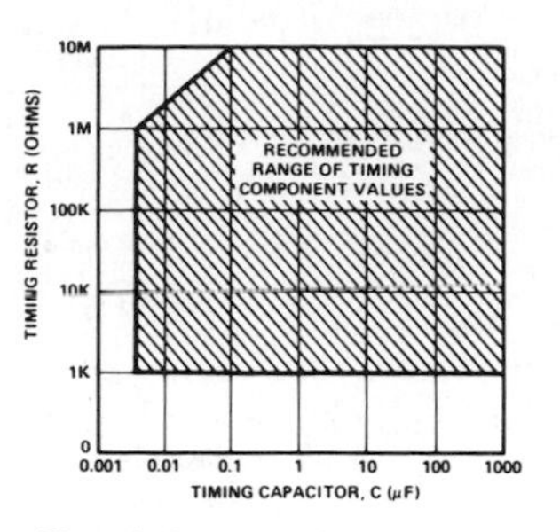

Figure 8. Recommended Range of Timing Component Values

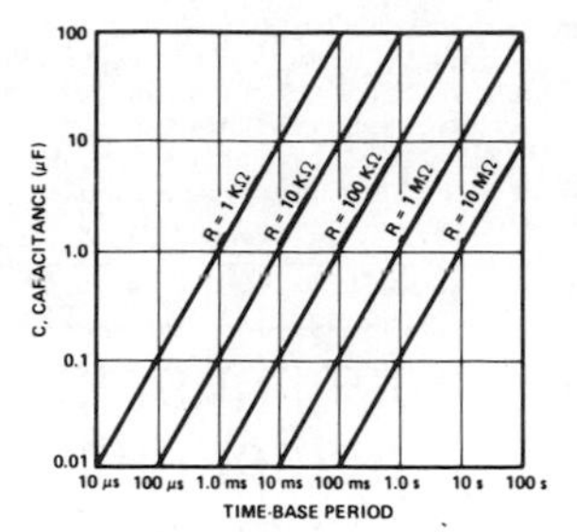

Figure 9. Time-Base Period, T, as a Function of External RC

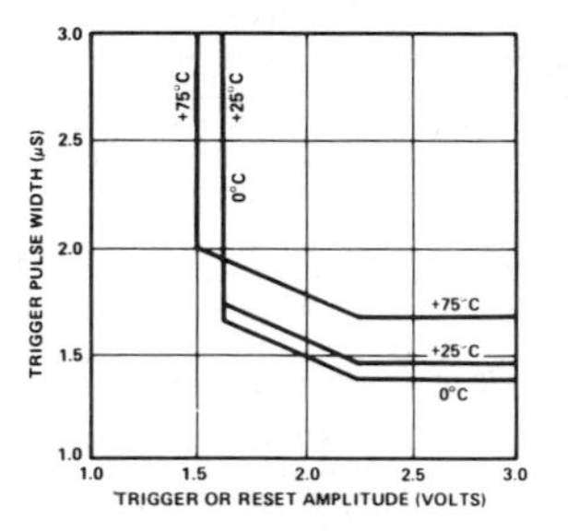

Figure 10. Minimum Trigger and Reset Pulse Widths at Pins 10 and 11

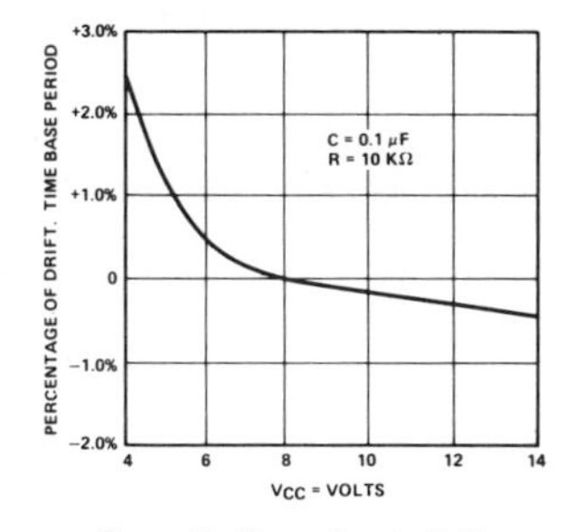

Figure 11. Power Supply Drift

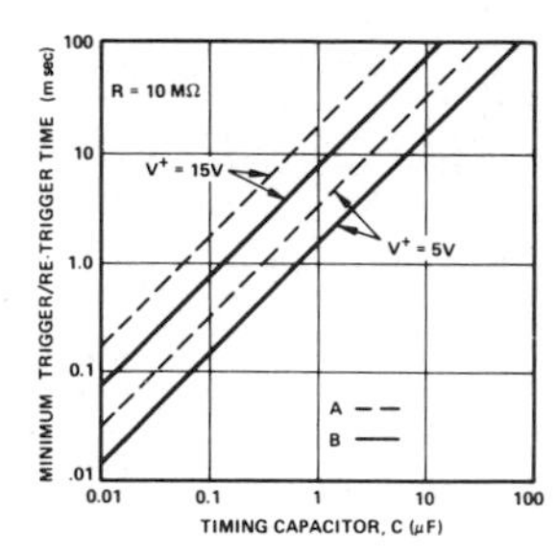

Figure 12.
A) Minimum Trigger Delay Time Subsequent to Application of Power
B) Minimum Re-trigger Time, Subsequent to a Reset Input

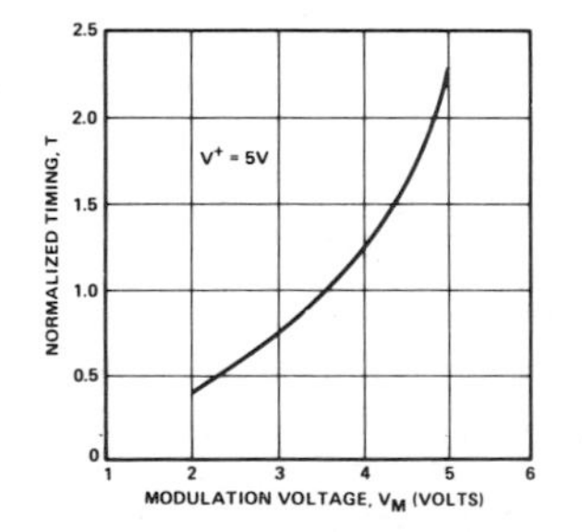

Figure 13. Normalized Change in Time-Base Period As a Function of Modulation Voltage at Pin 12

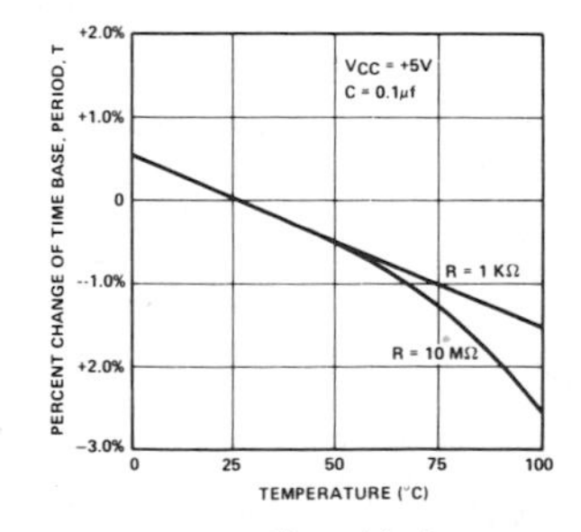

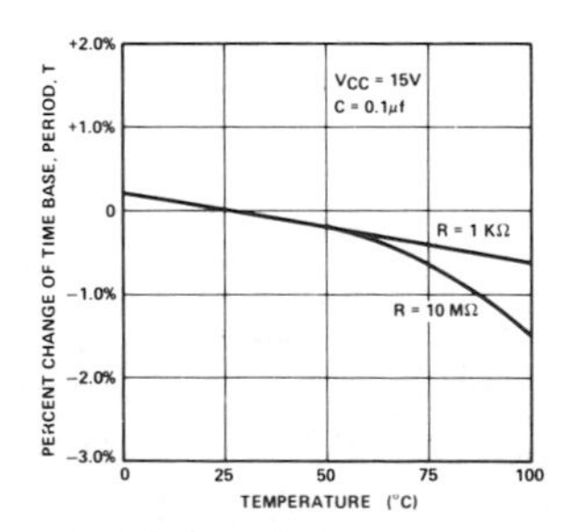

Figure 14. Temperature Drift of Time-Base Period, T

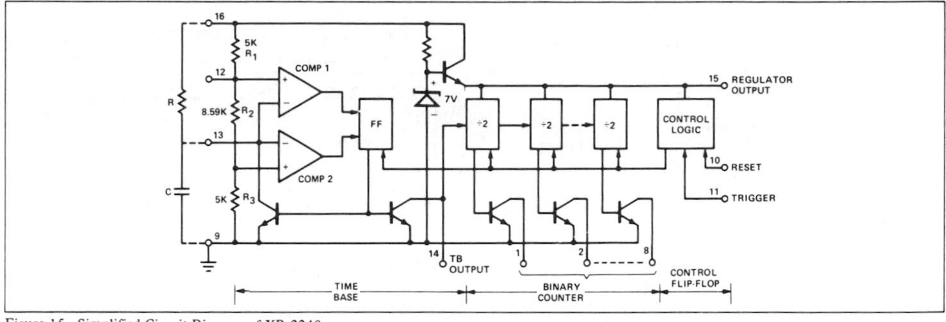

Figure 15. Simplified Circuit Diagram of XR-2240

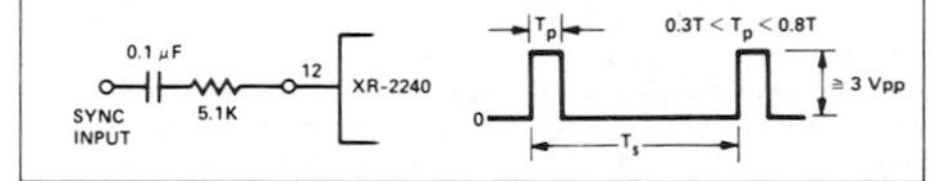

Figure 16. Operation with External Sync Signal.
(a) Circuit for Sync Input
(b) Recommended Sync Waveform

TIMING TERMINAL (PIN 13)

The time-base period T is determined by the external R-C network connected to this pin. When the time-base is triggered, the waveform at pin 13 is an exponential ramp with a period T = 1.0 RC.

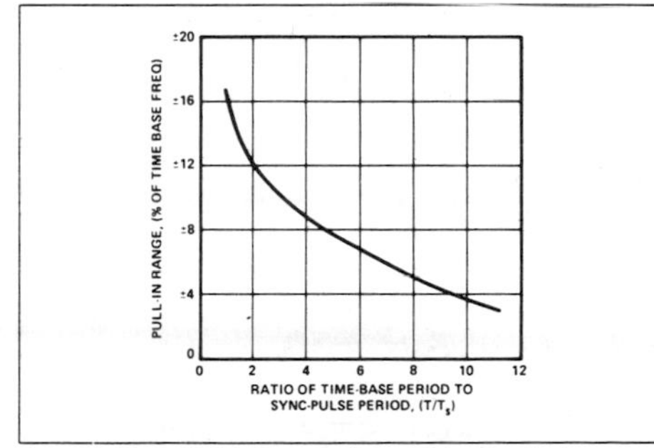

Figure 17. Typical Pull-In Range for Harmonic Synchronization

TIME-BASE OUTPUT (PIN 14)

Time-Base output is an open-collector type stage, as shown in Figure 15 and requires a 20 KΩ pull-up resistor to Pin 15 for proper operation of the circuit. At reset state, the time-base output is at "high" state. Subsequent to triggering, it produces a negative-going pulse train with a period T = RC, as shown in the diagram of Figure 5.

Time-base output is internally connected to the binary counter section and also serves as the input for the external clock signal when the circuit is operated with an external time-base.

The counter input triggers on the negative-going edge of the timing or clock pulses applied to pin 14. The trigger threshold for the counter section is ≈ +1.5 volts. The counter section can be disabled by clamping the voltage level at pin 14 to ground.

Note:
Under certain operating conditions such as high supply voltages ($V^+ > 7V$) and small values of timing capacitor ($C < 0.1$ μF) the pulse-width of the time-base output at pin 14 may be too narrow to trigger the counter section. This can be corrected by connecting a 300 pF capacitor from pin 14 to ground.

REGULATOR OUTPUT (PIN 15)

This terminal can serve as a V^+ supply to additional XR-2240 circuits when several timer circuits are cascaded (See Figure 20), to minimize power dissipation. For circuit operation with external clock, pin 15 can be used as the V^+ terminal to power-down the internal time-base and reduce power dissipation.

When the internal time-base is used with $V^+ \leq 4.5V$, pin 15 should be shorted to pin 16.

APPLICATIONS INFORMATION

PRECISION TIMING (Monostable Operation)

In precision timing applications, the XR-2240 is used in its monostable or "self-resetting" mode. The generalized circuit connection for this application is shown in Figure 18.

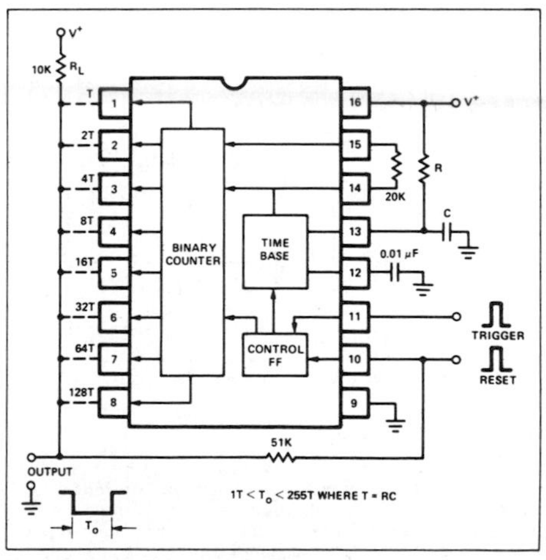

Figure 18. Circuit for Monostable Operation
(T_o = NRC where $1 \leq N \leq 255$)

The output is normally "high" and goes to "low" subsequent to a trigger input. It stays low for the time duration T_O and then returns to the high state. The duration of the timing cycle T_O is given as:

$$T_O = NT = NRC$$

where T = RC is the time-base period as set by the choice of timing components at pin 13 (See Figure 9). N is an integer in the range of:

$$1 \leq N \leq 255$$

as determined by the combination of counter outputs (pins 1 through 8) connected to the output bus, as described below.

PROGRAMMING OF COUNTER OUTPUTS: The binary counter outputs (pins 1 through 8) are open-collector type stages and can be shorted together to a common pull-up resistor to form a "wired-or" connection where the combined output will be "low" as long as any one of the outputs is low. In this manner, the time delays associated with each counter output can be summed by simply shorting them together to a common output bus as shown in Figure 18. For example, if only pin 6 is connected to the output and the rest left open, the total duration of the timing cycle, T_O, would be 32T. Similarly, if pins 1, 5, and 6 were shorted to the output bus, the total time delay would be $T_O = (1+16+32)\ T = 49T$. In this manner, by proper choice of counter terminals connected to the output bus, one can program the timing cycle to be: $1T \leq T_O \leq 255T$.

ULTRA-LONG DELAY GENERATION

Two XR-2240 units can be cascaded as shown in Figure 19 to generate extremely long time delays. In this application, the reset and the trigger terminals of both units are tied together and the time base of Unit 2 disabled. In this manner, the output would normally be high when the system is at reset. Upon application of a trigger input, the output would go to a low state and stays that way for a total of $(256)^2$ or 65,536 cycles of the time-base oscillator.

PROGRAMMING: Total timing cycle of two cascaded units can be programmed from $T_O = 256RC$ to $T_O = 65{,}536RC$ in 256 discrete steps by selectively shorting any one or the combination of the counter outputs from Unit 2 to the output bus.

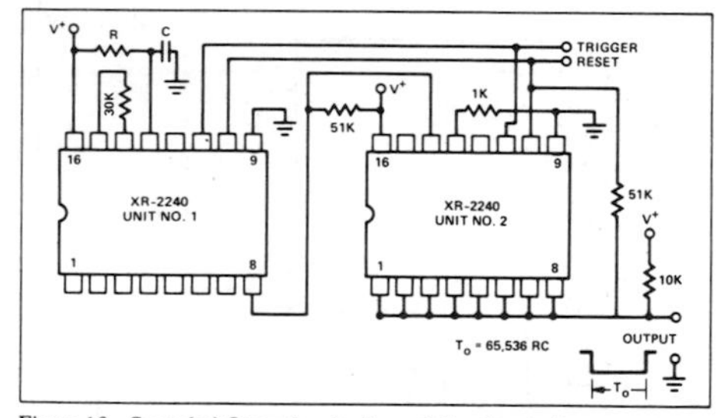

Figure 19. Cascaded Operation for Long Delay Generation

LOW-POWER OPERATION

In cascaded operation, the time-base section of Unit 2 can be powered down to reduce power consumption, by using the circuit connection of Figure 20. In this case, the V^+ terminal (pin 16) of Unit 2 is left open-circuited, and the second unit is powered from the regulator output of Unit 1, by connecting pin 15 of both units.

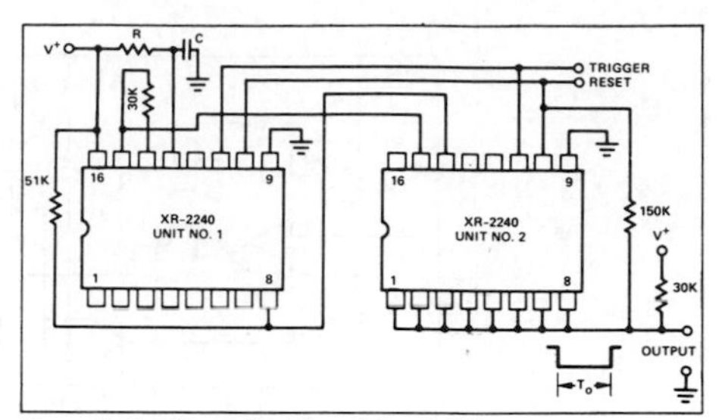

Figure 20. Low-Power Operation of Cascaded Timers

ASTABLE OPERATION

The XR-2240 can be operated in its astable or free-running mode by disconnecting the reset terminal (pin 10) from the counter outputs. Two typical circuit connections for this mode of operation are shown in Figure 21. In the circuit connection of Figure 21(a), the circuit operates in its free-running mode, with external trigger and reset signals. It will start counting and timing subsequent to a trigger input until an external reset pulse is applied. Upon application of a positive-going reset signal to pin 10, the circuit reverts back to its rest state. The circuit of Figure 21(a) is essentially the same as that of Figure 6, with the feedback switch S_1 open.

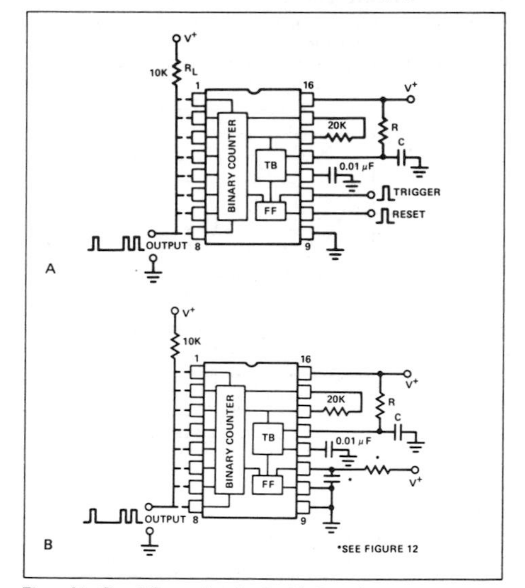

Figure 21. Circuit Connections for Astable Operation
(a) Operation with External Trigger and Reset Controls
(b) Free-running or Continuous Operation

The circuit of Figure 21(b) is designed for continuous operation. The circuit self-triggers automatically when the power supply is turned on, and continues to operate in its free-running mode indefinitely.

In astable or free-running operation, each of the counter outputs can be used individually as synchronized oscillators; or they can be interconnected to generate complex pulse patterns.

BINARY PATTERN GENERATION

In astable operation, as shown in Figure 21, the output of the XR-2240 appears as a complex pulse pattern. The waveform of the output pulse train can be determined directly from the timing diagram of Figure 5 which shows the phase relations between the counter outputs. Figure 22 shows some of these complex pulse patterns. The pulse pattern repeats itself at a rate equal to the period of the *highest* counter bit connected to the common output bus. The minimum pulse width contained in the pulse train is determined by the *lowest* counter bit connected to the output.

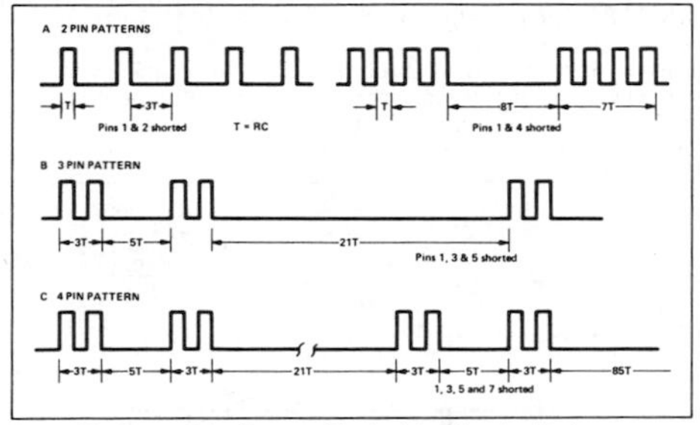

Figure 22. Binary Pulse Patterns Obtained by Shorting Various Counter Outputs

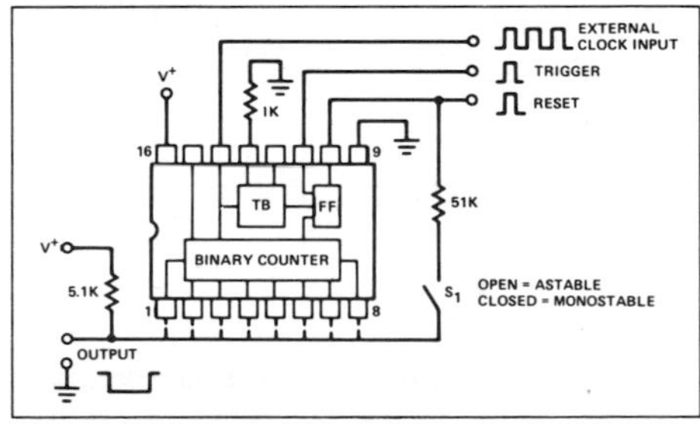

Figure 23. Operation with External Clock

OPERATION WITH EXTERNAL CLOCK

The XR-2240 can be operated with an external clock or time-base, by disabling the internal time-base oscillator and applying the external clock input to pin 14. The recommended circuit connection for this application is shown in Figure 23. The internal time-base can be de-activated by connecting a 1 KΩ resistor from pin 13 to ground. The counters are triggered on the negative-going edges of the external clock pulse. For proper operation, a minimum clock pulse amplitude of 3 volts is required. Minimum external clock pulse width must be $\geq 1\ \mu S$.

For operation with supply voltages of 6V or less, the internal time-base section can be powered down by open-circuiting pin 16 and connecting pin 15 to V^+. In this configuration, the internal time-base does not draw any current, and the over-all current drain is reduced by ≈3 mA.

FREQUENCY SYNTHESIZER

The programmable counter section of XR-2240 can be used to generate 255 discrete frequencies from a given time base setting using the circuit connection of Figure 24. The output of the circuit is a positive pulse train with a pulse width equal to T, and a period equal to (N+1) T where N is the programmed count in the counter.

The modulus N is the *total count* corresponding to the counter outputs connected to the output bus. Thus, for example, if pins 1, 3 and 4 are connected together to the output bus, the total count is: N=1+4+8=13; and the period of the output waveform is equal to (N+1) T or 14T. In this manner, 256 different frequencies can be synthesized from a given time-base setting.

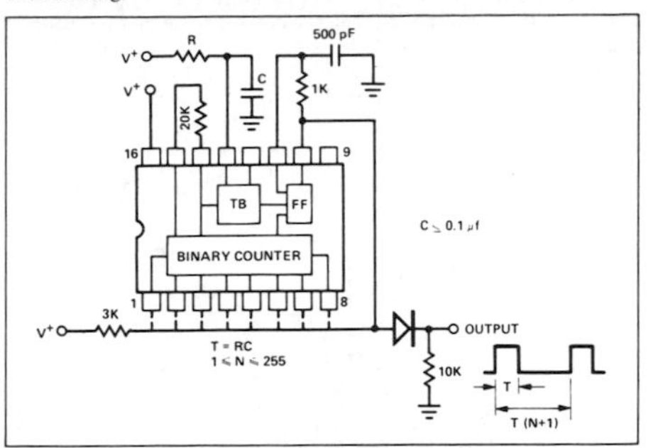

Figure 24. Frequency Synthesis from Internal Time-Base

SYNTHESIS WITH HARMONIC LOCKING: The harmonic synchronization property of the XR-2240 time-base can be used to generate a wide number of discrete frequencies from a given input reference frequency. The circuit connection for this application is shown in Figure 25. (See Figures 16 and 17 for external sync waveform and harmonic capture range.) If the the time base is synchronized to $(m)^{th}$ harmonic of input frequency where $1 \leq m \leq 10$, as described in the section on "Harmonic Synchronization", the frequency f_o of the output waveform in Figure 25 is related to the input reference frequency f_R as:

$$f_o = f_R \frac{m}{(N+1)}$$

where m is the harmonic number, and N is the programmed counter modulus. For a range of $1 \leq N \leq 255$, the circuit of Figure 25 can produce 2550 different frequencies from a single fixed reference.

One particular application of the circuit of Figure 25 is generating frequencies which are not harmonically related to a reference input. For example, by choosing the external R-C to set m = 10 and setting N = 5, one can obtain a 100 Hz output frequency synchronized to 60 Hz power line frequency.

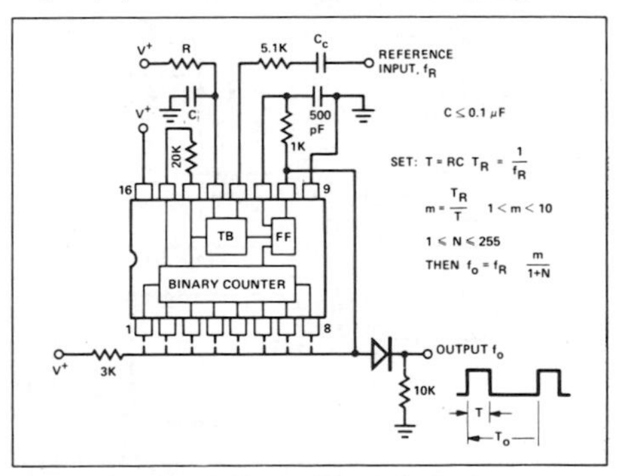

Figure 25. Frequency Synthesis by Harmonic Locking to an External Reference

STAIRCASE GENERATOR

The XR-2240 Timer/Counter can be interconnected with an external operational amplifier and a precision resistor ladder to form a staircase generator, as shown in Figure 26. Under reset condition, the output is low. When a trigger is applied, the op. amp. output goes to a high state and generates a negative going staircase of 256 equal steps. The time duration of each step is equal to the time-base period T. The staircase can be stopped at any desired level by applying a "disable" signal to pin 14, through a steering diode, as shown in Figure 26. The count is stopped when pin 14 is clamped at a voltage level less than 1.4V.

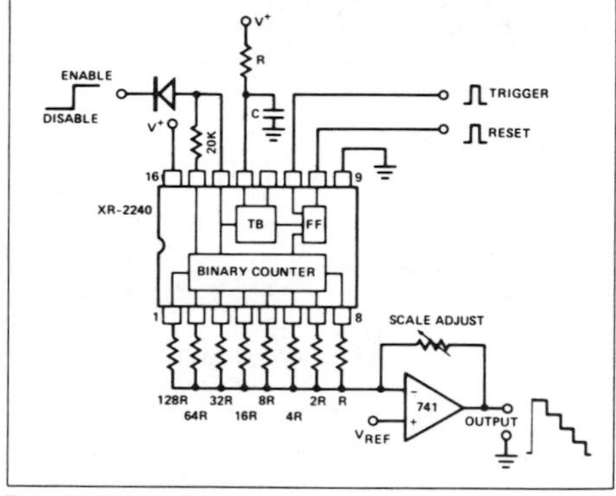

Figure 26. Staircase Generator

DIGITAL SAMPLE/HOLD

Figure 27 shows a digital sample and hold circuit using the XR-2240. The principle of operation of the circuit is similar to the staircase generator described in the previous section. When a "strobe" input is applied, the RC low-pass network between the reset and the trigger inputs of XR-2240 causes the timer to be first reset and then triggered by the same strobe input. This strobe input also sets the output of the bistable latch to a high state and activates the counter.

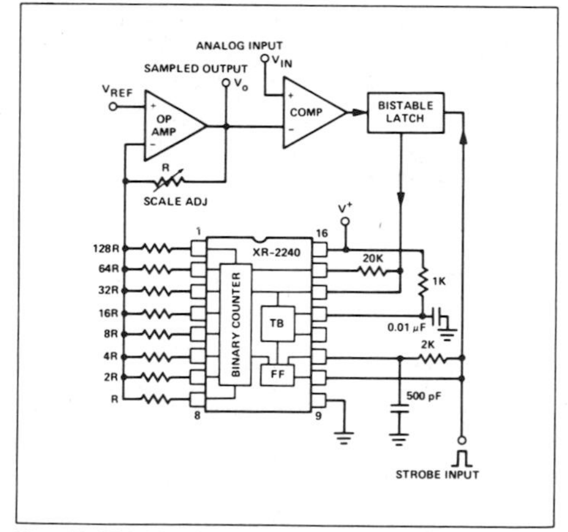

Figure 27. Digital Sample and Hold Circuit

The circuit generates a staircase voltage at the output of the op. amp. When the level of the staircase reaches that of the analog input to be sampled, comparator changes state, activates the bistable latch and stops the count. At this point, the voltage level at the op. amp. output corresponds to the sampled analog input. Once the input is sampled, it will be held until the next strobe signal. Minimum re-cycle time of the system is ≈ 6 msec.

ANALOG-TO-DIGITAL CONVERTER

Figure 28 shows a simple 8-bit A/D converter system using the XR-2240. The operation of the circuit is very similar to that described in connection with the digital sample/hold system of Figure 15. In the case of A/D conversion, the digital output is obtained in parallel format from the binary counter outputs, with the output at pin 8 corresponding to the most significant bit (MSB). The re-cycle time of the A/D converter is ≈ 6 msec.

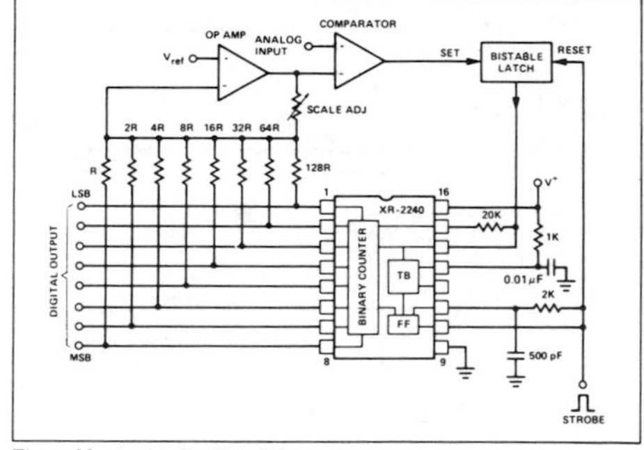

Figure 28. Analog-To-Digital Converter

ORDER INFORMATION

Part Number	Operating Temperature Range	Timing Error	Package
XR-2240M	−55°C to +125°C	2% max	Ceramic
XR-2240N	0°C to +75°C	2% max	Ceramic
XR-2240P	0°C to +75°C	2% max	Plastic
XR-2240CN	0°C to +75°C	5% max	Ceramic
XR-2240CP	0°C to +75°C	5% max	Plastic

Index

E

F

G

H

I

P

W

Z